Physics and Astrophysics

Physics and Astrophysics

Glimpses of the Progress

Authored by

Subal Kar

Former Professor and Head
Institute of Radio Physics and Electronics
University of Calcutta
Kolkata, India

CRC Press
Taylor & Francis Group
Boca Raton London New York

CRC Press is an imprint of the
Taylor & Francis Group, an **informa** business

CRC Press
Boca Raton and London

First edition published 2022
by CRC Press
6000 Broken Sound Parkway NW, Suite 300, Boca Raton, FL 33487-2742

and by CRC Press
2 Park Square, Milton Park, Abingdon, Oxon, OX14 4RN

ISBN: 978-1-0321-0527-7 (hbk)
ISBN: 978-1-032-21199-2 (pbk)
ISBN: 978-1-003-21572-1 (ebk)

DOI: 10.1201/9781003215721

Typeset in Palatino LT Std
by KnowledgeWorks Global Ltd.

Dedicated to all the inquisitive readers of the world

Contents

Foreword

Expanding the boundaries of our knowledge to understand the birth, existence, and extinction processes of the universe(s) presents one of the greatest challenges of all times. This effort includes the understanding of time, space, matter, energy, and life. This book provides us with a glimpse into the progress humankind has made from ancient times to today in understanding the intricacies of the macro- and micro-world through physics and astrophysics. The book starts with "The Beginning: Physics and Astrophysics in the Ancient Times Till the End of the 19th Century" and ends with: "Zooming Out to the Cosmic World of Astrophysics." Topics range widely, covering the development of physics and astrophysics from the days of Aristotle to Stephen Hawking in the 21st century and beyond. This book is unique in that it deals with all these wide-ranging topics in a compact volume and in easy to understand language while preserving the integrity of the scientific content. This presentation reflects the philosophical vision and value judgments of the author, who also has a passion for writing poetry expressing his feelings and philosophy. In the author's own words:

> Life is a receptacle of holy consciousness, inexpressible in worldly terms.
> Effulgent only in realization.
> Austere practice of life is to dedicate oneself in the honest search of the Eternal Being.
> Know that, in the purity of heart universal consciousness is self-effulgent:
> Making life truly a successful one.

Astronomy and cosmology have engaged the inhabitants of India for more than 5,000 years. The emphasis on 'Vidya', or knowledge, runs deep in this part of the world. This book is faithful to that tradition. It is a scientific-cum-philosophical-cum-popular book for readers of all ages. The present work is not a typical syllabus-based textbook. For students, faculty, and researchers, it will be a broad-based reference book that will refresh their fundamental knowledge of physics and astrophysics. Students/researchers today tend to turn away from the fundamental issues of physics/astrophysics. This book, with its wide coverage of scientific content but written in user friendly language, will surely instill a thirst for reviewing the fundamentals and awaken one's scientific spirit in its focus on the different discoveries of physics and astrophysics that have taken place beginning with ancient times. It promises to be an exciting book for intelligent nonspecialists and the public in general.

Much experimental verification of the theories of physics and astrophysics has been pursued in recent years. This includes the imaging of the black hole with the Event Horizon Telescope (EHT), the sensing of gravitational waves by the Laser Interferometer Gravitational Wave Observatory (LIGO), the discovery of Higgs boson in the Large Hadron Collider (LHC), and indication of physics beyond the standard model observed at Fermilab, LHC, and other research centers. These studies continue to generate the interest of students, faculty, researchers, and inquisitive general readers, and this compact book, which presents glimpses into the progress of physics and astrophysics, superbly caters to that need.

One reviewer commented that this book will be a fundamental pillar of physics and astrophysics, including their history of developments. Another reviewer observed that this book will remain relevant and useful as long as people remain inquisitive about physics and astrophysics.

The author's exceptional experience has fitted him well to assimilate the width and breadth of the topics included in this book. He was a Fulbright Scholar at Lawrence Berkeley National Laboratory, Berkeley, from 1999 to 2000, where he worked on Radio Frequency (RF) structures for the Muon Collider, which was a collaborative research program of Berkeley Lab, CA, conducted with Fermilab, Chicago and other US universities. He is one of the seven scientists of Berkeley Lab who were involved in MUCOOL RF Research and Development. The author visited CERN, in Geneva, in 2014, and the Cockcroft Institute Accelerator Facility in the UK in 2013. He also visited NASA Johnson Space Center in Houston, Texas, in 2004, at a time when I was working there as ST/Senior Scientist/Lead Technologist. The author's inquisitiveness and learning process gained through these experiences and many more have been instrumental in developing the contents of this book.

I have an interesting detail about the author to share with you. Earlier, we were both students at the Institute of Radio Physics and Electronics (INRAPHEL), University of Calcutta, India. This institute is well known for the contributions it has made through innovative R & D and teaching. Physics and astrophysics have been the center of the author's interest from the days when he was a BSc Physics (Honours) student at St. Edmunds' College, Shillong, Meghalaya, India. At INRAPHEL, the author was inspired by Prof. M. K. Das Gupta's discovery of the double radio source in Cygnus A, while Das Gupta was a PhD student at the Jodrell Bank Observatory in Manchester, UK. This motivated the author to vigorously pursue his self-study of astrophysics. The author served as the Head of INRAPHEL during 2009–2011. The author's long knowledge-journey has resulted in authoring this book.

I have served the space exploration Program of NASA at Johnson Space Center in Houston, Texas, for more than 54 years. The pursuit of understanding our universe(s) is in my DNA, and I believe that this book will surely help inject a new vigor and a clear vision into humankind's interest in physics and astrophysics.

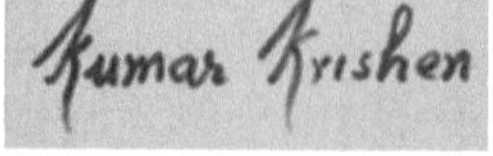

Dr. Kumar Krishen

ST/Senior Scientist/Chief Technologist, NASA Johnson Space Center (Formerly)

NASA Exceptional Service Medal Recipient

Fellow SDPS, Fellow and Distinguished Speaker, IETE

Preface

Physics and astrophysics are vast subjects. From the very beginning of the development of the history of science, these disciplines have germinated and grown side by side. With time, their interwoven nature has become more prominent. Although study of these two subjects together in a correlated way clarifies the concepts in both disciplines, a book written with this spirit in mind is practically unavailable in the market. In this book, my aim is to help the inquisitive reader to make the fascinating journey through different developments in physics and astrophysics from the time of Aristotle to Stephen Hawking and beyond.

Let us quickly glance through some of the applications of physics and astrophysics where similar physical principles have been used. The law of gravity applicable at Earth for falling bodies was also applied to arrive at gravitation for the movement of planets in the so-called Heaven. The mass-energy equivalence of relativity theory accounts for the space–time curvature that provides a new vision of gravitation in the cosmic scale, while the same relation is the basis of the generation of nuclear fission energy as well as the design of the nuclear bomb. Heisenberg's uncertainty principle that reigns in the micro-world of atomic and subatomic physics is also used with a subtle thoughtfulness in the design of the Laser Interferometer Gravitational-wave Observatory (LIGO), which in recent times has sensed the gravitational waves generated by the collision of two black holes in the far cosmic canvas. While journeying through the pages of this book, the reader will discover many more such applications of physical principles applied to both physics and astrophysics.

From ancient times starting with Aristotle, humans have sought to explain the physical phenomena around us. The functional nature of force, the motion of the celestial bodies, the shape of the Earth, the nature of space and time, and so forth are the topics that have received the greatest attention over time. From the 16th century onward, however, physics and astrophysics started taking a more logical shape with the contributions of Copernicus' heliocentric model of the universe, Galileo's laws of falling bodies, and Newton's laws of motion and universal gravitation. Just as Newton effected the first unification of the laws of Earth (through his insight into the falling of the apple) with the laws of Heavens (the motion of planets), J. C. Maxwell unified electricity and magnetism using the same framework of electromagnetism that has paved the way for our modern wireless technology of the Internet, smartphone, GPS, and the like. Faraday first introduced the concept of field in terms of lines of force between electric charges and magnetic poles and that has possibly modified the notion of force, even between two massive bodies. The latter was earlier thought to be a "ghost force." The last decade of the 19th century saw a lot of changes in the prevailing concepts of physics of that time, which ultimately ushered in a new dawn for further developments in physics and astrophysics in the 20th century. The aether prejudice that was so prevalent for a very long time was overcome by the Michelson–Morley experiment. Albert Einstein worked toward a "new physics" of the relativity theory, which has totally revolutionized our concept of space and time, mass and energy, gravitation, and so forth, completely changing our age-old views in both physics and astrophysics. Thus, the 20th century ushered in a new dawn for the progress of physics and astrophysics. All this material is the subject of Chapter 1.

The 20th century was truly a golden period of physics and astrophysics. At that time, two new terms were introduced to our science lexicon: relativistic mechanics and quantum mechanics. Chapter 2 discusses the intricate concepts of the theory of relativity and the theory of quantum mechanics, including their applications. The special and general theory of relativity was developed single-handedly by Einstein. With the advent of these theories, which he propounded in 1905 and 1915, respectively, a new philosophy of scientific thought was established. We have learned from these theories that "nothing in this universe can travel (relative to each other) faster than the velocity of light"; for bodies moving with a velocity comparable to the velocity of light, "time gets dilated" and "space get contracted," mass and energy are intimately and interchangeably related, and so forth.

Further, gravitation is not a ghost force; rather, it is due to the space–time curvature caused by the space–time being curved by the presence of a massive body. Space and time are both relative and not absolute, as it was thought prior to Einstein. In addition, space and time cannot be considered to be separate entities, and thus we have started seeing them as a single entity, the space–time. "Gravitational time dilation" is the lesson we learned from the general theory of relativity that is used even in correcting the atomic clocks installed in the satellites of our GPS system.

Quantum mechanics was developed by a number of physicists, most notably Niels Bohr, Louis de Broglie, Werner Heisenberg, Erwin Schrödinger, Paul Dirac, and Richard Feynman. Starting with Bohr's atomic model, we have come across the wave-particle duality of electrons propounded by de Broglie, Heisenberg's uncertainty principle, Schrödinger's wave mechanics, Dirac's consolidation of quantum mechanics, Feynman's diagram, and so forth. We have learned that there are three distinct chords in the heart of quantum mechanics: *granularity, relational,* and *indeterminism.* Although quantum mechanics is a weird subject, it has exceptionally good experimental support and the modern physics is totally based on this theory. In Chapter 2, in addition to relativity, all important topics related to quantum mechanics has been discussed including the Einstein–Podolsky–Rosen (EPR) paradox and quantum entanglement.

After the fundamental theories of relativity and quantum mechanics were formulated, a number of miscellaneous developments followed. The concept of an expanding universe followed from Einstein's theory of general relativity, which was subsequently supported experimentally by the Hubble team's astronomical observations in 1929. Karl Schwarzschild first indicated the possibility of the existence of a black hole as a solution of Einstein's field equation, which was accepted as a reality in the 1970s with astronomical observations. In 2019, the black hole was even imaged with an Event Horizon Telescope. The Big Bang theory, proposed by George Gamow in the 1940s as the theory of the origin of universe, was also established through observation of the cosmic microwave background (CMB) radiation in 1956. The possible existence of gravitational waves, as intimated by Einstein's relativity theory 100 years back, was established in 2016 by LIGO.

Unification efforts have also taken place beyond relativity and quantum mechanics, leading to Einstein's unified field theory, quantum field theory, and the theory of loop quantum gravity or loop theory. New thoughts such as the superstring theory—the so-called theory of everything—sought to unite all four diverse forces/fields found in nature into one comprehensive theory. However, the two competitive theories—superstring theory and loop theory—still lack any experimental support. Chapter 3 discusses all these issues, including antimatter, dark matter, and dark energy. The presence of antimatter has already been established in particle accelerators, and an initial map of dark matter was obtained in 2021 through the use of Dark Energy Spectroscopic Instrument (DESI).

Whether it be the smartphone in our hand or the galaxy in the vast expanse of the space, all are made of atoms of some kind or the other. Since the late 19th century, through the discovery of the electron by J. J. Thomson, we have known that atoms are not just minuscule billiard balls but, rather, they are composed of subatomic particles. From the mid-1950s, with the advent of particle accelerators or the atom smasher, our knowledge of the subatomic world has thoroughly changed, and we are now familiar with terms such as the photon, lepton, hadron, and quark, in addition to the already well-known electron, proton, and neutron. The particle accelerator at the Stanford Accelerator Center is a lepton-type linear particle accelerator, whereas that at CERN, Geneva, the Large Hadron Collider (LHC), is a hadron-type collider or particle accelerator. The last missing particle of the standard model of particle physics, Higgs boson, was discovered in 2012 at LHC. And today, in 2021, we are receiving exciting news about the possibility of physics beyond the standard model, as well as the possible existence of the fifth force. (Presently, we are familiar with only four forces of nature: gravitational, electromagnetic, strong and weak nuclear forces.) Chapter 4 discusses these issues in detail as they relate to the micro-world of the subatomic domain, including particle accelerators.

Looking at the sky above us and the cosmic canvas in general, the human mind has been awestruck from ancient times. In the daytime, on a clear day the sky is seen to be blue, the rising and setting Sun is found to be reddish in color, and so on. In the night time, the black cosmic canvas is filled with stars and various other celestial bodies such as the Milky Way galaxy, constellations, and many others. Chapter 5 begins our discussion of the sky and the physical reasons for the varieties of colors seen there, including the darkness of the night sky. It is followed by a description of our nearest star, the Sun, and the solar system in which our planet Earth is a member. A focus on the atmosphere of different planets helps us to understand why the Earth is the only planet where we find the life-forms that we see around us. While discussing the life cycle of stars, we cover the formation of red-giant, white-dwarf, neutron star, and the black hole, including novae and supernovae. Nebulae and galaxies are also discussed. Interestingly, we learn that our Sun and the solar system reside in the spiral arm of the Milky Way galaxy and that the Sun was born from the Crab nebula. The chapter also discusses the pulsating star (pulsar) and quasi-stellar star (quasar), and radio astronomy. The final section of Chapter 5 discusses the observable universe and multiverse, a concept that is supported by the most recent theories of physics but at the moment appears to be a favorite topic of science fiction writers.

Finally, I want to take the opportunity to acknowledge the kind and whole-hearted support that I have received from my wife, Rina, and my son, Sauradeep, while writing this book. The quality time that they have allowed me has given me the right space to be engrossed in deep thinking and to successfully complete this dream book of mine. I also extend my acknowledgment to the anonymous reviewers of the book proposal whose encouraging comments and fruitful suggestions have given me the courage to make last-minute improvements to the book. I am thankful to SRS Publishing Services, Puducherry, India, for their assistance with many of the illustrations in this book. I also express my heartfelt thanks to Ms. Aastha Sharma, Senior Editor, in whose hands this project started, and Ms. Jubi Borkakoti, Senior Editor, and her team at CRC Press, Taylor & Francis Group, who have taken this project forward. Further, I am highly thankful to Carly Cassano, Production Editor, Boca Raton, USA, Taylor & Francis Group, whose cooperation at the final stage of the book will always be remembered by me. I would like to thank the copyeditors too for doing their job excellently with patience and meticulousness. Last but not the least, my thank goes to Ashraf Reza, Project Manager at KnowledgeWorks Global Ltd (KGL), Noida, and his team for their untiring efforts in doing the typesetting and associated tasks for this book in order to make it a good product.

If this book is liked by the inquisitive readers, for whom it has been written, then all my humble efforts will be truly rewarded.

November 3, 2021
Kolkata: 700059, India.

Subal Kar
Former Professor and Head
Institute of Radio Physics and Electronics,
University of Calcutta, India

Author Bio

Subal Kar is former Professor and Head of the Institute of Radio Physics and Electronics, University of Calcutta, India. His field of specialization is microwave and terahertz (THz) engineering, metamaterials, and high-energy physics. Dr. Kar has three patents to his credit and has published a large number of research papers in peer-reviewed international journals. He was visiting scientist to various universities and institutes in the United States, Europe, and Asia, including Lawrence Berkeley National Laboratory, Oxford University, Cockcroft Institute, and Kyoto University. Dr. Kar has authored the textbook *Microwave Engineering—Fundamentals, Design and Applications* and has also contributed a number of chapters in books published by Elsevier, Springer Nature, and CRC Press. He is the recipient of the young scientist award of URSI and IEEE MTT, and the Fulbright award from the US Government.

1

The Beginning: Physics and Astrophysics from Ancient Times to the End of the 19th Century

Everything that exists had a beginning

1.1 Introduction

From ancient times, the human mind has been haunted by many fundamental questions. When did the universe start and how? What is its structure? Does the universe have a boundary? Who created it? Will it die? Such questions have been addressed throughout the ages in different ways by both mystics and scientists all over the world. The journey is still continuing at a full and even increased pace, especially by scientists. Our thirst for such knowledge has been quenched to some extent over the ages, but it appears that we are still at the tip of the iceberg and we have a long way to go. Perhaps the eternal joy of the thinking mind of humankind lies in journeying through such a never-ending path of endeavor in the quest for knowledge. In the words of the great American poet, Robert Frost: "And miles to go before I sleep, And miles to go before I sleep."

Knowledge is a process of an ever-widening horizon as we step through the stairs of vision and reason, which help us breathe the refreshing air of satisfaction that comes with leaping forward. Peter J. Ratcliffe, Nobel Laureate of medicine in 2019, made a simple but thoughtful comment: "Knowledge builds on knowledge." Isaac Newton's famous statement, "If I have seen further it is by standing on the shoulders of Giants," is worth mentioning here. In fact, our knowledge about everything that we see around us has developed exactly that way. Also, that our knowledge about anything is increasing means that the difference between the "knowable" and "earned knowledge" is decreasing. However, the knowable is literally infinite; thus, our journey in the thirst for knowledge continues—and probably will continue forever. One of my revered teachers in my college days asked me: "Can you tell me, Subal, what is the quotient between what I know and what I do not know?" Noting my silence and the inquisitiveness in my eyes, seeking for an answer, he politely but confidently answered: "It's a Big Zero." Now, I realize that this remark reflected not his modesty, but his honesty from the core of his heart as a great teacher. Here we can also call to mind Socrates' famous quote: "The only true wisdom is knowing you know nothing."

"Nature permits and Nature prohibits—you just have to understand the limits set by her." Herein lies the true knowledge, the meaningful realization of truth. If you understand this coded language of Mother Nature, you will be the master of philosophy, of the arts, of science—a "great master" true to its sense. Any great discovery of science is dependent on nature's permission and prohibition: Nothing is all encompassing. Certain aspects of nature have evolved into a particular discovery, but certain other aspects are yet

DOI: 10.1201/9781003215721-1

to be unveiled. The journey of knowledge thus proceeds with this possibly never-ending filling-of-the-gaps. This dalliance of Mother Nature is applicable to philosophy and the arts too, but science differs from other branches of knowledge because its permission and prohibition are to be established further by experimental verification. Since ancient times, knowledge of the unknown has taken us from the illusion of obscurity to the effulgence of reality, guided by the rational thinking and the realism of our journey through science.

Throughout the ages, thinkers have asked themselves this question: "What is reality?" They have produced a fascinating spectrum of responses. Is there more to reality than meets the vision of our bare eyes? "Yes," was Plato's answer over two millennia ago. According to Plato, what we humans call our everyday reality is just a limited and distorted (in Indian terms "illusive") representation of the true reality, and so we must free ourselves from our mental shackles to begin comprehending it. The most important lesson that modern physics has taught us about the ultimate nature of reality is that whatever it is, it is very different from what it seems to our gross sense perceptions. We have to use our scientific acumen to understand the reality with a sublime philosophical vision in order to visualize the true nature of reality. Thus, our journey in this book, while coursing through the fascinating history of the development of physics and astrophysics, is from the obscurity of our gross views obtained through our bare eyes of common sense to the effulgence of the finer understanding of the true nature of reality in terms of the objectivity of science and technology.

1.2 Physics and Astrophysics Developments from Ancient Times to Galileo

1.2.1 Introduction

If we go back some 2,620 years, say 600 B.C., Earth was known to humankind as just a patch of flat ground (as it indeed appears to be our gross vision through our eyes, barring the minor irregularities of the mountains and valleys). The sky overhead was known to hold the Earth—with tiny luminous objects (stars and planets) shining in it and two large disc-shaped shiny objects, one of which was visible in the day (the Sun) and the other at night (the Moon). The Earth was believed to be at the center of the universe, with the Sun, Moon, and others revolving around it.

1.2.2 Motion of Bodies and Miscellaneous Ancient Thoughts

The first systematic knowledge book of science known to us is *Physics*, written by Aristotle, the great Greek philosopher, scientist, and resident teacher of Alexander the Great. The book was not named after the discipline we now know as physics, but the said discipline received its name from Aristotle's book. Aristotle believed in two motions on the Earth—forced motion and natural motion. *Forced motion* was believed to be caused by a thrust or impulse and to end when the thrust/impulse ended. For example, a football moves when it is kicked, a breeze causes the branches of a tree to swing, a boat moves by itself (without being rowed) when placed in a flowing river—in all these cases, the thrust comes from leg, wind, and stream, respectively. The *natural motion* according to Aristotle is vertical—upward or downward. To throw a stone upward, one has to give it an upward thrust (force), and so it returns to Earth because, according to Aristotle, it wants to return to its *natural* level. All these accounts of the physical behavior of a body's motion on Earth given by Aristotle's physics may be a gross view, with only an approximate qualitative picture of the phenomena—which demanded significant correction as a result of the subsequent evolution of the knowledge domain of the subject of physics. But we cannot bluntly call them wrong—those were the seeds of rational thoughts. Even the physics of Newton later needed corrections in view of Einstein's theory of general relativity. Einstein's theory in turn failed near the singularity inside a black hole and needed a new set of laws called quantum gravity. Probably everything that we know today might one day be proved to be approximate or might need correction in the light of new concepts that might emerge with better understanding of physical phenomena. On the other hand, it may well be that Newton and Einstein will forever remain with us. There could be minor/major modifications of their theories as the evolution of knowledge demands, but the basic beauty, depth, and encompassing charm of their discoveries may well remain timeless for the inquisitive mind. The two basic questions that have forever changed the world of knowledge about the universe (in terms of classical physics) are: *If the apple falls, does the moon also fall?* (Newton) and *What would happen if I rode a beam of light?* (Einstein).

In addition to Aristotle, we must of course consider another great Greek thinker, Plato, disciple of Socrates and teacher of Aristotle. He understood from the works of Pythagoras, his predecessor famous for his theorem on triangles (taught in school days), that "mathematics (especially geometry) is the language best adapted to visualize the phenomena in the world around." Plato carved in the door of his school: "let not anyone enter here who is ignorant of geometry." Plato posed this momentous question to his disciples: "is it possible to find a mathematics that is able to describe and predict the movements of the planets" (the word "planet" means wanderer in Greek). Venus (the morning/evening star), Mars (the red planet), and Jupiter (the yellow planet) can be easily seen in the night sky, and their movements are little random, back and forth, among the other stars. In this connection, a comment from Galileo Galilei's *The Assayer* (1623) is in order: "Philosophy is written in this grand book, the universe, which stands continually open to our gaze. But the book cannot be understood unless one first learns to comprehend the language and read the characters in which it is written. It is written in the language of mathematics, and its characters are triangles, circles, and other geometric figures without which it is humanly impossible to understand a single word of it; without these one is wandering in a dark labyrinth."

Plato's analytical approach to scientific phenomena ultimately gifted humankind, after a few centuries' journeys by various astronomers like Aristarchus and Hipparchus, the famous book *Almagest* (in 135 A.D.), written by the Egyptian astronomer Claudius Ptolemaeus (known in the West as Ptolemy). In his book *Almagest*, Ptolemy summarized the results of work by the Greek astronomers in a masterly way.

From Pythagoras to Plato, including Neo-Platonism, thinkers in Greece and in Europe, in general, were very much influenced by the rich and highly developed spiritual wisdom as well as the scientific developments (especially in mathematics, medicine, and astronomy) of ancient India. The so-called Arabic numerals, the decimal system, the place value system for numerals, the use of zero, and the fundamental principles of algebra and geometry are distinctly of Indian origin and were carried to Europe by the Arabs. A few famous Indian mathematicians from ancient times were Āryabhaṭa, Brāhmagupta, Baudhāyana, and Śrīdharācāryya, while among the modern Indian mathematicians we may name Srinivasa Ramanujan, famous for his contribution to number theory and infinite series. All of us are familiar with the Pythagorean theorem of geometry: "the sum of the squares on the legs of a right triangle is equal to the square on the hypotenuse."

However, it is believed that Pythagoras came in touch with Brahmins in Persia, if not in India; and that has been confirmed by the German Indologist, Max Müller. Pythagoras possibly learned from Indian thought the 47th theorem of Euclidian geometry, which is found in the *Śulbasûtra* of Baudhāyana (800 B.C.). Leopold Van Schroeder, also a German Indologist, states that "the Pytogorous theorem had its origin in Śulbasûtra of Baudhāyana and Apastamba."

It is further believed that Pythagoras received his ideas regarding the science of music, the importance of numbers, and the existence of the fifth element from ancient Indian thought, all of which were then unknown in Greece and Egypt. In addition, Edward Royle, the British historian, feels that Hippocrates, the father of Western medicine, may have been indebted to India while writing his *Materia Medica.* As Royle notes, "we accepted the fundamental principles of medicine from Hindus." In the time of Alexander, "Hindu physicians and surgeons enjoyed a well-earned reputation for superior knowledge and skill." That India is "a parent civilization" is boldly declared by the French thinker Voltaire: "Everything came to us from the banks of the Ganges where the first Greeks traveled for nothing but knowledge." The supremacy of ancient Indian science over that of the rest of the world gradually started diminishing as the spirituality of mystics having higher strength and practice all over India in subsequent centuries started to assume the forefront in the journey of Indian civilization. From the time of Pythagoras and Democritus in Greece, India's science and philosophy filtered through Persia to Greece and even with the presence of Brahmins from India to Greece and visits of Grecian to India. In this way, Greece gradually started to become the cradle of metaphysics and science, and eventually formed the foundation of modern science in the Western world, while India shone as the spiritual mother of humankind.

1.2.3 The Static Earth Concept

Most ancient thought revolved around a very common (basically gross) observation that the Earth was static, while the Sun, Moon, and other planets went around the Earth, with innumerable little twinkling stars in the sky above, beyond which was possibly "The Heaven." Such thoughts were also prevalent with the Greeks, including Aristotle and Ptolemy. All the celestial bodies were supposed to move in the "void" with respect to time. Such fixed thoughts reigned supreme in ancient times and continued for a very long period of time thereafter, perhaps because they were also supported by the religious scriptures of ancient religions such as the Hindus and Hebrews.

Ptolemy elaborated on Aristotle's static Earth idea to a complete cosmological model of his time in which he assumed that Earth was fixed at the center of the universe surrounded by eight spheres of the Moon, the Sun, and five planets (Mercury, Venus, Mars, Jupiter, and Saturn) known at that time, as well as the stars (which apparently remain fixed relative to each other but rotate together across the sky), as shown in Fig. 1.1. What lies beyond that "observable universe" was not specifically mentioned, but in mystical terms in many parts of the world it was believed that beyond lay what was supposedly Heaven, the abode of the gods.

1.2.4 Geocentric to Heliocentric

We humans have long had a tendency toward hubris, arrogantly imagining ourselves at center stage, with everything revolving around us. But we have repeatedly been proven wrong. In reality, it is we who are revolving around the Sun, which is itself (including all

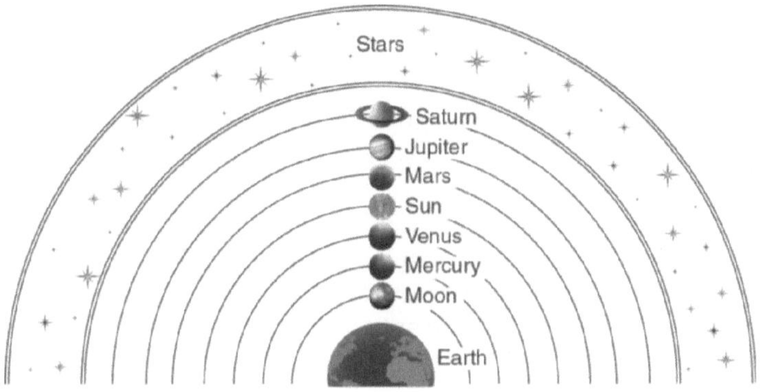

FIGURE 1.1
Representative diagram showing Ptolemy's geocentric model of the universe.

stars in our galaxy) revolving around the center of our own Milky Way galaxy and so forth. Modern astronomical observation shows that the Earth orbits around the Sun, with a relative velocity of 29.8 km/s, and the Sun orbits the Milky Way galaxy with a relative velocity of 230 km/s, while the Milky Way's relative velocity with respect to cosmic microwave background (CMB) radiation is 600 km/s.

Prior to Copernicus, the available model of the universe was that of Aristotle and Ptolemy; known as *geocentric* model. Aristotle believed in a stationary Earth, while the Sun, Moon, and other planets were believed to go around Earth in a circular orbit (which possibly was thought to be the most perfect orbit to follow). The revolt against Earth-centric (i.e., geocentric) ideas was first launched by the Polish priest/astronomer Nicholaus Copernicus, in 1514. The young Copernicus studied Ptolemy's *Almagset* of the geocentric (*ge* in Greek means Earth and *kentron* means at the center) universe and fell in love with it. He decided to spend his life devoted to astronomy, following in the footsteps of the great Ptolemy. However, he became the revolutionist who propounded a model in which the Sun was considered to be stationary at the center while all planets went around it in different circular paths. Copernicus anonymously circulated this revolutionary thinking in a handwritten book among a few (possibly with the fear that he would be branded a heretic by the Catholic Church). Today, such a model is known as the *heliocentric* (*helios* is the Greek word for the Sun) model of the solar (*sol* in Latin means Sun) system. Copernicus' *De revolutionibus orbium coelestium* (On the Revolutions of the Celestial Spheres) containing the concept of the heliocentric universe was published in 1543, on the very day of his death. In connection with the heliocentric model of universe propounded by Copernicus let us quote from his own writing: "As though seated on a royal throne, the sun governs the family of planets revolving around it."

Although his model wasn't completely correct, it formed a strong foundation for future scientists to build on and improve humankind's understanding of the motion of heavenly bodies. Indeed, subsequent astronomers extended Copernicus' work and have now established that there are many such solar systems across the vast universe and that we're far from the center of anything. Copernicus also suggested Earth's spinning rotation about its own axis (from west to east), just like a spinning top, which was experimentally established in 1851 by Léon Foucault using Foucault's pendulum. As per modern estimates, the speed of the spinning rotation of Earth about its own axis is 0.46 km/s. It may be noted that the diurnal motion of Earth that is responsible for the occurrence of day and night, as well as the observed change of position of stars in the sky, are caused by the spinning rotation of the Earth about its axis. On the other hand, the Earth's orbiting about the Sun results in the annual motion that causes the observed change of seasons in the Earth.

The heliocentric model of the universe was also proposed by the famous Indian astronomer and mathematician of the classical age of Indian mathematics and Indian astronomy, Āryabhaṭa (5th century A.D.), in his treatise *Arya-siddhanta*. However, Copernicus' model was scientifically a more definitive one. Āryabhaṭa studied at Nalanda University, India, an internationally famous university of that time founded in the 4th century A.D. However, that university was destroyed by Bakhtiyar Khalji (the invader in 1202 A.D.), including its very resourceful library that contained the treasure of ancient Indian science and astronomy. This event is reminiscent of the loss of the ancient world's single greatest archive of knowledge, the Library of Alexandria, Egypt, believed to be burned under Julius Caesar (in 48 B.C.).

1.2.5 Flat Earth or Round Earth?

Another concept about Earth was the flat Earth concept, which was a commonsensical view originating from ancient times and reigned supreme for eons but at the same time faced many peculiar questions. For example, if Earth is flat (with the exception of the mountains and oceans), then is it infinite in extent? And if it is finite, then what will happen when a person goes toward the edge of the Earth? Who is holding the Earth? To address this last question, it was supposed that the sky above us was a solid shield (as, indeed, it appears to be; also supported by the biblical term *firmament*) and that it came down to meet the Earth on all sides, as it appears to do even to our gross vision of the eyes. India's ancient Hindu thinkers placed the Earth on four pillars to get rid of its infinite extent in the so-called downward direction. But the question arose: "On what the pillars standing? On elephants! And on what the elephants standing? On a gigantic turtle! And the turtle? It swam in a gigantic ocean! And this ocean?" In short, though the assumption of a flat Earth seems to be commonsensical, this concept about the shape of the Earth inevitably had to face philosophical and scientific difficulties of the most serious sort.

Anyway, beginning in 350 B.C., the spherical Earth concept was firmly established, and since then this concept has been accepted all over the world. Aristotle's *On the Heavens* provided important logical information demonstrating that Earth is not a flat plate (which we have said was an earlier "commonsensical" notion) but round in shape. This is because during the lunar eclipse the shadow of Earth that falls on the Moon (when the Earth is between the Sun and the Moon) is circular in shape (only a sphere can cast a shadow of circular cross-section). Another common argument made by the Greeks in favor of the curved shape of Earth was that for an approaching sailing ship in the sea, the sail, and not the hull of the ship, first becomes visible over the horizon, while for a receding sailing ship the hull first disappears while its sail remains visible even at a long distance (see Fig. 1.2).

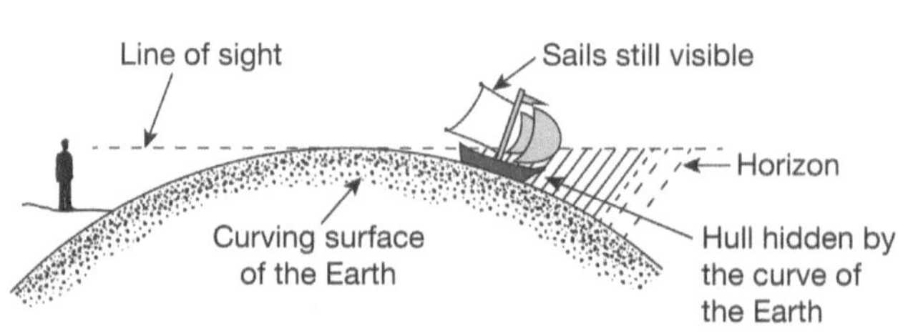

FIGURE 1.2
Hull-first disappearance of ship due to curvature of the Earth.

The safe return of Ferdinand Magellan, the Portuguese explorer who organized the Spanish expedition to the East Indies from 1519 to 1522, as well as other voyagers around the world, finally convinced people that the Earth's surface was curved back on itself into a sphere. However, our day-to-day work with Euclidian geometry, Pythagorean theorem, and trigonometric calculations needs the flat earth approximation, and it is correct in the sense that a small part of an arc is a straight line and that of a spherical surface is a plane. Thus, the workable area on the Earth's surface for all such calculations is practically a flat surface, and so we are happy with that. However, we cannot apply Pythagorean theorem if we need to find on the Earth's surface the hypotenuse of a triangle (!) that connects the center of the Earth with a pole as one apex of the triangle, while the other apex is the equator (as Earth's surface is curved). Thus, for truly macro applications, the Pythagorean theorem fails even on the curved Earth's surface. Similarly, we would subsequently see that in the truly macro-world (in gravitational scale) and the truly micro-world (in atomic scale), many of our common day-to-day experiences are no longer applicable; and a new world of realism becomes inevitable (see Chapter 2 on relativity and quantum mechanics).

The concept of spherical Earth at once put to rest any problem of an edge or end of the Earth, without introducing the concept of its infinite extent. A sphere has a surface of finite size but one without an end; it is finite but unbound. Also, with the spherical model of Earth, a number of notions got clarified with time; as, for example, the concept of down. Down is not a definite direction but is, rather, a relative one. In fact, we now define downward direction with reference to any point on Earth's surface as the one that points toward the center of the Earth. When we say that things naturally fall downward, we really mean to say that they naturally fall toward the center of the Earth (due to gravity). In that sense, nothing falls off the Earth, and people on the other side of the globe from us have no sense of standing upside down.

The spherical Earth concept may also be verified with a simple demonstrative experiment that we ourselves can conduct. Suppose that you are on the ground floor of a high-rise building and observe that the Sun has just set in the horizon. Take a swift elevator and go quickly to the 25th floor of the building to see if the Sun has set down the horizon (or make a quick phone call to your friend on that floor to report the setting of the Sun). Upon swiftly reaching the 25th floor, you or your friend (who is already on that floor) would find that the Sun is yet visible in the sky (though it will set soon). This catching of the sunset by you twice/or for your friend (already on the 25th floor) yet to see the first sunset is due to the round shape of the Earth. If our planet Earth were flat, once the Sun went over the edge from your first viewing position, it would be gone and your second viewing of sunset with increasing the height of viewing would not happen. Also, you on the ground floor and your friend on the 25th floor would have seen the sunset at the same time. The whole issue of your catching the sunset twice a few seconds/minutes later may be more prominent by taking the swift skylift elevator at the Burj Khalifa Hotel in Dubai (828 m high with 163 habitable floors, the world's current tallest building) or by sending up a drone with a camera on it.

1.2.6 The Nature of Space and Time

From the time of Aristotle to Newton until the end of the 19th century, it was believed that the vast space in which the Earth and all other celestial bodies were understood to move was flat and that the rules of Euclidian geometry (the geometry we all learned in our school days) were obeyed—parallel lines never meet. It was also a common scientific notion that "space is absolute." We all move through this space (including our Sun, planets, and the stars) in our own way and at our own speed. Regardless of our relative speed of motion and our mass, we experience the space in the same way and it is unaffected by our presence. Similarly, time was also understood to be absolute. That is, one could measure unambiguously the interval of time between two events regardless of the person measuring it or the place where it was measured—provided a good clock was used for time measurement. This is what most people also believe based on our commonsense observation. Thus, prior to Einstein, the absoluteness of space and time was the prevailing notion in physics and astrophysics.

But Einstein's relativity theory, when applied to the vastness of the cosmic world, put an end to the view of the flatness of space as well as to the absoluteness of space and of time. He combined space and time into a single entity called space–time, which is not flat but curved or warped by the mass/energy of the body residing in the space. The phenomena of time dilation and space contraction also became reasonable in the relativistic regime, as established in Einstein's special theory of relativity. Also, a clock is supposed to run faster when it is at a higher gravitational potential (farther from the source of gravitation, the Earth), as in an artificial satellite (say, in a GPS satellite) as per the general theory of relativity. However, it may be noted that the space–time remains absolutely flat in the world around us even today because the so-called space–time curvature makes little difference in normal situations of speeds (much less than the speed of light) and masses (much less than the mass of Sun, Earth, Moon, etc.) that we encounter in our day-to-day life. Also, the time dilation and space contraction phenomena of relativity do not affect day-to-day happenings around us in the terrestrial domain. But in further reaches of the universe, relativistic effects become significant to deal with the physics and astrophysics in the cosmic domain. (All of these new concepts of space and time and other relativistic fallouts are discussed in Chapters 2 and 3.)

1.2.7 Size of the Earth

Ancient data regarding the size of the round Earth used to be estimated by the ship-sail business using simple geometry. The Earth's radius is given approximately by $d^2/2h$, where d is the greatest distance at which the sail of the ship having a sail of height h can be seen. However, Eratosthenes of Cyrene, the Greek mathematician and geographer, obtained much more accurate data regarding the circumference of the Earth over 2,220 years ago by making clever use of angles. He knew that the Sun was straight overhead in the Egyptian city of Syene at noon on the summer solstice, but it was 7.2° south of straight overhead in Alexandria, located 794 km farther north. He therefore concluded that traveling 794 km corresponding to going 7.2° would give the circumference of the Earth as about: $794 \times 360°/7.2° \approx 39{,}700$ km (as going round the circumference of the Earth means going all around 360°)—which is remarkably close to the modern value of 40,000 km.

1.3 Galileo and the Beginning of Experimental Physics

The death-knell of the Aristotle/Ptolemy model of the universe came when Galileo Galilei of Italy observed the rotating satellites of Jupiter with his newly designed telescope (invented earlier in Holland). The telescope opened the myopic eyes of humankind into a world that now appeared to be vaster and more varied than ancient science could fathom. Based on his experimental observations, he concluded that everything need not have to orbit around the Earth (contrary to the thought of Aristotle and Ptolemy). Galileo publicly claimed in 1632 that Earth orbited the Sun, building on the Copernicus' model. However, the reward of his pathbreaking scientific declaration was to find himself under house arrest for committing heresy against the Catholic Church. A further step forward by Johannes Kepler, a German, following Copernicus, suggested that the planets follow an elliptical path, that is, an elongated circular path rather than a truly circular path around the Sun. Kepler was the assistant of a nobleman, Tycho Brahe, the Danish astronomer and a contemporary of Galileo, who spent his entire lifetime recording observations of the planets with the naked eye. Kepler later used Brahe's compiled data to formulate his three elegant laws of planetary motion that now go by the name *Kepler's laws*. He first showed that the only geometrical figure that would fit the observations of his mentor (Tycho Brahe) would be an ellipse (an elongated circle with major and minor axes and two foci instead of one center of a circle) for the orbits of the planets, with the Sun at one of the foci. Kepler's laws describe the movements of planets around the Sun with greater accuracy than any of the Greek forerunners.

Galileo is also credited with the Law of Falling Bodies; the first mathematical law discovered for Earthly bodies (which to that point had been reserved for movements of the planets, i.e., applicable to "The Heavens"). Galileo was unhappy with the available concept of his time, that of Aristotle, that a heavy body will fall faster than a light one due to a greater pull on the former toward the Earth. The Aristotelian tradition held that all the laws of the universe can be worked out (of course, qualitatively) on the basis of (gross) visual observation based on our day-to-day experience. In contrast, Galileo, for the first time in the history of humankind, performed an experiment designed to understand *how* objects move on the Earth when they are set free, that is, how they fall, and he also precisely measured their falling speed with respect to time. He adhered to the Pythagorean-Platonic vision that nature is understandable through reason and mathematics. Galileo confirmed through experiment that *each body increases its speed at the same rate when falling under the effect of Earth's attraction*; no matter what its weight is (he used two lead balls of different weights to perform his experiment). The result was momentous. Of course, a feather is found to fall slower than a lead ball. This is because the feather is slowed down by air resistance opposing its downward motion due to Earth's attraction. However, in vacuum they will also follow the same law of falling bodies, as was observed for two lead balls. In many books, it is mentioned that Galileo performed his experiment from the leaning tower of Pisa in Italy. Hopefully, this is just an apocryphal story because in his experiment he actually used very simple arrangements to roll balls of different weights down a smooth surface of single- and double-inclined planes (see Fig. 1.3).

The inference of the experimental results on falling bodies by Galileo may be summarized as follows. For motion of a body, we now come across the term *inertia*—meaning *resistance to change*. In other words, if the net external force is zero, a body at rest continues to remain at rest and a body in motion continues to move with uniform *velocity* (*speed, i.e., rate of change of displacement with a preferred direction,* a vector quantity). The state of rest and the state of uniform linear motion (with constant velocity) are equivalent; this insight into the law of motion of a body was missing in the Aristotelian view. Further, for a falling body under the effect of the pull of the Earth (named as the force of gravity by Newton), there is now a new term called *acceleration* (*rate of change of velocity with respect to time*) with which all bodies (small or big, i.e., irrespective of their weight) will fall, and this acceleration due to Earth's pull (gravity) Galileo estimated was 9.8 m/s^2. The law of falling bodies by Galileo can be expressed mathematically as $x = (1/2)gt^2$; where x is the displacement of the body,

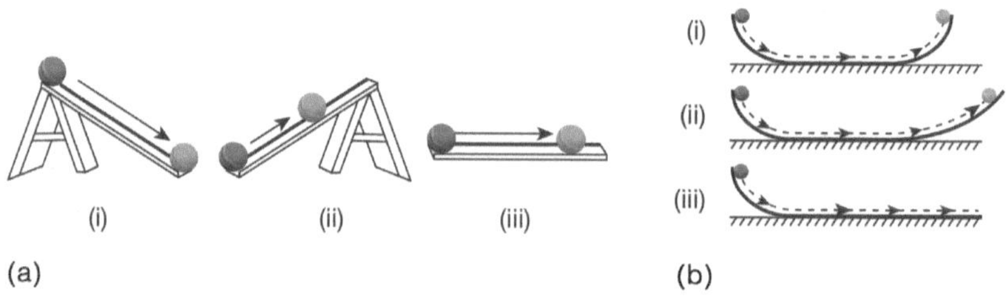

FIGURE 1.3
(a) Motion of ball in a single inclined and horizontal plane (i) accelerated, (ii) retarded, (iii) moves with constant velocity (neither accelerated nor retarded); (b) in (i) and (ii) the ball rises to the same height as the initial height, but in (ii) it travels a longer distance; (iii) when the slope of the second plane is zero, that is, horizontal, the motion of the ball never ceases, and it ideally travels an infinite distance (if the surface is frictionless).

t is the time taken to result in this displacement, and g (= 9.8 m/s^2) is the acceleration due to gravity. This in simple terms means that *all* bodies (irrespective of being small or big) will fall downward due to Earth's gravity covering a distance of 4.9 m in the first second, and then in 2 seconds it will cover a distance of 19.6 m and so on. In other words, for every second that an object falls under gravity, its velocity ($v = gt$) increases by 9.8 m per second.

Galileo's simple but revolutionary idea of falling bodies (obtained experimentally) dethroned Aristotelian mechanics, and the need for developing a new mechanics became essential. This task was accomplished almost singlehandedly by Sir Isaac Newton, the British scientific genius of the classical period of science. Newton, with the earnest request of Edmund Halley, after whom the famous Halley's Comet has been named, published his masterpiece, *Philosophiæ Naturalis Principia Mathematica* (Mathematical Principles of Natural Philosophy), often simply called *The Principia,* in 1686. There he stated his three laws of motion, including the universal law of gravitation (and many more scientific gems).

1.4 Newton's Laws of Motion and Universal Gravitation

1.4.1 Introduction

Like most breakthroughs in physics, Newton's laws of motion not only addressed the questions relating to the motion of bodies around us but also empowered human minds to realize that the laws of physics applicable in our Earth are equally applicable for the so-called heavenly bodies. His universal law of gravitation could satisfactorily answer a childish, yet serious, question of the day: Why does the Moon above the sky not fall to Earth just like a huge, dropped rock? Prior to Newton, the idea was that this happens because the laws of Heaven are different from those of our human Earth! Newtonian physics concerning the laws of motion and gravity not only set forth a revolution in the scientific thought of the day, but his concept spread like wildfire and was gradually applied to other topics of physics such as light, gases, liquids, solids, electricity, and magnetism. His concepts were extrapolated not only to the macrocosms of space, but also to the microcosms, finding that many properties of gases and other substances could be explained by applying Newton's laws of motion to the atoms they are made of. This scientific revolution ushered in the Industrial Revolution in the 18th century and ultimately led to modern-day space exploration and the information boom. Newton's law of gravitation can explain the cause of the tides, predict the occurrence of eclipses, simulate the launching of rockets from Mars to Neptune (and even today in making interstellar journey by Voyager II), and figure out when and how the spaceship will arrive at the desired destination. In this section, we discuss Newton's laws of motion and the universal law of gravitation.

1.4.2 Newton's Laws of Motion

Newton's first law of motion (also known as the law of inertia) is concerned with the inertia of rest and the inertia of motion of a body. It states that *a body at rest or in uniform motion would continue to remain in that particular state unless otherwise it is acted upon by a net external force* (i.e., an unbalanced force). A little elucidation will make the point clear. A body in uniform motion tends to continue at that state of motion until an opposing force reduces its velocity or brings it to rest. We observe this process when we apply the brakes to our moving car, which increases the frictional force between the wheel tire and the surface of the road to oppose the motion of the car. To *accelerate* the car, that is, to increase the rate of change of velocity of the car, we need to push the accelerator of the car harder (i.e., provide a net external force by the engine of the car in the direction of motion). A body at rest will start moving only when an external force tends to change its state of being in rest and forces it to go into the state of motion.

A practical experience while traveling on a bus or train may also be mentioned here to understand the effects of the inertia of motion and the inertia of rest, as stated above in regard to the first law of motion. Suppose we are traveling in a moving bus and the driver suddenly applies the brake to slow down the bus or to stop it, we are thrown forward (see Fig. 1.4a). This happens because with the moving bus, our whole body, including our feet, are in motion. When the driver suddenly applies the brake, our feet manage to stop with the bus due to friction between the floor of the bus and our feet in contact with it (allowing no relative motion between the two). However, the rest of the body continues to move (as our body is not a strictly rigid one as a whole, but it is deformable; that is, it allows some relative displacement between different parts of it). When the restoring muscular forces bring the body to rest, we come out of the initial jerk that has thrown our body forward.

A similar phenomenon happens when a stationary bus is suddenly started as the driver applies the accelerator; we get thrown backward by the jerk (see Fig. 1.4b). This is due to a sudden change from the inertia of rest to the inertia of motion.

Newton's first law of motion refers to the simple case in which the net external force on the body is zero. The *second law of motion* refers to the general situation when a net external

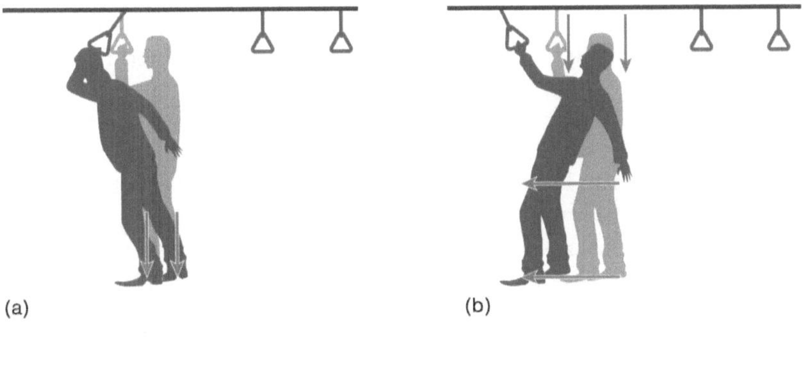

(a) (b)

FIGURE 1.4
Explanatory diagrams for the law of inertia. (a) The passenger is leaning forward when the bus suddenly stops by applying the brake; (b) the passenger is thrown backward when the bus suddenly starts by applying the accelerator.

force acts on the body; it relates to the net external force to the acceleration of the body. Mathematically, the second law of motion can be stated as: $F = ma$. Thus, if a body has a greater quantity of matter, that is, mass (m), its acceleration (a) will be less compared to a body having smaller mass when the same external force (F) is applied to both of the bodies. A familiar example is that a heavier car will have smaller acceleration (rate of change of velocity) compared to a lighter one, with both cars using an engine with the same power. For this reason, lighter cars have a small horsepower (1 hp = 750 Watts) engine, while heavier cars need to use an engine with larger horsepower. Further, for the same car to realize higher speed per unit time (i.e., acceleration), the engine needs to exert a larger force by having the driver press the accelerator harder.

Finally, we discuss *Newton's third law of motion*. Newton stated this law as "To every action, there is always an equal and opposite reaction." In fact, this crisp and beautiful wording made this sentence a commonly quotable one, sometimes even in a misleading way! The term *action and reaction* in this law may give the wrong impression that action precedes reaction; that is, if action is the cause then reaction is the effect. However, this is not the case in reality. According to Newtonian mechanics, force never occurs singly in nature; force is the mutual interaction between two bodies. Forces always occur in pairs. Further, the mutual forces between two bodies are always equal and opposite. There is no cause–effect relation implied in the third law; the force on a body A by the body B and the force on B by A acts at the same instant. With this reasoning, any one of them may be called action and the other one reaction. A practical experience exemplifying this third law of motion is the recoiling of a gun of mass m_2 when a bullet of mass m_1 is being shot from it. The Force F with which the bullet emerges from the gun with a velocity v_1 exerts the same force back to the gun in the opposite direction, as a result of which the gun hits the shoulder of the gun-shooter (against which the gun is normally held) with a velocity v_2. In the whole process, the momentum (mv) is said to be conserved, that is, $m_1v_1 = m_2v_2$.

1.4.3 A Comment by the Author

In order to state Galileo's law of falling bodies and Newton's laws of motion, we have used simple mathematical formulas (this will be done throughout this book wherever necessary, though in simple terms). Such elementary analytical visualization of scientific laws, in addition to the philosophical ideas embedded in its physics, is essential to feel the objectivity of scientific discoveries. If we dissociate mathematics from physics or vice versa, we do not have true science. Physics and mathematics may be thought to be, respectively, the Adam and Eve of science. Mathematics is the implicit language of science that gives its analytical expression, while the philosophical insight is contained in the physics of the phenomena and the two together make the complete physical science.

A beautiful comment regarding the relation of mathematics with science has been made by the physics laureate Chen Ning Pang, who received Nobel Prize in Physics in 1957 at the age of 31 along with Tsung Dao Lee, "for their penetrating investigation of the so-called parity laws which has led to important discovery in the elementary particles." Pang commented: "The scientific equations we seek are the poetry of nature."

1.4.4 The Gravitation and the Newton's Universal Law of Gravitation

The term "Gravity" or "gravitation," derived from the Latin word *gravitas* meaning *weight*, is a natural phenomenon by which all things with mass or energy—including planets, stars, galaxies, and even light—are brought toward (or gravitate toward) one another.

According to Newton's law of gravitation, it is understood to be a "ghost force" acting between bodies having mass while according to Einstein's concept of gravitation, based on general theory of relativity, it is caused by space–time curvature resulted around a body due to its mass/energy. On the Earth, gravity gives weight to physical objects, and the Moon's gravity causes the ocean tides on Earth. The gravitational attraction of the primordial gaseous matter in the universe caused it to begin coalescing and forming stars and caused the stars to group together into galaxies. Thus, gravity is responsible for many of the large-scale structures in the universe. Gravity/gravitation has an infinite range over which it acts, although its effect become weaker as objects get further away. It may be noted that gravity is one of the four fundamental forces in nature; however, in the standard model of particle physics, it is not included as gravity is understood to be practically negligible at the atomic level.

Newton's most important contribution to knowledge about the universe is the Universal Law of Gravitation (though there is a mild debate about whether Isaac Newton or Robert Hooke is to be credited for that accomplishment). It is on the basis of this law that we can understand why and how the planets go around the Sun and also the trajectories of our space vehicles going to the Moon and different planets, including those entering interstellar space. In school textbooks, the background of this discovery is again apocryphal. As the story goes, Newton was pensively sitting in the garden when he saw an apple falling from the tree, whereupon the concept about gravitation immediately flashed in his mind. Whether the story be fiction or fact, Newton's law of gravitation is definitely the foundation for classical physics. Newton included this law in his famous book *The Principia*. He discovered the laws of gravity in 1687, and as noted earlier, these laws were radically revised by Einstein in 1915 with his general theory of relativity (which remains unchallenged to this day; these are discussed in Chapter 2).

With Galileo's law of falling bodies and Kepler's laws of planetary motion in the backdrop, Newton was seriously pondering the *reason* for planetary motion (that happens in the Heavens!) and whether it had any connection with the acceleration due to gravity g obtained by Galileo from his law of falling bodies in the Earth. Newton imagined that a *Little Moon* might be above the highest mountain peak, Mount Everest, in addition to the real Moon orbiting around the Earth. According to Newton, either of the Moon will experience a *centripetal force* (a force that acts toward the center of Earth and makes a body follow a curved path rather than a straight path) exerted by the Earth's gravitation. The centripetal force gives rise to a centripetal acceleration (rate of change of velocity) of the body (little moon or the real moon). The value of this centripetal acceleration of the body was calculated to be exactly equal to the value of $g = 9.8\ \text{m/s}^2$ obtained from the law of falling bodies by Galileo. This was an exciting, yet *natural,* coincidence! *The force that causes the Little Moon to turn around the Earth in its orbit must be the same as that which causes objects to fall to the ground on Earth's surface.* Newton used this inductive reasoning based on empirical observations. In other words, the new vision that human knowledge gained is that the same pull of gravity that pulled an apple down from a tree (when it is let loose) kept the Earth in motion around the Sun; the age of reason in true scientific sense thus dawned.

For planetary motion, Newton proposed the inverse square law of variation of the gravitational force between the two bodies: $F = G(m_1 m_2 / r^2)$, where $G = 6.67 \times 10^{-11}\ \text{Nm}^2/\text{kg}^2$ is the universal gravitational constant, m_1 and m_2 are the mass of two bodies, and r is the distance between them. The universal law of gravitation mentioned in Newton's *Principia* is the following: "Everybody in the universe attracts every other body with a force which is directly proportional to the product of their masses and inversely proportional to the square of the distance between them." The value of G was determined by the English Scientist Henry Cavendish in 1798. Knowledge of the value of G combined with the value of acceleration due to gravity (g) and the radius of the Earth (R_E) allowed Cavendish to find the value of the mass of Earth (M_E). This explains the popular statement about Cavendish: "Cavendish weighted the Earth."

1.4.5 Weightlessness Experienced by Astronauts in Space Flights

The acceleration due to gravity, g, has a maximum value of 9.8 m/s^2 on the surface of the Earth, but its value decreases as we go above or below the surface of the Earth and when we go from polar to equatorial region. This follows from Newton's universal law of gravitation. Further, $g_{\text{polar}} > g_{\text{equitorial}}$ since the Earth is slightly bulging at the equator and flattened at the poles (i.e., the polar radius of the Earth is smaller than its equatorial radius). The weightlessness experienced by astronauts in manned space flights orbiting around the Earth is due to the so-called zero gravity experienced by the astronauts (being far away from the Earth's surface g reduces to practically zero value), and they just float inside the space vehicle (Fig. 1.5).

1.4.6 Fallout of Newtonian Gravitational Law and the Laws of Motion

Newton's gravitational law was an enormous intellectual triumph. When combined with his laws of motion, it explained the orbits of the planets around the Sun and Moon(s) around the planets, the ebb and flow of ocean tides, and the fall of rocks; and it taught Newton and his 17th-century compatriots how to weigh the Earth and the Sun.

FIGURE 1.5
Weightlessness experienced by astronauts in a space vehicle orbiting around the Earth [1].

The Newtonian mechanics and the Newton's law of gravitation, suddenly after millennia, could give shape to an immensely new and clear vision of the universe. With the legacy of Pythagoras and the Alexandrian astronomers' mathematical physics, Newton established that the laws in the so-called Heaven were being followed equally well on the Earth; there is no such "natural level" for things (as presumed by Aristotle). The universe is a vast (infinite) space studded with planets and stars, without limit and without any center. Within it, material bodies run free and straight, unless a force (gravitation), generated by another body, deviates/deflects them.

Until the mid-1600s the Church and scholars of the day believed in two distinct types of laws: The laws governing the Heavens (a perfect and harmonious one) and mortals on the Earth living under the physical laws (the coarse and vulgar ones). Anyone going against this view could be put to death or imprisoned by the Church. Giordano Bruno, an Italian philosopher and mathematician, was burned alive in 1600 in Rome for speculating that our Sun was just another star and concluding that "there are then innumerable Suns." Three decades later, in 1633, Galileo Galilei was forced to repudiate his scientific finding "that the Earth moved around the Sun" because it was heretical, being against Church teachings; however, during his trial, he is said to have muttered under his breath, "But the Earth does move!" After the trial, the court found Galileo "vehemently suspect of heresy" and sentenced him to imprisonment. He was kept under house arrest until his death in 1642.

Nonetheless, humankind arrived at a new dawn of civilized conscience when the young Isaac Newton discovered the law of gravitation applicable to the so-called heavenly bodies and established that the same force was responsible as gravity for falling bodies in the Earth—thus unifying the laws of Heaven and Earth. After Newton's discovery, the motion of the entire solar system could be calculated with almost perfect accuracy.

Newton's discovery of the laws of mechanics and universal gravity had such a momentous outcome that for the next two centuries no one questioned it, for it had proved to be powerful beyond all expectations. The entire technology of the 19th century and that of our modern world too rests largely on Newton's formula. Be it the building of bridges, skyscrapers, trains, engines, and hydraulic systems, including the flying of airplanes; be it the prediction of the existence of a new planet (in fact, the dwarf planet Pluto was discovered this way) or the sending of spaceships to Mars or to Saturn, including interstellar voyages, none would have been possible without Newtonian mechanics and calculations of gravitation with his "Little Moon."

1.4.7 The Bentley Paradox

With the discovery of the universal law of gravitation, Newton was elated because his theory was found to be capable of establishing that the motion of objects on Earth due to gravity can also be applied to the universe in general. While he was basking in the fame that *The Principia* brought him in 1687, five years later an English scholar, critic, and theologian, Richard Bentley, stumped Newton by asking a simple question for which even Newton had no plausible answer. Bentley pointed out that if the universe was finite, then all the stars would collapse into themselves due to the attractive force of gravity, and our night sky would have been an absolute scene of carnage. On the other hand, if it were infinite, then the force on any object, tugging left or right, would also be infinite, and therefore the stars should be ripped to shreds in a fiery cataclysm. This is what is known as the *Bentley paradox*. Newton resorted to a patchwork type of answer to address the paradox. He assumed an infinite universe that is perfectly uniform, and thus he reasoned that a star that experiences an infinite force from the right also experiences the same amount of force from the left, with the forces nullifying each other and resulting in a stable stellar system.

This escape route taken by Newton from the Bentley paradox was not completely flawless. If what Newton said was correct, then the universe would simply become analogous to a house of cards; a little jitter would collapse the whole system. Newton's further but feeble response was to resort to a "divine power" that prevented his house of cards from collapsing. His non-plausible answer was that "a continual miracle is needed to prevent the Sun and the fixed stars from rushing together through gravity." However, with the advent of Einstein's general theory of relativity, gravitation no longer remained a "ghost force" (as supposed by Newton), but it was henceforth understood to be due to the curvature of space–time around a body caused by all the matter or energy of the body under consideration. For example, the mass/energy of the Sun curves the space–time around it, and thus the Earth and other planets experience the gravitational effect due to Sun. This concept of gravitation as per the general relativity of Einstein, in addition to the expanding nature of the universe proved through Hubble's astronomical observations, finally settled the problem of the Bentley paradox.

1.4.8 Gravitation: Is It a Force or Is It Due to Space–Time Curvature?

The word that appears time and again in Newton's intellectual framework of mechanics and gravitation is *force*. It is the *net external force* that makes the motion of a body possible, and the *force of gravity* causes a body to fall on the surface of Earth and the planets to orbit around the Sun, satellites to orbit around the planets, and so forth. However, one thing bothered Newton most just after he propounded his theory of gravitation: How, he asked, does the Earth manage to attract the Moon, which is so distant! How can the Sun attract the Earth without coming in contact with it? The force (rather, the "ghost force") of gravity as described by Newton is an "action at a distance"; its effects on faraway objects are instantaneous, no matter the distance. In an exchange of letters with R. Bentley, however, Newton stated that it was "inconceivable that inanimate brute matter should, without the intervention of something else which is not material, operate upon and affect matter, and have an effect upon it, without mutual contact." He further mentions his utter reservation about this very attraction at a distance that he himself introduced. "That gravity should be innate, inherent and essential to matter, so that one body may act upon another at a distance through a vacuum, without the mediation of anything else, by and through which their action and force may be conveyed from one to another, is to me so great an absurdity, that I believe no man who has in philosophical matters a competent faculty of thinking, can ever fall into it. Gravity must be caused by an agent acting constantly according to certain laws, but whether this agent is material or immaterial, I have left to the consideration of my readers."

It is the characteristic of a genius to be aware of the limitations of his own findings, and he modestly acknowledges the same. In Newton's own words: "I do not know what I may appear to the world, but to myself I am only a little boy playing in the sea-shore and trying to collect a smoother pebble than ordinary—whilst the vast ocean of truth lay all undiscovered before me." This clearly indicates that all that he had learned only made him feel how little he knew in comparison to what remained to be known.

The "reader" who could intuitively visualize Newton's *unfound agent* responsible for the "ghost force" to act at a distance was another Briton—Michael Faraday. Not with the study of gravitational attraction between bodies but while pondering the force between electric charges and magnetic poles, in 1854, Faraday was the first who intuitively introduced the concept of *field*, an English term. He introduced fields as properties of space (even when devoid of matter) having physical effects mediating between electric charges or magnetic poles via the so-called lines of force (details of which will be discussed in Section 1.5). He argued against "action at a distance" and proposed that interactions between objects occur via space-filling "lines of force." This description of fields remains unchallenged even to this day.

The greatest visionary scientist of them all, Albert Einstein, proposed a truly revolutionary idea of *space–time curvature* to understand the gravitational phenomena with the help of his general theory of relativity (which is discussed in Chapter 2). Einstein's new concept of gravitation allowed better estimates for a number of unanswered issues such as the perihelion shift of the planet Mercury in our solar system, deflection of starlight by the Sun, the strong gravitational pull by the black hole, and so on.

From Newton's universal gravitation, it is known that the Moon's gravitational force results in a tidal force that squeezes and stretches the Earth's oceans which changes every few hours in response to Earth's diurnal motion (spinning motion around its own axis drawn from north to south). Squeezing is in the transverse direction (perpendicular to the Moon's direction), while stretching is in a longitudinal direction (both toward and away from the Moon), the Moon's direction being in the horizontal direction (Fig. 1.6). On the basis of the general relativity theory, the same issue can be addressed in terms of the space–time curvature produced around Earth by Moon's gravity. However, whether we use the Newtonian gravitational force approach or Einstein's space–time curvature approach, the Moon is found to produce tides roughly 1 m in size, so the difference between high tide and low tide in the oceans is typically 2 m. It is important to note here that Einstein's initial effort to introduce gravity in the general theory of relativity started with the concept of tidal gravity for freely falling bodies and was finally formulated with the enlightenment received from the four-dimensional space–time concept of the physicist Hermann Minkowski and the mathematician David Hilbert's mathematical treatment of space–time curvature, along with the Riemannian geometry of curved surfaces. (All of these issues are discussed in detail in Chapter 2.)

Although Newton's universal law of gravitation, first presented in 1687, had gone unchallenged for more than 200 years, one pesky observation threatened to derail the unchallangeable law: The motion of the innermost planet of the solar system, Mercury. Kepler described

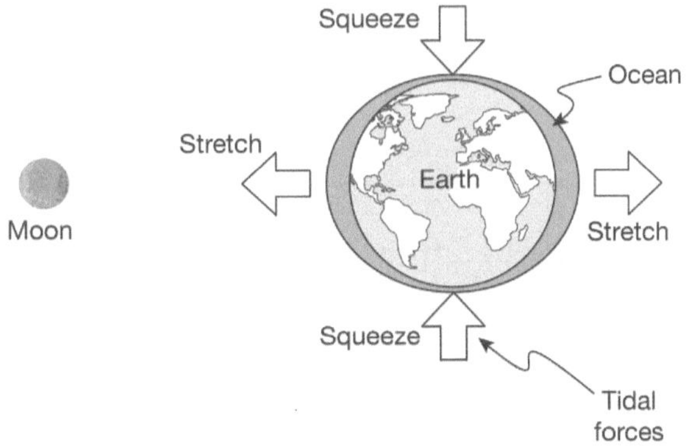

FIGURE 1.6
Ocean tide in the Earth due to the Moon's gravitationl effect.

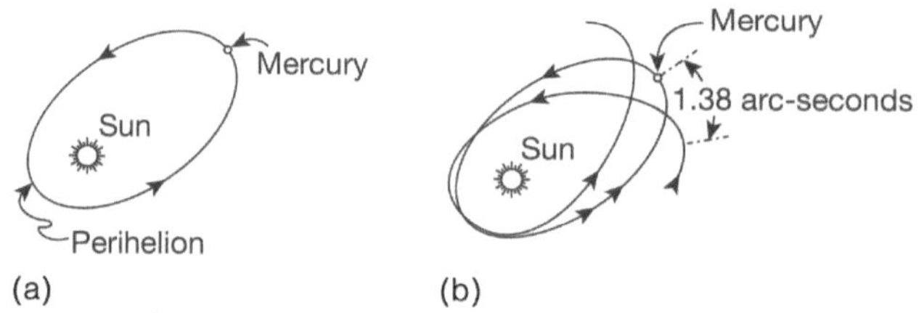

FIGURE 1.7
The perihelion shift of the planet Mercury (a) orbit according to Kepler and (b) actual orbit showing perihelion shift as observed experimentally and supported by Einstein's relativity theory.

the orbit of Mercury as an ellipse, with the Sun at one of the foci (Fig. 1.7a). However, by the late 1800s, astronmers had deduced from their experimental observations that Mercury's orbit was not quite elliptical. After each trip around its orbit, Mercury fails by a tiny amount to return to the same point at which it started. This failure can be described as a shift, with each orbit in the location of Mercury's closest point to the Sun, causing a shift of its perihelion. Astronomers measured a perihelion shift of 1.38 seconds of arc or arc-second during each orbit (Fig. 1.7b). Newton's law of universal gravitaton could account for 1.28 arc-seconds of this 1.38 arc-second shift. But Einstein's concept of gravitation in terms of general relativity could account for exactly a 1.38 arc-second shift of perihelion of the planet Mercury. The words 'perihelion' and 'aphelion' come from the Greek words *peri* meaning close, and *apo* meaning far, and *helios* means the Sun. They are used in astronomy to refer to the closest and farthest points of the orbits of any object revolving around the Sun.

However, Newtonian physical laws appeal to our everyday notion of space and time in our terrestrial backdrop. When gravity is not very strong, the Newtonian flat space–time paradigm and gravitation as a "force entity" works perfectly as in the domain of everyday life and in most understandings of the solar system. But in the exotic parts of the distant universe, such as pulsars, quasars, and black holes where gravity is exceptionally strong, we have to resort to the curved space–time paradigm of Einstein's general theory of relativity. In other words, we can say that Newtonian physical laws are the limiting case of Einstein's relativity theory, the latter being a universal law, a meta-principle, especially for the macro-world. But even today when studying gravity in the solar system, physicists often switch back and forth with impunity between the Newtonian flat space–time paradigm and Einstein's curved space–time paradigm (as the predictions from both are almost identical, except for the perihelion shift of the planet Mercury), using at any time whichever strikes their fancy or seems to be more insightful.

1.5 Faraday and Maxwell's Electromagnetism

1.5.1 Introduction

Both electricity and magnetism have been known for more than 2,000 years. In the past, electricity and magnetism were treated as separate subjects. Electricity dealt with just rubbing a glass rod with amber (which could attract pieces of paper), batteries, lightning, and so on, while magnetism showed the attraction/repulsion of poles, deflection of the compass needle, and so on. The name *electricity* comes from the Greek word *electron* meaning amber, and magnetism is from the Greek term *magnetislithos* or lodestone, which showed the properties of attraction/repulsion. The study of electricity and magnetism progressed at a slow pace during the 18th century. In the beginning of the 19th century, Faraday was working in a London laboratory with bobbins, needles, iron cage, and the like, in an effort to explore how electric and magnetic things attract and repel. Interestingly, Faraday was an impoverished Londoner without formal education. He was an inquisitive learner and, also, a voracious reader of scientific books that he used to read while working in a book-bindery in his early formative years. Responding to his extreme eagerness to work in his laboratory, Humphrey Davy, the inventor of Davy's safety lamp used in coal mines, allowed Faraday to work in his laboratory, initially as a bottle washer. With his scientific acumen and competence, Faraday soon gained Davy's confidence, ultimately becoming the most brilliant experimenter of 19th-century physics and its great visionary. As Davy in his later life gracefully commented: "My best discovery is Michael Faraday."

1.5.2 Faraday's Contribution

Faraday had an exceptionally clear vision. Without knowing mathematics, he wrote one of the best physics books ever written, *The Forces of Matter*, which was virtually devoid of equations. With his mind's eye, he tried to understand or, more specifically, he could "see" the force that acts between electric charges and magnetic poles. Charles Coulomb in 1785 devised the mathematical formulation for the force between electric charges q_1 and q_2: $F = k|q_1 q_2|/r^2$; here $k = 9 \times 10^9$ Nm^2C^{-2} (the Coulomb Constant), and r is the distance between the two charges. However, Faraday intuitively visualized the physical nature of this force and gave its pictorial representation, from which we now have the concepts of field and lines of force (or field-lines) acting between electric charges and magnetic poles. According to Faraday's visualization, between two electric charges or magnetic poles there are invisible bundles of infinitely thin lines (forming a cobweb) that fill the space between and around the charges/poles. He termed these lines the lines of force because, in some way, they carry the force. That is, they transmit the electric and magnetic forces from one body to another, as if they were cable lines through which the pulling (for opposite ones) or pushing (for like ones) of the electric charges/magnetic poles takes place (Fig. 1.8).

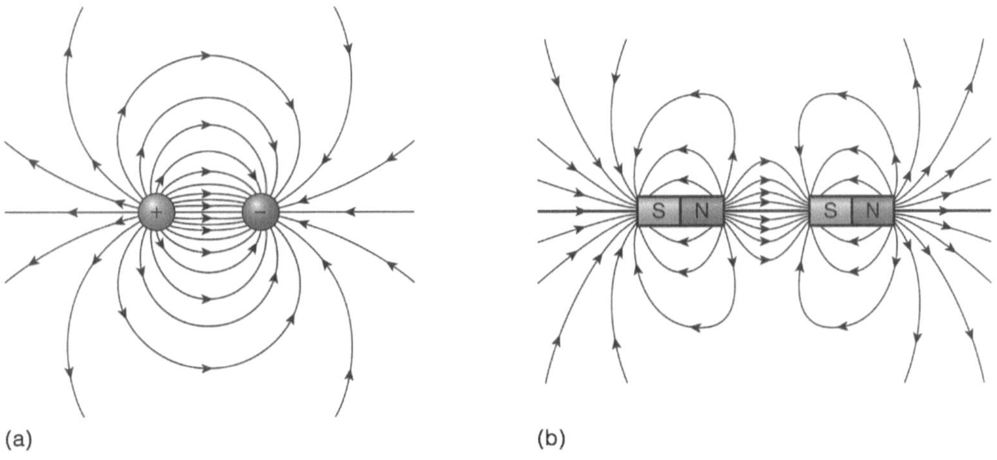

FIGURE 1.8
Lines of force between (a) electric charges and (b) magnetic poles.

1.5.3 J. C. Maxwell's Contribution

James Clerk Maxwell was a wealthy Scottish aristocrat and became one of the greatest mathematicians of the 19th century. Though a wide gulf of difference separated Faraday and Maxwell in both their social origin and their intellectual style, their work ultimately led to the development of a new branch of scientific knowledge known as *electromagnetism.* Faraday propounded the concept of electromagnetic induction on the basis of his experimentation on the movement of magnets inside and out of a current carrying conducting coil. Maxwell with his mathematical genius and physical insight unified the then available experimental results and physical concepts of electricity and magnetism and proposed the now famous *Maxwell's equations of electromagnetism.* He used Faraday's law of electromagnetic induction as well as the work of Hans Christian Oersted, Carl Friedrich Gauss, and Andre-Marie Ampère, together with his own introduction of the so-called displacement current (in contrast to the conduction current that flows via a conductor which we are familiar with in our day-to-day household use of light and the fan).

By the first half of 1800, the rigid separation between electricity and magnetism was breaking down as scientists started to realize that vibrating/time-varying electric fields could create magnetic fields and vice versa. For example, just by shoving a bar magnet into a coil of wire, one can generate a small electric current in the wire. Thus, a changing magnetic field is capable of generating an electric field. Similarly, one can reverse the demonstration just by running an electric current through the wire when a magnetic field will be produced around the coil. This same principle illustrating that the changing electric field can produce magnetic fields, and vice versa, is used for generation of electricity in hydroelectric/thermal/nuclear power stations.

Until 1860, the coupling of electric and magnetic field was not clearly understood and also could not be mathematically related. But an obscure 30-year-old Scottish physicist and gifted mathematician at Cambridge University, James Clerk Maxwell took the challenge to combine the electric and magnetic fields in a coupled framework—henceforth known as electromagnetism. Through mathematical calculations and physical visualization, he argued that it might be possible for electric and magnetic fields to vibrate together in precise synchronization, creating one another. This chain of vibrating/oscillating electric and magnetic fields was able to generate a wave called the *electromagnetic wave* (Fig. 1.9). This figure shows that the electric and magnetic field vibration takes place in the *x-y* plane (called the wavefront), perpendicular to the direction of propagation, *z*. In the wavefront, the electric and magnetic fields, E_x and H_y, are also mutually perpendicular to each other. This type of

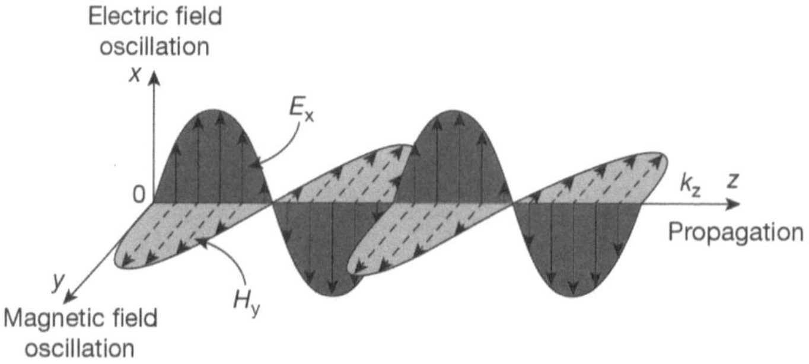

FIGURE 1.9
Oscillating electric and magnetic field vibrating synchronusly with eachother generates a propagating electromagnetic field.

electromagnetic wave propagation in free space is known as transverse electromagnetic wave (TEM) propagation; that is, both the electric and magnetic fields are perpendicular (transverse) to the direction of propagation of the electromagnetic wave.

Maxwell described the physical visualization and mathematical foundation of electromagnetism and electromagnetic wave in terms of his four equations of electrodynamics, known as Maxwell's equations, notably:

$$\nabla \cdot \vec{D} = \rho \text{(Gauss' law)}$$

$$\nabla \cdot \vec{B} = 0 \text{(non} - \text{existent magnetic monopole)}$$

$$\nabla \times \vec{E} = -\frac{\partial \vec{B}}{\partial t} \text{(rotational electric field, Faraday's law)}$$

$$\nabla \times \vec{H} = \vec{J} + \frac{\partial \vec{D}}{\partial t} \text{(rotational magnetic field, Ampère's law modified by Maxwell)}$$

In the above equations, $\vec{D}(=\varepsilon\vec{E})$ is the electric displacement, and $\vec{B}(=\mu\vec{H})$ is the magnetic flux density, where ε and μ, respectively, represent the permittivity and permeability of the medium. (In simple terms, the permittivity of a medium/material is a measure of how easily the electric field lines can pass through the medium/material, and permeability is the similar term applicable to magnetic field.) E and H are, respectively, electric and magnetic field intensities, and J is the current density, while ρ is the electric charge density in the medium/material. The symbol ∇ is called the *del* (or *nabla*) operator. The del operator is a vector operator whose dot (·) product is the divergence of the physical vector quantity (E or H), while the cross (x) product is the rotation/curl of the physical vector quantity.

The first Maxwell's equation represents *Gauss' law* for electric charges, which states that the total outward electric flux (i.e., divergence) through any surface surrounding charges is equal to the amount of charge being enclosed by the surface. The second Maxwell's equation, which is commonly known as Gauss' law for the magnetic field, is very well known to us from our common experience, showing that there are no isolated magnetic poles. In other words, the magnetic field is said to be *solenoidal* (i.e., the lines of magnetic flux close upon themselves between the south and north poles of the magnet, having the appearance of a tube or pipe, meaning the Greek word *sol-een*), which means the nonexistence of the magnetic monopole. The last two equations are known as Maxwell's curl equations, of which the first curl equation with the electric field comes from *Faraday's law*. The second curl equation (i.e., the fourth Maxwell's equation) is a modification of *Ampère's circuital law*, in which $\partial\vec{D}/\partial t$ represents a current, called the displacement current (which flows in the dielectric material of a capacitor unlike the conduction current, which is known to flow in a conductor). Ampère's circuital law states that the circulation of the magnetic flux density in a nonmagnetic medium around any closed path is proportional to the total current (J) flowing through the surface bounded by the path. It may be seen that in the time-varying case (unlike the static case, in which the physical quantity has no time variation), both the electric and magnetic fields become *rotational*; that is, they have nonzero curl. From our day-to-day experience, it may be observed that a whirlpool (which is created when water flows vigorously from the tap into the waste hole of a wash basin) is the physical example of curl, while a faucet (from which water rushes out in a pipe) is the example of divergence.

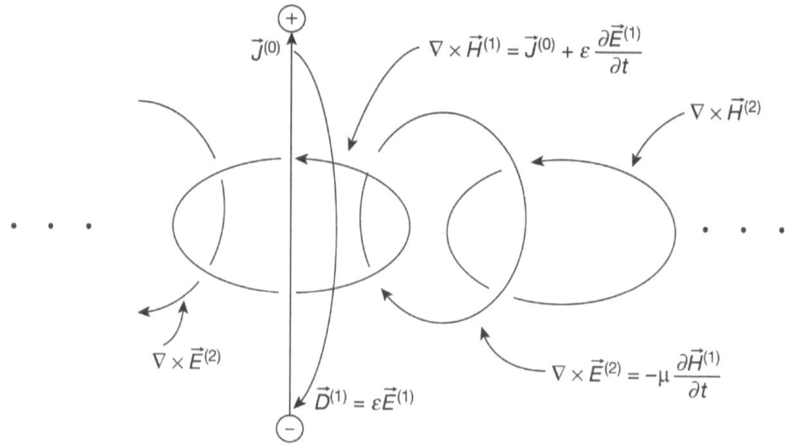

FIGURE 1.10
Conceptual representation of the propagation of electromagnetic waves based on Maxwell's two curl equations.

It may be noted that the circulation (i.e., curl) of a vector may exist, even when the divergence of the vector is zero (e.g., compare the second and fourth Maxwell's equations for magnetic field).

Maxwell's equations implied the existence of *electromagnetic waves* (which Maxwell established from his two curl equations), a phenomenon whereby electric and magnetic fields propagate from one spatial point to another (Fig. 1.10), at a finite speed and turned out to be the speed of light. Maxwell's wave equation thus established that light is simply a wave rippling through a medium via the electromagnetic field (which Einstein in 1905 showed to be aided by the flow of the quantum of particles called photons; in fact, in one second, 10^{20} photons aid our vision of an object by visible light). Action at a distance (as was implicit in Newton's law of gravitation) was thus conclusively refuted. The theory of classical electromagnetism was completed in 1862 with Maxwell's equations.

These equations appeared in Maxwell's *A Treatise on Electricity and Magnetism*, published in 1873. The universal importance and impressive beauty of Maxwell's equations led Ludwig Boltzmann to ask: "Was it God who wrote these equations?" Maxwell's equations are so unique that they do not need relativistic correction (Newtonian mechanics, on the other hand, does need that correction). In fact, they also provided Einstein with the clue he needed to propound his special theory of relativity.

Maxwell's equations are equally applicable for metamaterials or left-handed materials (LHM). These are human-made artificial materials of metal-dielectric composites that can mimic the natural or right-handed material (RHM) with artificially fabricated electric and magnetic atoms. First, metamaterial was made in 2001 in University of California, San Diego (UCSD). Such artificial material can give counterintuitive phenomena like the reversal of Snell's law, the reversed Doppler Effect, and the reversed Čerenkov Effect, and they are popularly known to make things invisible like Harry Potter's invisibility cloak. Incidentally, the first plasmonic metamaterial in India was successfully made and tested by our group at Calcutta University in collaboration with SAMEER, Kolkata Centre, and BARC, Mumbai, in 2009, refer to: http://www.nature.com/nindia/2009/090820/full/nindia.2009.273.html.

Maxwell's equations play the same role in electromagnetism as Newton's equations of motion and the law of gravitation in mechanics and planetary motions. In 1931, during the centenary celebration of Maxwell's birth, Max Planck commented: "This (Maxwell's) theory remains for all time one of the greatest triumphs of human intellectual endeavor." And Einstein commented: "The work of Maxwell was the most profound and the most fruitful that physics has experienced since the time of Newton." Most scientists agree that the 19th-century physicist James Clerk Maxwell, had the greatest influence on 20th-century science and can be given the same ranking as Newton and Einstein for his accomplishments in the field of electromagnetism.

Electromagnetic force between electric charges or magnetic poles is one of the four fundamental forces of nature: Gravitational force, electromagnetic force, weak nuclear force, and strong nuclear force. Most of the phenomena that occur around us can be understood in terms of electromagnetism or electromagnetic force. It is that force which makes chemical forces possible that hold together matter forming solid bodies, atoms in molecules, and also electrons in atoms. Even the forces describing processes occurring in the cells of organisms have their origin in electromagnetic force. This force operates in the neurons of our brain and governs our processing of the information that helps us to perceive the world around us and so on.

Like gravitational force, the electromagnetic force also acts over large distances and does not need any intervening medium. However, electromagnetic force is enormously stronger than gravitational force. The electric force between two protons, for example, is 10^{36} (trillion, trillion, trillion) times the gravitational force between them, for a fixed distance. It may further be noted that the Coulomb force ($F = k|q_1q_2|/r^2$) between two electric charges q_1 and q_2, and gravitational force ($F = G(m_1m_2/r^2)$) between two masses m_1 and m_2 follow a similar inverse square law; the force decreases as the square power of distance (r) between the two charges or the two masses. But gravitational force has only one sign (it is always attractive), whereas the Coulomb force can have both signs (attractive and repulsive, depending on the charges q_1 and q_2 being one positive and one negative, or both positive/negative), thereby allowing the possible cancellation of electric forces. This is why gravity, despite being a much weaker force, can be a more dominating and pervasive force in nature.

1.5.4 Fallout of Maxwell and Faraday's Electromagnetism

The most important and exciting outcome of Maxwell's equation is the prediction of wave-like disturbances in the electromagnetic field, giving rise to an electromagnetic wave that travels at a fixed speed, the speed of light. In 1864, Maxwell wrote: "This velocity is so nearly that of light that it seems we have strong reason to conclude that light itself (including radiant heat and other radiations) is an electromagnetic disturbance in the form of waves propagated through the electromagnetic field according to electromagnetic laws." The fact that this light travels at a finite but very high speed was first discovered by Danish astronomer Ole Christensen Roemer in 1676 by observing the eclipses of the Moons of Jupiter (by Jupiter), though his measured value was not very accurate (1,400,000 miles per second). Hippolyte Fizeauin 1849, through his interferometric measurement, estimated the value of the velocity of light c to be 313,300 km/s (194,700 mi/s), which is within 5% of the correct value of c = 299,792,458 m/s in vacuum presently used by international agreement. Note that a meter is defined as the lengths of the path traveled by light in a vacuum during a time interval of 1/299,792,458 second. Further, it is also the speed at which all massless particles like photon and field perturbations travel in a vacuum, including electromagnetic radiation (of which light is a small range in the frequency spectrum) and gravitational waves.

In this context, it is worthwhile to discuss the reason for using this particular value of the speed of light (c) universally for all precise calculations in physics and astrophysics. We all know that in modern physics the value of the speed of light c in a vacuum is an important cornerstone in every other formula, a common example being Einstein's mass-energy conversion formula, $E = mc^2$. The actual value of c used for all these purposes is $c = 299{,}792{,}458$ m/s (though in our school and college days, we were habituated to use $c = 3 \times 10^8$ m/s). This is because the velocity of light, c, directly relates to the *fine structure constant* (α), a fundamental constant of physics. Our model of the universe has chosen the fine structure constant to be 1/137 (approximately 0.007) and nothing else. If its value is changed, it will interfere with numerous processes taking place in our universe. In other words, if the speed of light in vacuum is taken as anything other than the one mentioned above, in that case all the laws of physics and the whole universe would be surprisingly different from the one we are currently living in. Even though quantum field theory indicates that vacuum is not absolutely an empty space and that in fact vacuum does have some quantum fluctuations, which might change the value of c by a very small amount, even then physicists/astrophysicists cannot compromise with the value of c in vacuum from its fixed and long-used value of $c = 299{,}792{,}458$ m/s.

Let us discuss here a few words about the "fine structure constant" mentioned earlier. The universe we live in is controlled by a magic number called the fine structure constant (α), with a value which nearly equals 1/137 or 1/137.03599913 (=0.007297351), to be precise. The equation for α is dependent on Planck's constant (h), velocity of light (c), and electronic charge (e); and as the units of h, c, and e cancel each other, α becomes a "pure number." The number α quantifies the gap in the "fine structure" of the spectral lines of the hydrogen atom (from which it derives its name), that was first introduced by Arnold Sommerfeld in 1916 and its value has been measured precisely by a number of experiments. Appearing at the intersection of such key areas of physics as relativity, electromagnetism, and quantum mechanics is what gives 1/137 its allure. The so-called magic number (α) preoccupied the mind of many great physicists, including the Nobel laureate Wolflong Pauli who was obsessed with it his whole life and commented: "When I die my first question to the Devil will be: What is the meaning of the fine structure constant?"

From Maxwell's equation we now understand that the electromagnetic wave is constituted of mutually connected time-varying electric and magnetic fields curling (rotating) around each other and propagates in free space with the velocity of light. Yes, all the modern communication technology we use in our daily life—radio, television, mobile and fixed-line telephone, satellites, GPS, radar, Wi-Fi, internet—all these systems use electromagnetic wave predicted by Maxwell's equations. The antenna/aerial we use in all these electronic communication systems acts as a transducer to convert electromagnetic signal to electrical signal and vice versa.

The insight behind the electromagnetic waves predicted by Maxwell's equations establishes that Faraday's so-called lines of force or field-lines can tremble and undulate, just like the waves of the sea. His computation of the undulation of Faraday's lines turned out to be exactly equal to the velocity of light. Thus, Faraday's physical vision of the electric and magnetic field in terms of lines of force and Maxwell's mathematical genius of the prediction and calculation of the velocity of electromagnetic wave produced the brilliant 19th-century discovery that *light is an electromagnetic wave*! The light that allows us to "see" something near to us or at a distance does not take place without anything mediating between the observer and the object. The mediating agent is this rapid vibration of the spiderweb of Faraday lines. When we see some object, it is through this lake of vibrating lines that transport the image of the object to us. Now, what is the color that we see in the world around us? It is simply caused by a different frequency (the speed of oscillation) of the electromagnetic wave (the spiderweb of Faraday lines through which the interwoven electric and magnetic field acts) that the light is. If the wave vibrates more slowly, the light is redder. If it vibrates a little more rapidly, it is bluer. Slow vibration means lower frequency or larger wavelength (the distance between one wave crest and the next). The color we perceive is the psychophysical reaction of the nerve signal generated by the receptors of our eyes, which distinguish electromagnetic waves of different frequencies.

The marvel of Maxwell and Faraday's work on electromagnetism comes from the application point of view when we find that Maxwell's equations are equally applicable for the situation when the waves vibrate more slowly than light (as for infrared, microwaves, and radio waves) and also for waves vibrating more rapidly than light (as for ultraviolet, X-rays, and gamma rays), all of them being electromagnetic waves. If the wavelength of these waves is a meter or more, we call them radio waves. Shorter waves are known as microwaves (having a few centimeter wavelengths) or infrared (with wavelength of the order of ten thousandth of a centimeter). Visible light has a wavelength of between only forty and eighty millionth of a centimeter. Even shorter wavelengths are known as X-rays and gamma rays (see Fig. 3.4, Chapter 3). Just for a comparison, X-rays have wavelengths 10,000 times shorter than those of light, while radio waves have wavelengths a million times larger than those of light. X-rays typically behave like high-energy particles (photons) and are thus most easily detected with Geiger counters in which the X-ray photons hit atoms, knocking electrons off them.

Radio waves almost always behave like waves of electric and magnetic force and are thus most easily detected with wire or metal aerials/antennas in which the waves' oscillating electric force pushes electrons up and down, thereby creating oscillating signals in a radio receiver attached to the antenna. The electric current with which our light bulb and fan works is caused by the shaking of an electric charge *here* (at the switch), producing the so-called transient, which produces a wave (a propagating disturbance) at the speed of an electromagnetic wave (i.e., with the speed of light) and drives an electric current *there* (at the bulb or tube and the fan). Is it not the marvel of electromagnetism whose founding fathers were Maxwell and Faraday? By the way, it may be noted that although we describe electric current as the flow of free electrons, that does not mean that electrons physically move with its so-called drift velocity from the switch to the bulb or fan. This is because if we calculate the typical drift velocity of free electrons in a conductor carrying an electric current, it will come out as only a few meters per hour!

Only a few years later, the electromagnetic waves that Maxwell had theoretically anticipated were experimentally generated in 1889 by Heinrich Hertz, a German professor of physics and a gifted experientialist. For example, he generated electromagnetic waves nearly 66 cm wavelength with his spark-gap generator. Hertz was a very modest person; after the discovery of electromagnetic waves made possible by his experiment, he commented: "This is just an experiment that proves Maestro Maxwell was right, we just have demonstrated these mysterious electromagnetic waves that we cannot see with the naked eye" (except visible light; none of the electromagnetic waves like radio wave, microwave, infrared, ultraviolet, X-ray, or γ-ray can be seen). Hertz also proved experimentally that these invisible electromagnetic waves (with a 66 cm wavelength), the so-called wireless waves, show reflection and refraction, and, most importantly they traveled at the same speed as that of light, as Maxwell predicted.

A few years later, the Italian-born engineer Guglielmo Marconi, working in England (in his own words as "an amateur in radio," though this was far from truth), foresaw the business side of wireless telegraphy. He was the first to transmit intelligible signals by means of Hertzian waves (around 350 m wavelength) via radio-telegraphic communication. That is, communication was achieved not by electric currents along wires as in ordinary telegraphs, but by radiating electromagnetic waves (hence the name "radio") through free space (from transmitting antenna to receiving antenna). On December 12, 1901, Marconi received the radio signal (corresponding to the Morse code "S" of telegraphy) at Signal Hall in Newfoundland, which was sent by his team workers located in Cornwall, England, 1,700 miles away on the other side of the Atlantic Ocean. History has credited him with the invention of an early form of radio (i.e., wireless) telegraphy, and in 1909 he won the Nobel Prize in Physics jointly with Karl Ferdinand Braun "in recognition of their contributions to the development of wireless telegraphy." Thus, Maxwell's prediction of electromagnetic waves and Hertz's generation of such waves with a simple experimental arrangement ultimately evolved into the vast industry of wireless technology—with radio, television, mobile phone, GPS available for daily use, including radar for both civilian and military use.

Even our present-day teleconferencing via Skype and our smartphone all became possible with this electromagnetic wireless technology. In our everyday experience, when we tune in to our favorite FM radio station, say, 92.7 MHz on the dial of our radio set, thanks to Maxwell's electromagnetic visualization, the electric and magnetic fields interwoven in that radio wave are turning into each other at the rate of 92.7 million times per second! Incidentally, I completed my BTech., MTech., PhD (Tech.) from the Institute of Radio Physics and Electronics, University of Calcutta, India. The founder of this Institute, Professor Sisir Kumar Mitra, was a pioneer in radio research in India, in whose name one of the craters in the Moon has been named—the Mitra Crater. This lunar-impact crater is attached to the western outer rim of the larger crater Mach, on the far side of the Moon.

In this connection, during 1895–1897, Jagadish Chandra Bose of Calcutta, India, also did some pioneering research work on the properties of radio (wireless) signals like reflection, refraction, and polarization (similar to light waves) at millimeter-wave frequency (60 GHz, i.e., having 5 mm wavelength) during 1895–1897. Bose's work in radio (in his own words "invisible light") was specifically directed toward studying the nature of the phenomenon of such waves and its applicability for remote control, but with no known attempt to develop any radio communication system as such. He made remarkable progress in his research of remote wireless signaling. In November 1895, he presented a public demonstration at Calcutta Town Hall—igniting gunpowder and ringing a bell at a distance using a millimeter-wave signal with his radio equipment. He also presented the instrument used for wireless experimentation in the Friday discourse of the Royal Society, UK, in 1897 in the presence of giants such as Lord Kelvin. The Institution of Electrical and Electronics Engineers (IEEE) named him as one of the fathers of radio science. Bose was the first to use a metal-semiconductor junction to detect 5 mm waves, and he used various other millimeter-wave components in his radio equipment resembling the present-day horn antenna, waveguide, and the like. Nevill Francis Mott, a Nobel Laureate in 1977 known for his own contributions in solid-state electronics, remarked that "J.C. Bose was at least 60 years ahead of his time." Bose subsequently switched over to research on plant physiology and made a number of pioneering discoveries in that domain. He used his own invention, the Crescograph, to measure plant response to various stimuli and thereby scientifically proved the parallelism between animal and plant tissues.

1.6 A New Dawn in the Making for the 20th Century

1.6.1 Introduction

By the late 19th century, all the disparate phenomena of the physical universe could be explained beautifully by a handful of simple laws: Newton's laws of motion, including his universal law of gravitation, and Maxwell's electromagnetic field equations. Newton's laws helped to understand the motion of car on the road, the orbiting of planets around the Sun, ebb and flow of ocean tides and so forth. Similarly, by manipulating the simple set of electric and magnetic laws and Maxwell's electromagnetic field equations, physicists could explain lightning, radio waves, propagation, reflection, diffraction of light, and so forth. Fame and fortune awaited those who could harness the Newtonian laws and Maxwell's electromagnetism for technology. By manipulating the Newtonian laws of heat, James Watt discovered how to convert a primitive steam engine devised by others into the practical device that came to bear his name. Similarly, by using Joseph Henry's understanding of the laws of electricity and magnetism, Samuel Morse devised his profitable version of the telegraph. Marconi, benefiting from the work of Hertz and Maxwell, succeeded in making the first transatlantic radio telegraphic communication in 1901.

1.6.2 The Aether (Ether) Prejudice till the Late 1990s

So far so good. However, the physics theories of the late 19th century were prejudiced by the Newtonian concept of absolute space and absolute time (which was also Aristotle's concept). With the prevalent idea that sound waves are possible because air vibrates (sound cannot propagate through a vacuum) and ocean waves are possible because water ripples—that is, they require a "medium" to transmit its wave motions—so light was also understood to require a medium, the "luminiferous aether" (commonly known as ether), to transmit its wave motions. During that period of time, it was supposed that everything around us, including vacuum, was filled with ether—an all-pervading elastic medium. Even the electromagnetic waves predicted by Maxwell supposedly describe the motion of waves through the ether (a hypothetical medium introduced by Christiaan Huygens to explain the wave nature of light that can propagate even through vacuum).

According to Newton's concept of absolute space, a person at rest in the so-called absolute space will measure the same speed of light in all directions. But if one is in motion, say, in the northward direction, then he or she should see the northward-propagating light slowed down and the southward-propagating light speeded up, just as a person on a northbound train sees northward-flying birds slowed and southward-flying birds speeded up. In reality, birds of any particular species move at the same speed by beating their wings, but it is the air through which they are flying that regulates their flight speed. Similarly, for light it was understood that ether (which is at rest in the absolute space according to the Newtonian concept) regulated the propagation of the speed of light. Thus, anyone at rest will measure the same speed of light in all directions, while anyone in motion will measure different light speeds. By beating its electric and magnetic fields against the ether, light always propagates at the same universal speed through the ether, regardless of its propagation direction. Now, Earth moves through absolute space around the Sun in one direction in January, and then in the opposite direction 6 months later, in June. Correspondingly, the speed of light is expected to be different in different seasons—though the difference is very small (about one part in 10,000) because the Earth moves very slowly as compared to light.

Newton's foundational concept of absolute space stood firm, producing one scientific triumph after another for more than two centuries; from the domain of the Heavens to the domain of Earth; there was no sign of a crack in that solid foundation. However, in 1881, Albert Michelson's experiment conducted to verify the speed of light with respect to the motion of the Earth in two different directions crumbled the whole edifice of Newtonian physical laws.

1.6.3 The Michelson-Morley Experiment—The Concept of Space Contraction

In the later part of the 19th century, scientists began to ponder the possibility that if the ether was real, at least its effect might be detected with a precision measurement. Albert Michelson, a 28-year-old American, using his self-invented experimental technique known as Michelson's interferometry, observed no variation in the speed of light with direction. In his initial experiment in 1881, the speed of light turned out to be the same in all directions and in all seasons. This experiment was repeated with much higher precision in his work with a chemist, Edward Morley, in 1887. The Michelson-Morley experiment of 1887 looked for the effects of an ether wind that would blow past the Earth as it moves in its orbit around the Sun. Light should appear to move more quickly when the ether is blowing toward us and more slowly when it is blowing away from us. Michelson and Morley found no such effect. Michelson reacted with a mixture of elation at his discovery and dismay at the possible consequences of this experimental result.

Heinrich Weber (a physics teacher and staunch believer and expert in Newtonian physics, about whom Einstein once commented that Weber's lectures were 50 years out of date) and most other physicists of the 1890s reacted with skepticism to the results of the Michelson-Morley experiment. Yes, it is easy to be skeptical, but to open up one's mind and to react with wisdom to a new but challenging outcome of a discovery is difficult. The Irish physicist George FitzGerald, however, the first to accept the Michelson-Morley experiment at its face value, speculated about its implications. His explanation of the null result was that the ether wind caused objects to compress. If objects (including all measuring devices) got smaller in exact proportion to the intensity of the ether wind, then the speed of light would always appear the same: When the wind is blowing toward us, our ruler will shrink, and when the wind is blowing away from us, it will appear to expand.

Hendrik Lorentz, Henri Poincare, Joseph Larmor, and others between 1896 and 1905 were also investing the problem with Maxwell's law of electromagnetism in two frames of reference, one at rest and another moving with relative speed. They observed that in the former frame of reference (i.e., the one at rest), the magnetic field satisfies the observed facts that magnetic field lines have no ends, i.e., they make closed loops, but in the later frame of reference (i.e., one moving with relative speed), a few field lines may get cut by the motion when we rely on Newton's notion of absolute space. Lorentz, Poincare, and Larmor solved this problem mathematically, contrary to the Newtonian precepts, by assuming that all moving objects get contracted along their direction of motion by precisely the amount that Fitzgerald needed to explain the Michelson-Morley experiment. However, Lorentz, Poincare, and Larmor were still prejudiced in favor of the concept of the ether and also had no idea about the so-called time dilation introduced later by Einstein in his special theory of relativity.

1.6.4 Birth of Einstein's Special Theory of Relativity

It was Albert Einstein, the patent examiner second class in the Swiss patent office at Bern, who was pondering over some anomalies observed when the Newtonian mechanics is applied to Maxwell's electrodynamics for bodies moving close to the velocity of light and was able to propound his special theory of relativity in 1905. He may not have been aware of the work of Fitzgerald, Lorentz, Poincare, and Larmor. In addition, he did not take any interest in the experimental results of Michelson and Morley, for he believed that experimental results are not very important to constructing the foundation of the "new physics" that he was contemplating. He relied on his innate intuition as to how things ought to behave. His conviction that the universe loves simplicity and beauty, and his willingness to be guided by this conviction, even if it meant destroying the foundations of Newtonian physics, led him—with a clarity of thought that others could not match—to his new description of space and time. In a true sense, Einstein was not a destroyer of the old foundations of physics; rather, he was a great creator, offering a new foundation that would replace the old one—a foundation that turned out to be in far more accord with the real picture of the universe.

Einstein totally rejected the Newtonian physical laws and proposed a revolutionary viewpoint: "there is no such thing as absolute space, there is also no such thing as absolute time. And as for the ether—it does not exist." He further presumed that the "speed of light is absolute and nothing can go faster than light." By making such a revolutionary presumption, he concluded that space and time are relative (in contrast, Newton assumed that space and time are absolute), and hence he found speed of light to be relative. Einstein further assumed that no one state of motion is to be preferred over any other; all states of motion must be equal, in light of the physical laws in the so-called inertial frame of reference. Einstein published his seminal, yet somewhat mundane titled article, "On the Electrodynamics of Moving Bodies" (which made Maxwell's equations consistent and beautiful in all inertial frames of references) published in *Annalen der Physik* in 1905. This formed the foundation for the special theory of relativity. However, Einstein reinvented some of the concepts and possibly was influenced by some of Lorentz and Fitzgerald's mathematics, but he showed that no ether was required, and so he discarded the belief that empty space cannot be truly empty. It may also be noted that the Michelson-Morley experiment did not, by itself, lead to special relativity, but for Einstein it was an important part of figuring out how the universe really works—which he ultimately established with his theory of relativity.

One of the central issues presented by the special theory of relativity is that in all (i.e., in different) inertial frames of reference, the laws of physics must assume the same mathematical and logical forms. "Different frames of reference" mean different states of motion. The inertial frames of reference are the special frames of reference that move under their own inertia, neither pushed nor pulled by any external force (including gravity). Thus, they continue always onward in the same state of uniform motion with which they began. In so doing, Einstein idealized our universe as one in which there is no gravity. If gravity is present, then the reference frame will no longer remain an inertial frame of reference because the mass of Earth, for example, will pull the frame of reference with the force of gravity (an external force). There is no way whatsoever to shield the reference frame from gravity's pull. Thus, the relativity theory he propounded in 1905 was called the *special theory of relativity* ("special" because it correctly describes the universe only in those special situations where gravity is unimportant or not considered). Extreme idealizations like this are central to progress in physics. Thus, Einstein first gained intellectual control over an idealized universe without gravity, and then he turned to the more difficult task of understanding the nature of space and time in our real, gravity-endowed universe. This task eventually led him to conclude that gravity curves/warps space and time—the golden fruit of the *general theory of relativity* ("general" because it generalized the relativity theory to include the effect of gravity). Einstein was a true genius. His intuition as a physicist was prodigious. Thus it was that in the beginning of the 20th century humankind was gifted with the most beautiful of theories, one such is the theory of relativity, which is discussed in Chapter 2.

2

The Golden Period: Two Master Strokes of the 20th Century—Relativity and Quantum Mechanics

Golden period of everything becomes most productive and has a lasting value

2.1 Introduction

Prior to the 20th century, the scientific knowledge needed to understand the mysteries of the universe was based on Newtonian mechanics and Maxwell's electromagnetism. Toward the end of the 19th century, scientists believed that they were close to producing a complete description of the universe. But stepping into the next century, two absolutely new terms appeared in the scientific lexicon: *relativistic mechanics* and *quantum mechanics*. These concepts have completely changed our world view henceforth. *Relativity* uncovered the secrets of space and time, energy, gravity, and dwells primarily in the cosmic world describing such exotic phenomena as warping or curvature of space–time, black holes, and Big Bang. Einstein's masterpiece, general relativity, sought to answer questions such as: Is there a beginning and an end to time? What is the farthest point in the universe? What lies beyond the farthest point? What happened at the moment of creation? And so forth. *Quantum mechanics*, on the other hand, is a theory of matter of the micro-world, describing the atomic and subatomic physics by uniting the dual concepts of waves and particles. Inventions such as the transistor, laser, and electron microscope and even our modern understanding of the periodic table of chemical elements are based on quantum mechanics. Quantum mechanics tries to answer the questions precisely opposite to those answered by relativity, for example: What is the smallest object in the universe? Can matter be divided into smaller and smaller pieces without limit? And so on. Quantum mechanics can be used to calculate only the behavior of atoms and subatomic particles, not the large-scale behavior of the cosmic world like the galaxies, expanding universe and so forth.

Einstein pioneered the first theory—relativity—and concentrated his efforts on understanding the nature of gravity and light. The foundations for understanding the nature of matter at the micro level, however, were laid by the second theory, quantum mechanics, which governs subatomic phenomena. It was primarily started by Niels Bohr and expanded on by Louis de Broglie, Warner Heisenberg, Erwin Schrödinger, Paul Dirac, and others.

DOI: 10.1201/9781003215721-2

2.2 Relativistic Mechanics

2.2.1 Introduction

With deep introspection into the anomaly between the Newtonian mechanics and Maxwell's electromagnetism regarding space and time, Einstein with his stroke of brilliance and a visionary's physical insight formulated the most important theory of the 20th century—*the theory of relativity or the relativistic mechanics.* According to Newtonian mechanics, the pulse of time beat uniformly throughout the cosmos. A clock on the Earth beat at the same rate as a clock on the Moon. Maxwell's equation, however, contained the clue that under certain circumstances one of the clocks would slow down. Scientists overlooked this strange (and seemingly absurd!) phenomenon in Maxwell's equation for nearly half a century. Only in 1905 did the then totally unknown young (26-year-old) Albert Einstein, a technical expert second class of the patent office at Bern, Switzerland, propose his revolutionary concept of the *special theory of relativity.* When Einstein first submitted his paper for publication, it was rejected by the reviewers, who were schooled in the notion of Newtonian physics. However, Einstein's outlandish but correct vision was ultimately published a year later. When evidenced with experimental support, the scientific community realized that the ideas in Einstein's paper contained a stroke of genius. The special theory of relativity demonstrated that space and time are interwoven into a single continuum known as space–time. The theory not only united space and time, but it also established the concept of the interconnection between mass and energy (Einstein's famous equation: $E = mc^2$)—one being the manifestation of the other.

2.2.2 The New Archimedes Is Born

While working at the Swiss patent office, Einstein found a quiet refuge and plenty of time to think over the queries he had harbored from childhood, especially regarding the mystery of light. The most important question that bothered him at the age of 16 was a thought experiment: What would a beam of light look like if one could race next to it at the speed of light (c = 1,86,000 miles per second or 3×10^8 m/s—light can travel around the Earth seven times in one second!). In terms of Newtonian mechanics (where velocity/speed is always considered with respect to the speed of something else), the speed of light with respect to the observer in motion will be the difference of the two velocities (the velocity of the observer and the velocity of light). When the observer moves at or close to the velocity of light, the light will thus appear to be *frozen* in time with respect to the observer, and the observer will see the stationary waves of electric and magnetic fields (that comprise light, as per Maxwell's electromagnetic theory). But this contradicts the prediction of Maxwell's equation that whatever the observer's velocity, he or she would measure a constant velocity (c) of light. Perhaps Maxwell himself realized this principle but was unable to probe the issue further since he died soon after proposing his theory, at the young age of only 48.

Einstein, however, grasped the singular importance of this fact, for he could visualize that this meant *we must change our notion of space and time.* Einstein's special theory of relativity overturned the notion of space and time that had dominated scientific thinking for thousands of years. Like all scientific geniuses, Einstein could foresee the underlying unifying symmetry in Mother Nature that links seemingly dissimilar entities—in his case, space and time, as well as mass and energy. Just as Newton could find the unification of terrestrial and celestial physics with his universal law of gravitation and Maxwell's electromagnetism united electricity and magnetism, so too was Einstein able to unite space and time in his relativity theory. For his part, Einstein modestly acknowledged the importance of Maxwell's theory and the mystery of light addressed by the theory in his statement "Special theory of relativity owes its origin to Maxwell's equations of the electromagnetic field." When the *Annalen der Physik* published Einstein's article on the special theory of relativity, the apparent contradiction between the equations of Maxwell and Newtonian physics was found to be addressed with extreme elegance: there was a momentous impact on the world of physics. One account of this impact goes like this: An austere professor of physics of Krakow University, Poland, came out of his dimly lighted study room, waving around Einstein's article and screaming, "The new Archimedes is born!"

Archimedes of Syracuse (3rd century B.C.), the greatest scientist of ancient times, a believer of rational thinking and a retaliator to those who remain satisfied with their own ignorance and belief that mysteries of the universe are *intrinsically* inaccessible to human thought. In his *Psammites* (The Sand Reckoner), Archimedes proclaims: "Some think, O King Hiero, that the grains of sand cannot be counted"—but he counted the grains of sand of the universe! In this work, Archimedes developed a new system of numbering (based on the power of the myriad, ten thousand) resembling our exponentials, which makes it possible to deal with very large numbers and shows its power by counting how many grains of sand there are not just on the seashore but in the entire universe; (In his count, 8×10^{63} grains of sand were required to fill the universe.). Archimedes had the zeal and bravery to oppose anything that he considered not rationally correct. In view of his opening statement in *Psammites*, we may mention the then prevailing and revered thought expressed in the book of *Ecclesiasticus* (which Roman Catholics consider to be part of the holy Bible).

Ecclesiasticus opens with the astounding questions: "Who can number the sand of the sea, and the drops of rain, and the days of eternity? Who can find out the height of the Heaven, and the breadth of the Earth, and the deep, and wisdom?" These thoughts of holy Bible are surely philosophically illuminating and poetically sublime, but regarding "infinite" science has its own view. The modern science establishes the principle that *there is nothing called infinity*. The cosmos is enormously vast, but surely it is a finite one. This message of modern cosmology resonates with Archimedes' rebellious thought. Unfortunately, centuries have passed, and the text of Ecclesiasticus, along with the rest of the Bible, can be found in countless homes, while only a few read Archimedes' text. (Archimedes was slaughtered by the Roman soldiers during the Roman invasion of Syracuse, the ancient Greek city and the last proud remnant of Magna Grecia to fall to the Romans.)

Archimedes is most famous for discovering the law of hydrostatics, which is popularly known as Archimedes' principle: a body immersed in fluid loses weight equal to the weight of the amount of fluid it displaces. In the layperson's terms, it can be called the law of flotation, which helps us to understand why even though a ship is heavy it floats easily in the sea water. A popular story about Archimedes goes like this: When Archimedes envisioned that physical law of nature, he emerged from his bath with unbound joy shouting Eureka, that is, "I got it!"

2.2.3 Most Beautiful of Theories

Lev Landau, the former Soviet Union's outstanding theoretical physicist, remarked that the general relativity is the most beautiful of theories. What is the general theory of relativity? Einstein himself was not fully satisfied with his special theory of relativity, even though

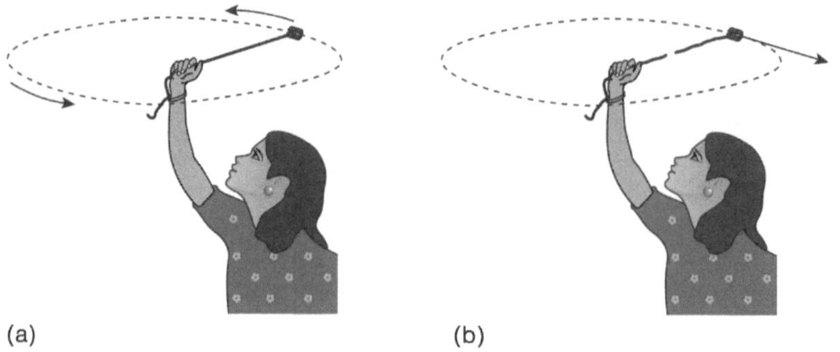

FIGURE 2.1
(a) Spinning of the yo-yo with the string follows the arrowed circular path; (b) the motion of the yo-yo will be along the straight path with the arrow marking when the string snaps.

it brought him worldwide acclaim and offers from innumerable universities. Einstein was bothered by the fact that the special theory of relativity was unable to accommodate the concept of gravitation, Newtonian or otherwise, as acceleration was not included in the special theory of relativity. Acceleration is caused by a net external force, but the special theory of relativity is based on the assumption of inertial frames of reference and is insulated from all forms of external forces, including gravity.

Let us first address the following question: If the Sun suddenly disappears, what will happen to the Earth or any other planet that revolves around it? According to Newtonian mechanics, the answer is that they will just fly into the deep space wildly in no time! This notion may be understood through a very common example of spinning a yo-yo with a string attached to it. If we spin the yo-yo in a circle around our head by pulling with a force on its string and if the string gets snapped for one reason or another, the yo-yo will move in a straight line (instead of the circular path) with a constant speed (Fig. 2.1). This is because the force with which we were pulling it with the attached string was deflecting it from this straight-line motion to go in a circular path. In the solar system, it's not a string but the Sun's gravity—the "ghost force" according to the Newtonian concept—which keeps the planets in respective circular orbits around the Sun. If the Sun suddenly disappears, the Sun's gravity is no longer there. Thus, according to classical Newtonian mechanics, the Earth will start moving in a straight line instead of in a circular orbit and will instantaneously get lost in deep space.

To Einstein this was unacceptable; nothing, including gravity, could go faster than the speed of light. It should take at least 8 minutes (to be more precise 8 minutes 20 seconds) for Earth to be wobbled out of its orbit in the solar system and be lost in deep space (because the time taken by light from the Sun to reach the Earth is approximately 8 minutes). For this reason, if the Sun were suddenly to wink out of existence, the space–time around it would go from being curved to being flat (as its mass/energy that caused the curvature of the space–time around it according to the general theory of relativity would be absent). But that transformation is not instantaneous, as the space–time is a dynamic fabric (according to Einstein's theory of relativity). Hence, that transformation has to occur in some snapping motion that would send very large ripples—the gravitational waves through the universe via the space–time fabric, just like outward propagating ripples in a pond. These gravitational waves will propagate with the velocity of light, and hence approximately for 8 minutes the Earth will still continue orbiting (the Sun?), after which it will be lost in deep space. Thus, through his relativity theory, Einstein felt that Newton's theory of gravitation must be incapable of understanding the above-mentioned problem in the right perspective (which makes no reference to the speed of light). But according to Maxwell's electromagnetism and his own special theory of relativity, the ultimate limiting velocity in the universe is the velocity of light. *The motion of anything in this universe faster than the velocity of light is impossible.* Thus, a new theory was required to deal with the issue of gravity as laid down in the Newtonian theory of gravitation and to make his relativity theory compatible with the problem of gravity. This led Einstein to propose his general theory of relativity.

Newton himself felt it *absurd,* as mentioned earlier in Chapter 1, that a force could act between bodies that were not in touch with each other and were so far apart (say Earth and the Moon or Earth and the Sun). However, a possible clue regarding the physical nature of such a force or field for magnetic poles and electric charges (and not for gravity of course) was visualized by Michael Faraday in terms of his *lines of force* acting between electric charges or magnetic poles. It may thus be thought that gravitational force might act with the gravitational field, but what would be its physical nature and how to develop the mathematical basis for that (just as Maxwell developed the mathematical basis of the electromagnetic field) made Einstein restless to find a genuine solution. Einstein immersed himself in the problem and with unceasing effort, after 10 years, in 1915, came up with the grand solution: The *general theory of relativity.*

2.2.4 Gravitation as per the General Theory of Relativity

According to Einstein's general theory of relativity, gravitation is not due to any ghost force, as Newton mentioned in his theory of gravitation but, rather, to the curving or warping of space–time around a body caused by its mass/energy. This notion may be better understood with an example that visualizes the physical perspective of such a statement. Imagine a rubber sheet (a child's trampoline, for example) in the form of a net with edges held high in the air by long poles and a cricket ball placed in the center of it (Fig. 2.2a). It is quite natural that the ball will press down into the fabric of the trampoline, as a result of which the trampoline net will dimple/sink to some extent, depending on the elasticity of the trampoline net and the weight of the ball. Now let us consider a small marble that is moving along the surface of the curved (sunken) net. It is easy to understand that the marble, instead of traveling in a straight path, will travel in a circular orbit around the depression caused by the ball at the center of the trampoline net.

As per the Newtonian gravitation principle, one can imagine that an invisible "force," a "ghost force," is acting between the marble and the ball (Fig. 2.1a). But according to Einstein's general relativity concept, the curvature/warpage (Fig. 2.2b) of space–time (trampoline net) by mass/energy of the Sun (ball) causes the Earth (marble) to move in a circular orbit with the Sun (ball) at the center of the space–time (trampoline net). This is the concept of gravitation as per Einstein that results in the apparent force balance in the solar system. Hence, according to the general theory of relativity, our whole concept of visualizing gravitation is thoroughly revolutionized, and now our understanding is that gravity is not a ghost force at all, but the bending/curving/warpage of space–time caused by the presence of mass/energy (of the Sun for the example above). If the ball (Sun) is suddenly removed from the trampoline net (space–time fabric), the undulation caused by its removal will propagate like a ripple (which is the gravitational wave in the cosmic canvas) along the surface of the net (space–time). A fraction of a second later (in the case of the Sun–Earth combination, it is 8 minutes), the ripple (the gravitational wave) will hit the marble (i.e., the Earth), and the marble's (i.e., the Earth's) course of motion will be altered. This may be the physical picture of the problem of what would happen if the Sun were suddenly to disappear from the solar system. The waves of gravitation (i.e., the *gravitational waves*), traveling

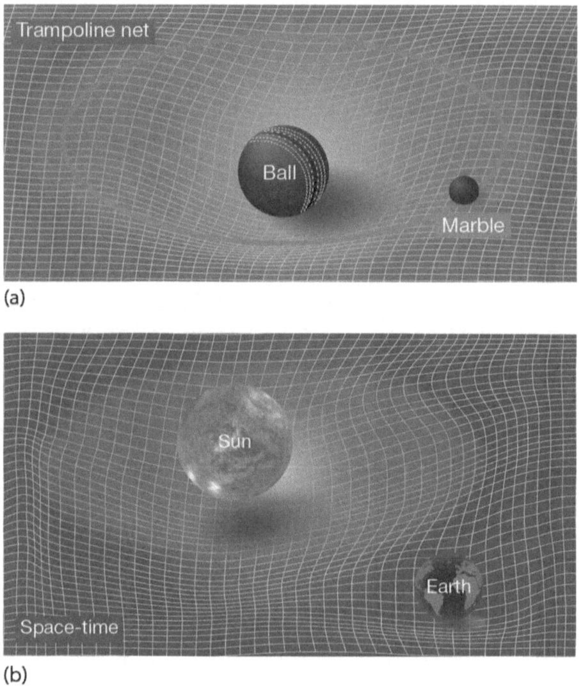

FIGURE 2.2
(a) A ball at the center of the trampoline net with a marble moving along the surface of the curved trampoline; (b) the curvature of space–time by the mass/energy of the Sun causing the gravitation to make the Earth revolve around the Sun.

at the speed of light (just like electromagnetic waves) would take 8 minutes to reach the Earth after the Sun disappeared. The theory of gravitation and the theory of relativity now become compatible with the concept of the general theory of relativity.

2.2.5 Experimental Proof in Support of General Theory of Relativity

Einstein's general theory of relativity was initially greeted with skepticism (as the revolutionary thoughts behind the special theory of relativity were already difficult to digest) by scientists the world over. It was predicted from the general theory of relativity that light rays would bend on passing a massive object, so that the apparent position of the stars would seem to shift slightly compared to the actual position if their light passed near the Sun. As per general relativity, Einstein calculated that the starlight passing close by the Sun should be deflected gravitationally by an angle of 1.75 arc-second.

After Einstein set forth his general theory of relativity in 1915, there were a few chances to test it: In 1916, when the outbreak of World War I interfered with a test; in 1918, when attempted observations were defeated by cloud cover; and on May 29, 1919, when the first successful test took place and the theory of gravitation changed forever in favor of Einstein. Sir Arthur Eddington, a famous British astronomer, masterminded an expedition that involved two teams, one in Brazil and one in Africa, to photograph and measure stellar positions during one of the 20th century's longest total solar eclipses: nearly 7 minutes (actually 6 minutes and 51 seconds) in duration. The results of those observations were compelling and profound. The bending was measured to be exactly

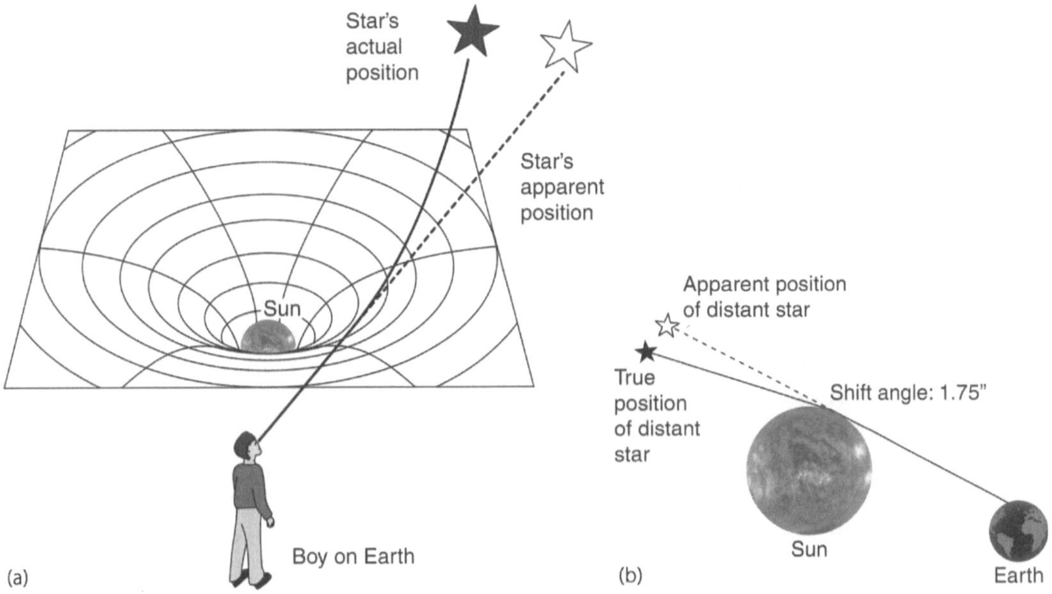

FIGURE 2.3
Bending of starlight due to the curving/warping of space–time in the vicinity of the Sun caused by its gravity: (a) as to an observer; (b) depiction of the theoretical result of relativity that matches with measured data.

1.75 arc-second; the value Einstein had predicted (but almost double the value predicted by Newtonian gravitation). This dramatically proved that Einstein's theory made a correct prediction; the light beam's path from the stars (like matter) should become bent when passing by the Sun (Fig. 2.3). This means that the Sun's vast mass/energy could somehow warp or curve the space–time, causing the starlight around the Sun to bend; this simple experimental result verified a great theoretical prediction.

The 1919 eclipse experiment established the first experimental foundation of the general theory of relativity, catapulting Einstein into fame. Newspapers around the world trumpeted this tremendous success. Even today, some of the world's best science writers are still publishing wonderful books on this remarkable achievement, which confirmed the prediction of general relativity first propounded by Einstein in 1915. Today general relativity is so well understood that it is used to weigh galaxies and locate distant planets according to how they bend light. Since light bends around heavy objects, we can see even some stars that are actually located behind the Sun, as shown in Fig. 2.3b.

Johann Soldner in 1804 had also pointed out that Newtonian gravity predicts that starlight will bend around a massive object, as Newton had already observed in 1704 in his Queries No. 1 in *Opticks*. However, that value of bending (0.87 arc-second) is half the value that is predicted from Einstein's general theory of relativity and that is proved by experimental observation. In all tests of Einstein's general theory of relativity, the theory has been borne out; not a single observation has disproved it. Einstein was so certain about the correctness of the principle and the equations he propounded for the general theory of relativity that he was not surprised by the results of the eclipse experiment. That year a student asked Einstein what his response would have been if the experiment had failed. Einstein's cool answer was: "Then I should have been sorry for the dear Lord, but the theory is correct."

A question that naturally arises is, how does gravity affect photons (that is, bend light that is made of photons) if photons have no mass? It is true that the photon has no mass (see Table 4.2, Chapter 4), and it is also true that we see light to get bent around bodies with large mass due to gravity—the phenomenon of the bending of starlight while passing close by the Sun. This is not because the gravity of the massive object pulls on the photons directly; rather, it is because the massive object curves the space–time around it significantly through which the photons travel. According to the theory of general relativity, all objects with mass warp (curve) space–time around them; the more massive an object, the more pronounced the warping it causes. When photons travel through the region near a massive object (the Sun, for example), which has caused significant warping of space–time around it, the otherwise straight path of travel of starlight (the photons) appears to follow a curved (bent) path because the space–time through which they are traveling is curved.

Today the distortion of light (electromagnetic wave) caused by gravity can be measured even in the laboratory without having to send light beams close by the Sun. In 1959 and again in 1965, Harvard professor Robert Pound and his colleagues showed that when gamma rays (a form of electromagnetic radiation) traveled a distance of 74 feet from the top of a building to the bottom, the force of gravity shifted their wavelength by an extremely small but finite amount (one part in hundred trillion)—the amount that is expected from Einstein's theory.

On April 20, 2004, NASA launched the so-called gravity probe (Fig. 2.4), a satellite-based relativity gyroscope experiment that verifies the prediction of Einstein's general relativity that a massive body warps the space–time around it. This Earth-orbiting satellite was placed at 650 km altitude, crossing directly above the poles, and carried four gyroscopes. With very precise measurement of the tiny changes in the spin of the four gyroscopes, the space–time curvature near the Earth (i.e., the *geodetic effect* caused by the mass of the Earth) was measured; initially with an accuracy of 1%. This accuracy was further improved to one better than 0.5% of the value predicted by the general theory of relativity.

Einstein's general theory of relativity recently passed another test. In 2020, a team of astronomers using the European Southern Observatory's Very Large Telescope, with nearly three decades of painstaking monitoring, reported observing a subtle shift in the orbit of the closest known star S2 to the supermassive black hole Sagittarius A^* (the superscripted sign [asterisk {*}] is used for a black hole) at the center of the Milky Way galaxy—and the movement matches Einstein's theory precisely. As per prediction from general relativity, the elliptical 16-year orbit of S2 would, with time, experience the so-called Schwarzschild

FIGURE 2.4
Geodetic effect (space–time curvature caused by Earth) measured with a gravity probe [2].

precession to trace out a Spirograph-like flower pattern of its orbit in space. That finding has been supported by recent astronomical observations. However, on the basis of Newton's classic description of gravity, S2 should continue along exactly the same path through space as it followed on its previous orbit. But it did not, thereby once again proving Einstein to be correct.

2.2.6 Newtonian Mechanics or Einstein's Relativity?

With the appearance of Eisenstein's relativity theory, do we need to abandon the Newtonian laws of physics? The answer is no, not necessarily. The domain of validity of Newton's laws, however, is restricted to the situations when the relative speeds of the objects are very small compared to the speed of light and the mass of the bodies concerned is not stupendously high (as of the black hole or so). Thus, Newton's laws are still valid and used extensively in everyday life in most fields of science and technology. We do not need to pay attention to the "time dilation" (of relativity) when planning an airplane trip, and engineers do not worry about length contraction (of relativity) when designing an airplane, for dilation, and contraction issues of relativity are far too small to be of any concern in our everyday life. But in particle accelerators where the particles (electrons/protons, etc.) move close to the velocity of light, in exotic phenomena of pulsars, quasars, and the black hole in distant parts of the universe, everywhere we need to resort to Einstein's relativity theory (as the relative velocities related to these cosmic phenomena are comparable to the velocity of light and the mass of the celestial bodies concerned results in significant space–time curvature). But even Einstein's relativity theory fails deep inside a black hole (at the so-called black hole singularity) and in the Big Bang, from which the universe was born and the beginning of time is reckoned. In such cases we may need to resort to the so-called quantum gravity (which is still a lofty theoretical framework understood and appreciated by only a few physicists; discussed in Chapter 3).

2.2.7 Comment by a Literature Laureate

Literature Nobel Laureate George Bernard Shaw, the Irish playwright famous for *Man and Superman* (1902), *Pygmalion* (1912), and *Saint Joan* (1923), among many works, while paying tribute to Physics Laureate Albert Einstein at a speech at the Savoy Hotel in London, on October 27, 1930, commented: "Ptolemy made a universe which lasted 1,400 years. Newton also made a universe, which have lasted 300 years. Einstein has made a universe and I can't tell you how long that will last." The latest image of the black hole taken by the Event Horizon Telescope on April 10, 2019, indicates that Einstein's general theory of relativity from which the prediction of the black hole was made is perhaps the correct one.

2.3 Quantum Mechanics

2.3.1 Introduction

By 1916, the scientific community was complacent that our knowledge about the universe is satisfactorily explained by the relativity theory (both the special and general theory propounded by Einstein) that takes care of space and time, including matter and energy, with a new vision of the concept of gravity. The classical theories of Newton and Einstein formed the bedrock on which the entire superstructure of physics rested until the 1920s. However, the discovery of quantum mechanics soon made clear that the fundamental physical reality in the micro-world cannot be addressed even with the most serious candidate of the macro-world, the relativity theory.

The two invaluable jewels of the 20th century—relativity and quantum mechanics—thoroughly revolutionized our concept of the physical world, respectively, in terms of the space–time continuum in the macro-world (i.e., in the cosmic domain) and wave–particle duality with in-built uncertainty and probabilistic nature in the micro-world (i.e., in the atomic and subatomic domains). We have already observed that the theory of relativity (both the special and the general one) was single-handedly developed by Einstein during 1905–1915 (and, of course, the mathematical foundations of great mathematicians like Hendrik Lorentz, Bernhard Riemann, and David Hilbert were extremely helpful for him in doing that). However, quantum mechanics was developed through the efforts and brilliance of a number of the great physicists of the 20th century, most notably Niels Bohr, Louis de Broglie, Werner Heisenberg, Erwin Schrödinger, Paul Dirac, Richard Feynman, and Max Born.

Toward the end of the 19th century, it was generally believed that the description of the laws of nature was in its final stage. But in the late 1800s and early 1900s, the scientific world was thrown into turmoil by a number of exciting new experiments such as the photoelectric effect, black body radiation, and radiation spectra (discrete emission lines) of the hydrogen atom. These experiments challenged the scientific knowledge of that time to account for those experimental observations with a satisfactory theoretical explanation. Thus, Mother Nature had the right to "laugh" at the overconfidence of those people who thought that they had uncovered all of the secrets it hid in its vast and mysterious storehouse. It was against this backdrop that quantum physics was born: Dealing with the behavior (the "mechanics") of particles of matter ("quanta")—in other words, the physics for the micro-world in the effort to obtain a plausible answer to all the questions related to those seemingly unaddressable experimental results (by the classical physics) and to open up an absolutely new way of understanding the science of the micro-world. In comparison with classical physics, quantum mechanics considers the properties of matter on a deeper and more fundamental level.

The development of quantum physics and the so-called quantum mechanics can be divided into three stages: (1) the seeding stage (end of 19th century–1912), the age of quantum physics based on the work of Max Planck and Albert Einstein; (2) Niels Bohr's quantum model of the atom (1913–1922); and (3) the firm establishment of quantum physics or more specifically the quantum mechanics by Louis de Broglie, Werner Heisenberg, Erwin Schrödinger, Paul Dirac, and others (1923–1927).

The consequences of quantum mechanics are all around us. The equations of quantum theory and their effects are used daily in a wide variety of fields: By physicists, engineers, chemists, and biologists. Quantum mechanics is capable of explaining conductivity in metal, semiconducting effects in Si, GaAs, and so on, that led to the invention of the transistor by quantum physicists William Shockley, John Bardeen, and Walter Brattain of Bell Laboratory in 1948. These physicists shared the Nobel Prize in Physics in 1956 "for their researches on semiconductors and their discovery of the transistor effect." We all know that the transistor is used to amplify or switch electronic signals and power. This discovery revolutionized electronics and made it possible to make cheaper and smaller devices than the old vacuum tube devices such as the triode and pentode valves.

John Bardeen received the Nobel Prize in Physics twice: in 1956, for co-inventing the transistor, and in 1972, for developing the theory of superconductivity. The superconductivity is a phenomenon which can only be explained by quantum mechanics. The superconducting phenomenon was first discovered on April 8, 1911 by the Dutch scientist Heike Kamerlingh Onnes who was studying the resistance of solid mercury at cryogenic temperature using liquid helium as refrigerant. At the temperature of 4.2 K, he observed that the resistance abruptly disappeared. In the same experiment, he also observed the superfluid transition of helium at 2.2 K, without recognizing its significance. Following this discovery, many scientists including Walther Meissner and Robert Ochsenfeld (1933), Brain David Josephson (1962) have contributed in many aspects of the phenomenon of superconductivity. However, the complete microscopic theory of superconductivity was finally proposed in 1957 by John Bardeen, Leon Cooper, and John Robert Schrieffer, known as the BCS theory—which earned them the 1972 Nobel Prize in Physics.

The superconductivity is about how extremely cold metals are able to conduct electricity with great efficiency (with practically no resistance offered to the flow of current). Superconductors have paved the way for the development of NMR spectroscopy and hence its medical subtool MRI is used for getting images of inner organs made of soft tissues (in contrast, X-ray imaging being used for imaging hard tissues or bones). Cryogenically cooled superconducting coils are also used to realize very high magnetic field (with large electric current and having practically no heating-up) required in circular particle accelerators. In fact the LHC of CERN, Geneva has used 23 km of superconducting magnets around its 27 km circumference of the collider. The Niobium–Titanium (NbTi) wires of the electromagnet's coil are kept at a cryogenic temperature of 1.9 K (–273.3°C) using liquid helium. However, for commercial use in mobile phones, laptops, and also for possible low-cost transmission and distribution of electricity, we would need superconductors operating at room temperature and at normal atmospheric pressure. A 2020 research on superconductors have reported the realization of its operation at 58°F but the superconducting material need to be squeezed to 267 gigapascals, or more than 2 million times the Earth's atmospheric pressure. If the temperature and pressure issues of superconductors get resolved then we might see a dramatic transformation in electrical technology too, making a second electrical revolution—the first electrical revolution, we all know, was brought about by Faraday in 1831 with his electrical generator.

In the absence of the transistor and the subsequent development of integrated circuits, modern electronic instruments like radio, television, computers, and the smartphone would not have been possible. Quantum mechanics led to the development of the laser, which we use for many industrial applications and optical communications, including medical applications such as eye surgery. Laser discs have changed the way stereo recording is done. Our children and grandchildren would probably watch three-dimensional television based on laser technology in their living rooms. Unquestionably, the success of quantum mechanics has altered the foundations of industry, commerce, and medical science in a way that was unimaginable in the beginning of the 20th century. But ironically, quantum mechanics, which seems so definitive and clear-cut in its practical applications, actually is based on uncertainties, probability, and philosophically bizarre ideas. In short, quantum mechanics dropped a bomb on the world of physics. Here, let us quote Niels Bohr in this respect: "Anyone who is not shocked by quantum theory; has not understood it." And Richard Feynman, who more than anyone has known how to juggle quantum theory, wrote: "I think I can state that nobody really understands quantum mechanics." "Quantum mechanics," he notes, "describes nature as absurd from the point of view of common sense and yet it fully agrees with experiment. So I hope you can accept nature as she is—Absurd." However, I think that the strangeness or the so-called obscurity of quantum mechanics, if any, is due to the limited capacity of our imagination. The limitations of our intuitive vision may make it appear as strange and crazy; in reality, its triumph of efficacy is stupendous!

2.3.2 The Formative Stage of the Quantum Concept

2.3.2.1 Max Planck's Contribution

The physicists of the 1890s were mystified by the so-called black body radiation—the electromagnetic radiation emitted by a body when it is heated. The term *black body* stands for an idealized opaque and nonreflecting body that emits radiation over all wavelengths. The experimental results (Fig. 2.5a) indicate that at a given temperature T the intensity distribution of electromagnetic radiation over wavelengths increases very sharply, passes through a maximum, and then falls gradually to zero. Further, with the increase of temperature, the maximum should shift toward lower wavelengths (a piece of iron when heated appears red at 500°C but bluish at 1000°C, as the wavelength decreases from red to blue; see Figs. 2.5a and 3.4). In 1900, Lord Rayleigh and James Jeans tried to apply classical mechanics to the problem of black body radiation and observed that the black body emits electromagnetic radiation, having a distribution over different wavelengths/frequencies given by: $B_\nu(T) = (8\pi\nu^2/c^3)k_BT$, where B is the spectral radiance or radiation intensity of the black body, ν is the frequency, c is the speed of light in the medium, k_B is the Boltzmann Constant, and T is the absolute temperature of the body. Although the Rayleigh–Jeans law worked fine for higher wavelengths/lower frequencies (10^5 GHz), it utterly failed for lower wavelengths/higher frequencies. At higher frequencies, it diverges as the square power of frequency, that is, ν^2. This divergence at higher frequencies when the radiation intensity approaches infinity at higher frequencies is called the ultraviolet catastrophe or the Rayleigh–Jeans catastrophe (Fig. 2.5b). Thus, if we proceed with the classical approach for black body radiation, then looking toward the Sun we will be blinded by the γ-rays

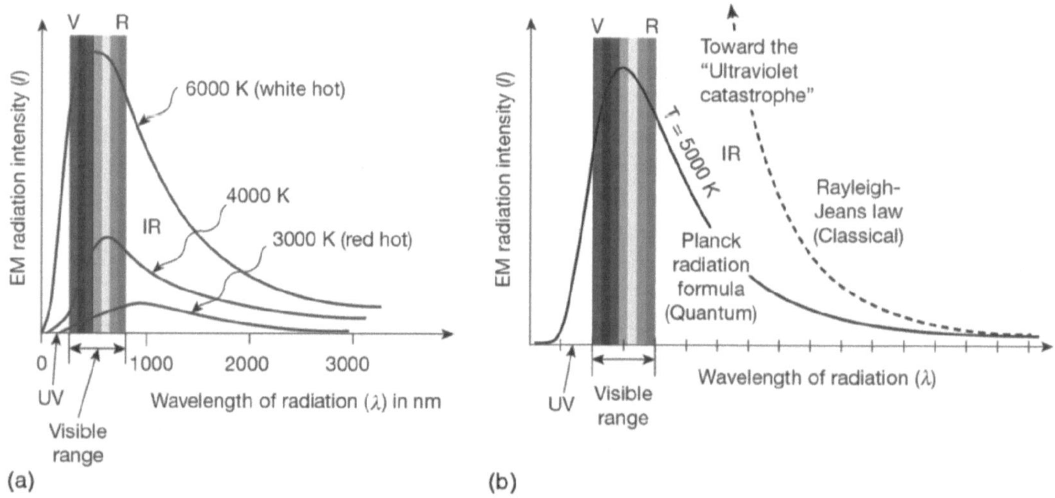

FIGURE 2.5
Black body radiation characteristics: (a) experimental results; (b) theoretical model curves by Rayleigh and Jeans (classical approach) and Planck (quantum approach).

(highest frequency and highly ionizing radiation) of the electromagnetic spectrum, which is practically not the case.

On December 14, 1900, the generally accepted date for the birth of quantum theory, Planck presented his theoretical results (Fig. 2.5b) for the black body treated as an ensemble of harmonic oscillators. With considerable reluctance, he postulated that electromagnetic radiation emitted by a black body can take place only in *quantized* form (an elementary unit or packet of energy) given by: $E = h\nu$, where h is known as Planck's Constant (which Planck termed the *elementary quantum of action*) and ν is the frequency (on the basis of which the color of light is determined) of the emitted radiation. According to his hypothesis, the emission of radiation from the black body can take place only in equal portions ("quanta") of energy $h\nu$, proportional to the frequency ν of vibrations of a single oscillator of the black body. It is exactly this hypothesis about energy quanta that led to the agreement of theory with the experimentally observed results and the elimination of the so-called ultraviolet catastrophe (compare Figs. 2.5a and 2.5b). This could explain why a bar of steel when heated to high temperatures becomes red hot at first and eventually becomes white hot, or it may explain why lava is red hot when it spews from an erupting volcano, or why the Sun emits light in the visible and ultraviolet region of the frequency spectrum. Thus, we are saved by the quantum nature of radiation, which says that light is made of particles (photons), as Planck assumed (and which is the *reality*).

Thus, Planck while searching for a suitable theoretical justification of the experimental result of black body radiation proposed that the radiation was not entirely wavelike, as the classical physicists of that time thought, but that the energy transfer takes place in the radiation process in definite and discrete energy packets, quanta of energy $h\nu$. Why Planck made such a leap-taking but ad hoc assumption is not known; however, in his own language he noted: "a purely formal assumption ... actually I did not think much about it. . . . " The physics community in Planck's time reacted with intense skepticism to his novel idea that light was not continuous but granular. The idea that light could be thought to be chopped into "quanta" that act like particles was considered preposterous at the time. As Planck could not offer a good justification for his assumption of energy quantization, his contemporaries did not take this energy quantization idea seriously until Einstein invoked a similar assumption to explain the photoelectric effect. Henceforth, the famous equation for quantized nature of electromagnetic radiation, $E = h\nu$, came to be known as *Planck's equation* or *Planck–Einstein's quantization equation*. Anyway, the famous Planck's Constant followed soon after when Planck tried to match the experimental results with the adjustable parameter h, the Planck's Constant, in his theoretical proposition of the quantum nature of electromagnetic radiation from a black body.

Planck's quantum theory for black body radiation with the radiation emitted as specific quanta of energy now accounts for the experimental observations of the above-mentioned specificity of red-hot and white-hot colors of the heated steel bar and for radiation from the Sun in the visible and ultraviolet region of the electromagnetic spectrum. However, it is realizable why in our common daily experience we cannot visualize the "granular" nature of radiation (i.e., the quantum effects) because the size of each such energy packet (quanta) is incredibly tiny (determined by the Planck's Constant: $h = 6.63 \times 10^{-34}$ J-s, which is astronomically small). This particular leap-taking assumption made by Planck is regarded as the birth signal of quantum physics and the greatest intellectual accomplishment of his career. Planck published his work, "On the Law of Distribution of Energy in the Normal Spectrum," based on quantum analysis of black body radiation in 1901 in *Annalen der*

TABLE 2.1

Planck Units in Modern Physics and Astrophysics

Quantity	Mathematical Expression	Value
Length (m)	$l_p = \sqrt{(\hbar G/c^3)}$	1.62×10^{-35}
Time (s)	$t_p = \sqrt{(\hbar G/c^5)}$	5.39×10^{-44}
Mass (kg)	$m_p = \sqrt{(\hbar c/G)}$	2.18×10^{-8}
Temperature (K)	$T_p = \sqrt{(\hbar c^5/Gk_B^2)}$	1.42×10^{32}

where ħ ("h bar" = h/2π) = 1.054571818 × 10^{−34} J-s is known as reduced Planck's Constant.

Physik. According to his analysis, the radiation intensity of electromagnetic energy from a black body is given by Planck's radiation law:

$$B_\nu(\nu,T) = \frac{8\pi\nu^2}{c^3}\frac{h\nu}{e^{(h\nu/k_BT)} - 1}$$

As shown in this equation, at lower frequencies when $h\nu \ll k_BT$, the exponential quantity $e^{(h\nu/k_BT)}$becomes $1 + (h\nu/k_BT)$. Hence, Planck's law reduces to the Rayleigh–Jeans law in the low frequency regime.

Planck's Constant is the "unit of quantization." Its presence in any expression implicitly indicates the quantum-mechanical nature of the expression; however, the converse is not true. It would be incorrect to attempt, as is sometimes done, to reduce the whole "essence" of quantum mechanics to the presence of Planck's Constant. In reality, quantum mechanics needs many more intricate details of the micro-world, which will be understood on the basis of an in-depth understanding of quantum mechanics later developed by de Broglie, Heisenberg, Dirac, Schrödinger, and others.

As shown in Table 2.1, there are Planck units like Planck length, Planck time, Planck mass, and Planck temperature, based on the fundamental or natural constants c (velocity of light), G (universal gravitational constant), h (Planck's Constant), and k_B (Boltzmann Constant).

One might ask: What is the use of such astronomically small or large units which are not at all conceivable in our remotest experience or knowledge whatsoever? Yes, it is true that Planck units have no practical application. No car odometer will be calibrated in Planck lengths, no stopwatch will tick off Planck times, and no thermometer will ever give temperatures as a teeny, tiny fraction of the Planck value. These numbers only tell us the limits of physics as we currently know it and may be even the limit of physics as it could ever be known—therein lies the importance of Planck units in physics and astrophysics.

2.3.2.2 Einstein's Contribution

In 1905, in one of his epochal papers, Einstein provided the explanation for the photoelectric effect in terms of Planck's quantum theory. The photoelectric effect was, however, first observed by Heinrich Hertz in 1887 while doing his electromagnetic wave experiment. Hertz observed that high-voltage sparks across the detector loop were enhanced when the emitter plate was illuminated by ultraviolet light from an arc lamp. Light shining on the metal surface somehow facilitated the escape of free charged particles, which we now know as electrons. However, Hertz did not proceed any further with understanding of the phenomena. Unlike Planck, who was a reluctant, almost timid revolutionary, and whose temperament was a typical 19th-century physicist; Einstein struck out boldly in new directions with his quantum theory of photoelectric effect. Einstein asked what happens when a particle of light (the light quanta—termed *photon*, the Greek word for light and a term coined by Gilbert N. Lewis in 1926) strikes a metal. If light were a particle (granular) obeying Planck's theory, then it should bounce the electrons out of some atoms of the metal and generate electricity (the so-called photoelectricity). Einstein, using Planck's Constant, calculated the energy of the ejected electrons using the formula $E = (h\nu - p)$; where p is a constant that depends on the characteristic of the metal. Today's solar cells used for generating clean electric power use the principle of photoelectric effect (to be more specific, photovoltaic effect), and the image sensors of the digital camera use this effect for its operation. The particle (i.e., corpuscular) nature of light in terms of the quantum concept was first established by Einstein. However, Newton also proposed the corpuscular nature of light but with a different perspective (classical approach) in his *Opticks*. In Newton's own words: "Are not the rays of light very small bodies emitted from shining substances?"

It did not take much time for experimental physicists to verify Planck's and Einstein's equations. Planck won the Nobel Prize in Physics in 1918 "in recognition of the services he rendered to the advancement of Physics by his discovery of energy quanta." Einstein won the Nobel Prize in Physics in 1921 for the photoelectric effect (in addition to his epoch-making contribution in formulating the relativity theory) "for his services to Theoretical Physics and especially for his discovery of the law of the photoelectric effect." However, after a large number of unsuccessful nominations, Einstein finally received the 1921 Nobel Prize in Physics (neither uniquely for his relativity theory nor for his light quanta).

2.3.3 Journey Toward Understanding the Micro-world—The Quantum Mechanics

The atomic model of Niels Bohr, who adopted the solar system like model of the atom (first proposed by Ernest Rutherford in 1911), introduced the concept of stationary orbits or quantum states from which radiation of photon can take place when electron transition occurs. This can be considered to be the beginning of the journey toward developing the *new quantum theory* and the subsequent development of *quantum mechanics*. With the help of his atomic model, Bohr was highly successful in accounting for the radiation spectrum of the hydrogen atom. However, his atomic model was based on analytical derivations that were a mixture of Newton's classical mechanics and Planck's quantum concept. Louis de Broglie henceforth introduced a radically new idea known as the *wave–particle duality* of micro-particles of matter like the electron and proton, drawing an analogy with the wave-photon nature of light, which could justify the existence of the stationary or quantum orbits of Bohr's atomic model. This can be seen as the true beginning of quantum mechanics. This development was followed by the *uncertainty principle* of Werner Karl Heisenberg, the *wave mechanics* of Erwin Schrödinger, and the consolidation of new quantum theory by Paul Dirac, who logically put together the *granular, relational,* and *indeterminacy* properties, which are in the genetic code of quantum mechanics. Others like Richard Feynman and Max Born also made significant contributions to the development of quantum mechanics. Details of their contributions are discussed in Section 2.7.

Throughout the 1930s, 1940s, and 1950s, quantum mechanics could be seen as an unstoppable Mack truck barreling down the highway, flattening all problems that had puzzled physicists for centuries. However, we should emphasize that quantum mechanics worked only when physicists used it to analyze micro-world phenomena where the velocity of particles is much lower than the speed of light. For example, the electrons in the hydrogen atom typically travel at speeds one hundred times less than the velocity of light. When attempts were made to include special relativity, the Mack truck of quantum mechanics encountered a tough brick wall too difficult to crack. If nature would have created atoms where the electrons traveled at velocities close to the velocity of light, special relativity would have dominated even in the micro-world, making quantum mechanics less successful.

On Earth natural phenomena rarely approach the speed of light, so quantum mechanics is triumphant in explaining everyday phenomena such as the transistor, laser, superconductivity, and so on. But when we need to analyze the properties of ultra-fast and high-energy particles in the cosmos (or in modern particle accelerators), quantum mechanics can no longer ignore relativity. In such situations, quantum mechanics must be married to relativity. Thus, the effort to develop the theory of quantum gravity and many more interesting theories emerged within the domain of modern physics, the details of which are discussed in Chapter 3.

2.4 Albert Einstein's Relativity—A New Cosmic Vision

2.4.1 Introduction

In the history of science, no scientist has ever attained the same popularity as Einstein. His fame far transcends the boundaries of physics: he is known not only to professional scientists and the students of science but also to people whose interests are remote from science. This popularity stemmed largely from the fact that Einstein's work played a revolutionary role in the development of physical knowledge of the universe and, moreover, touched on the most profound problems of the scientific world outlook with which all thinking persons are concerned. Einstein's scientific creativity and insights into the physical world made an enormous impact on the development of 20th-century philosophical thought, enriching almost all branches of knowledge related to science in some form or other.

Einstein had a longstanding liking for philosophy. He read the works of Democritus, Aristotle, Plato, La Mettrie, Spinoza, Berkeley, Hume, Mach, Kant, Russell, and others, but his thoughts were not shadowed by any single philosophical system. He believed that "the critical thinking of the physicists and intricate nature of modern physics cannot cope with its problems without philosophical vision." Einstein's special and general relativity theories played a major role in altering the scientific view of the universe (in terms of the space–time continuum and the origin of gravitation) and held a prominent position among the outstanding attainments of modern scientific thoughts. The picture of the universe founded on these theories is radically different from that of the classical physics that has come down to us from Aristotle to Newton. The impact of Einstein's scientific creativity on scientific thinking is enormous and revolutionary. Einstein worked out new standards for scientific knowledge, which further developed the Copernican tradition of rejecting anthropomorphic self-obviousness. It has enabled scientists to revise the traditional views and conceptions of the structure of the universe, revealing deep and close ties between philosophy and natural science. For this reason, neither physicists nor philosophers were indifferent to Einstein's work on relativity; both were attracted by its illuminating novelty.

Natural scientists saw the relativity theory as the resolution of the inner contradiction of classical mechanics (of Newton) and electrodynamics (of Maxwell), while dialectical materialists regarded it as a natural scientific confirmation of the matter and its attributes reflected in modern philosophical thoughts.

2.4.2 Einstein's Religious Thought

Einstein saw no reason to resort to religious dogma in explaining the mysterious phenomena in the universe around us. He believed that religion was historical in nature, emerging as it did at a certain stage in the development of human society. According to him, "A man who is thoroughly convinced of the universal operation of the law of causation cannot for a moment entertain the idea of a being who interferes in the course of events—provided, of course, that he takes the hypothesis of causality really seriously." A God who rewards and punishes was inconceivable to him for the simple reason that human actions are determined by necessity, external and internal, so that in God's eyes he cannot be responsible, any more than an inanimate object is responsible for the motions it undergoes. In a letter written just a year before his death, Einstein dismissed God in religion as "the incarnation of primitive superstition." In another letter written in 1954 to the German philosopher Erich Gutkind, Einstein wrote: "The word of God is for me nothing but the expression and product of human weakness, the Bible a collection of venerable but still rather primitive legends."

I strongly believe that Einstein was definitely not against the ethical values found in the religious teachings of Buddha, Jesus, and others that promote peace and universal understanding. Despite Einstein's apparently negative attitude toward the so-called official religions and the idea of God, he took refuge in the "cosmic religion"—a rapturous amazement at the harmony of natural law. According to Einstein, "cosmic religious feeling. . . can give rise to no definite notion of a God and no theology—it merely inspires a human being (especially a scientist) to perceive the loftiness and marvelous order of the universe." Further, he noted that "I believe in Spinoza's God, who reveals himself in the harmony of all that exists, not in a God who concerns himself with the fate and doings of mankind." It may be mentioned herewith that in Hindu philosophical scripture *Ṛig Veda* and also in the Pyramid texts of ancient Egypt, we respectively come across such terms as "Ritam" and "Re," which mean Universal Harmony—and the human soul is in incessant search of that unique rhythm of the universal one (possibly Einstein's "cosmic religion" was in search of that effulgence of universal harmony). Perhaps with this cosmic religion in mind, Einstein made a significant comment: "Science without religion is lame, religion without science is blind." However, this famous aphorism has also been the source of endless debate between believers and nonbelievers of religion who want to claim the greatest scientist of the 20th century as their own.

2.4.3 Albert the Rebel

Albert was a rebel even from his early age. His parents left him in Germany to attend the high school there, but he found the German school system too rigid and militaristic (like the stony-faced Prussian soldiers). Unable to stand the authoritative hardness of the school, he abandoned his studies in Germany. He joined his parents in Pavia, Italy, and spent his time loafing. Later he went to Switzerland for his studies, initially failing to gain admission to Zurich Polytechnic. While completing his studies there, he struggled to survive by part-time tutoring since he was rejected for a teaching position in the universities.

In 1902, through the recommendation of the father of his close friend, he got a humble job as a technical expert second class in the patent office in Bern, Switzerland, to support his wife and newborn child. Although Einstein was overqualified for the job, he accepted it as it served both his economic and intellectual purposes surprisingly well: The job gave him ample time to think and work independently in the isolation and quietness afforded by the patent office. Further, his job required him to isolate key ideas from the inventors that taught him how to think in terms of physical pictures and zero in unerringly on the fundamental ideas that make a theory work in practice.

2.4.4 Albert the Genius

From childhood, the mystery of the universe captivated Einstein. His most profound and fascinating experience came from his encounters with the unknown: "It is enough for me to make amazed surmises about these mysteries and to attempt humbly to form a limited impression in my mind of the perfect structure of all that exists." Einstein had the unique capacity to imagine how the universe might be constructed, and like Faraday, he was able to "see" all these novel ideas in his mind's vision. Afterward, the equation for him came to substantiate in analytical language the vision of physical reality that flashed in his inner mind. For Einstein, relativity was not merely a collection of equations: It was a mental image of the universe arduously translated into equations.

There is an interesting anecdote about how the concept of time dilation flashed in his mind. He was sitting in the Zurich railway station where there were a number of wall clocks on the walls of the platform. Suddenly the following thought flashed in his mind: If one of the clocks were to be placed in a frame of reference that was moving close to the velocity of light, would that clock register a different time with respect to the other clocks on the wall of the platform? This was reportedly the seed thought for his special theory of relativity—just as the falling of an apple from a tree was the seed thought for Newton's law of universal gravitation.

While working in the Bern patent office, Einstein sent four research articles to the scientific journal *Annalen der Physik,* each of which was worthy of a Nobel Prize or even higher recognition. The first article presented a new vision of the so-called Brownian molecular motion that deals with the motion of microscopically visible particles suspended in liquid; this work provided empirical evidence of the reality of an atom. The second article focused on the photoelectric effect and stated that luminous energy can be emitted or absorbed only in a discrete amount called light quanta or photon. This was a complete picture of the theory of photoelectricity (first observed by Hertz in 1887), which was given a firm theoretical foundation through Einstein's theory of photoelectricity. The third paper reconciled Maxwell's equations for electricity and magnetism with the laws of mechanics by introducing major changes to Newtonian mechanics close to the speed of light. This was Einstein's *special theory of relativity.* The fourth paper concerned the mass–energy equivalence in which Einstein deduced what is arguably the most famous of all equations in the history of humankind: $E = mc^2$.

2.4.5 Similarities of the Lives of a Genius and the Great Scientists

The life of a genius always bears some marked difference from that of the ordinary individual. If we look at Newton during his early 20s, we find some interesting similarities with Einstein's life during that period of their life. When Newton was a 23-year-old student at Cambridge University, he was sent home because the dreaded black plague of 1665 was sweeping the land, and thus the government closed down most of the universities and other institutions in Europe to maintain social distancing during the pandemic. As a result, Newton spent the years between 1665 and 1667 with his mother in the country house. With plenty of time on his hands, Newton thought deeply about the motion of objects that fell to the Earth and the planets that revolved around the Sun. And in a stroke of brilliance, he conceived his famous theory of universal gravitation. During this period, he also completed some base work on calculus, experiments on the prism that paved the way for the foundation of optics.

Einstein's relativity theory was also propounded when he was just 26-year-old, and that was also when he was working alone toward achieving a "new physics" in the lonesome environment of the patent office as a technical expert second class in the first decade of 1900. He was unaware of (or not interested in noting) the latest developments in physics during that period. Instead, he was engrossed in his revolutionary thinking about a so-called new physics. In fact, Einstein's legacy lies in his leap beyond the horizon and his ability to think outside of the box, though with reason.

Most great scientists seem to perform their landmark work when they are between 20 and 35 years of age. Louis de Broglie, for example, proposed the famous wave–particle duality of quantum mechanics at the age of 22 when he was a PhD student, whereas Paul Dirac contributed significantly to the foundation of quantum mechanics at the age of 25. De Broglie was awarded the Nobel Prize in 1929 "for his discovery of the wave nature of electrons," while Dirac received the Nobel Prize jointly with Erwin Schrödinger in 1933 "for the discovery of new productive forms of atomic theory." Heisenberg was only 24 years old when he published his theory of quantum mechanics featuring uncertainty principle for which he was awarded Nobel Prize in 1932. "for the creation of quantum mechanics, the application of which has, inter alia, led to the discovery of the allotropic forms of hydrogen." Brian Josephson proposed the Josephson effect (leading to superconductivity) in 1962 when he was a 22-year-old PhD student at Cambridge University, while Leo Esaki, at the age of 34, invented the Esaki diode (tunnel diode). Both Josephson and Esaki received the Nobel Prize in Physics in 1973 "for their experimental discoveries regarding tunnelling phenomena in semiconductors and superconductors, respectively." And, the youngest medicine laureate is Fredrik Banting who discovered insulin (used to control blood sugar in people who have diabetes); he was just 32 years old when received the prize. Many such examples may be found in the history of science.

2.4.6 Einstein's Relativity Theory

Let us understand the significance of the term *relativity* in relation to the theories of relativity propounded by Einstein. The term was known from the days of Aristotle, who was the first to emphasize that we only perceive *relative* speed. Speed is not a property of an object of its own: It is the property of the motion of an object *with respect to another object*. Galileo's observation in 1639 that a falling object behaves the same way on a moving ship as it does in a motionless building also points toward the concept of relativity. On land, we talk of our speed with respect to the Earth, while on a ship we talk about our speed with respect to the ship. Galileo understood that this was the reason why even though the Earth moves with respect to the Sun, we do not feel the movement. However, the Galilean relativity is applicable only to mechanical phenomena. Einstein did reformulate Galileo's relativity to deal with the bizarre things that happen at near-light speed, where time slows down and space gets compressed. Einstein, the father of modern relativity theory, generalized it to apply to all processes of nature, including electromagnetic ones (electromagnetism was not known at the time of Galileo, and light was supposed to have infinite (!) velocity). Einstein's relativity is concerned with the relativity of motion of one frame of reference with respect to the other, moving with uniform velocity (for special theory of relativity), unaffected by any external force including gravity and with uniform acceleration (for the general theory of relativity), where the external force is that of gravity—assuming an invariant velocity of light in both the frames of reference and the velocity of light being the highest possible velocity in the universe with which a body can move with respect to another.

Einstein's relativity theory marked a major breakthrough in science, for it subjected all the age-old views of space and time to a complete overhaul. Einstein's relativity theory deals with the so-called four-dimensional space–time continuum, time dilation, and space contraction, a new concept of simultaneity, warping of space–time, and so on. Through the relativity theory, we are now able to decipher a complex range of mysterious celestial phenomena such as the slowing down of clocks near massive bodies, the bending of starlight by the Sun, the black hole, expansion of the universe, the Big Bang, and so forth.

Many of us may be surprised to learn that Einstein did not use the term *relativity* for his coveted theory of the new physics. Rather, his paper on the special theory of relativity of 1905 was entitled "On the Electrodynamics of Moving Bodies." In fact, he hated the term *relativity*, preferring *invariance theory* (because the laws of physics look the same to all observers; there is nothing "relative" about it). However, Max Planck in 1906 was the first to use the term *relative theory* for Einstein's work in order to emphasize how the theory uses the principle of relativity used by Galileo back in 1639 but with a different perspective. The foundation of the whole theory is based on considering/weighing the velocity of a body relative to the velocity of light. If comparable, the theory is applicable, otherwise it is not. In the discussion section of the same paper by Planck, Alfred Bucherer for the first time used the expression "theory of relativity" (German: *Relativitätstheorie*). Anyway, the reason for prefixing "special" and "general" for the theories propounded in 1905 and 1915 by Einstein is discussed in Chapter 1. Although the special and general theories of relativity were Einstein's brainchild, while developing the mathematical part of these theories time and again he was helped by his old college friend of ETH (Swiss Federal Institute of Technology in Zürich), Marcel Grossmann. In addition, the mathematical foundations of great mathematicians like Riemann and others helped Einstein to form the analytical basis of his relativity theory.

2.5 Special Theory of Relativity

2.5.1 Introduction

The special theory of relativity has two fundamental postulates: (1) the laws of physics are the same in all inertial frames of reference; and (2) the speed of light (c) in vacuum is the same in all inertial reference frames regardless of the motion of the observer or source. A frame of reference is said to be inertial if a body in it moves uniformly (not exposed to any external force—in other words, it is nonaccelerating). To be clearer, the inertial frame of reference is a gravity-free frame of reference and has to be very small compared to the distance over which the strength and direction of gravity change. The first postulate was known long before Einstein; in fact, in one of the corollaries to the laws of motion, Newton indicated this, and in 1904 in Poincare's works such a statement was also mentioned precisely in relation to relativity. However, the second postulate was unique with Einstein's special theory of relativity—which also asserts that the motion of anything in this universe faster than the velocity of light is impossible.

2.5.2 Time Dilation and Space Contraction

The second postulate may stem from an interesting and implicit prediction from the electromagnetic equations of Maxwell, which indicate that all observers should measure the same speed of light, no matter how fast they are moving. In other words, a person on the Earth and a person on a speeding space vehicle will measure the same velocity (1,86,000 miles per second or 3×10^5 km/s) for the light beam. Maxwell's equations do not admit stationary waves as a solution; that is, light waves can never be seen at rest. On the basis of this prediction of Maxwell's equation, Einstein showed that time does not pass at the same rate for everyone. A fast-moving observer measures time passing more slowly than a (relatively) stationary observer would. This is called the *time dilation* of the special relativity theory. In this visualization, a clock placed in a moving frame of reference would beat more slowly than a clock placed in a relatively stationary frame of reference. Also, any measuring stick in the moving frame of reference would shrink in length. In other words, a fast-moving object appears shorter along the direction of motion relative to a slow-moving one. This is known as *space contraction*. Both time dilation and space contraction effects are very subtle until the object travels close to the speed of light. The consequence of special relativity that time must slow down (time dilation) and length must contract (space contraction) for speeding bodies seems to violate common sense. This is only because common sense deals with occurrences that are far removed from the speed of light. However, in the cosmic world where the velocity of bodies becomes comparable to the velocity of light, these are dominant effects. Even on a terrestrial scale, if we can measure very small time differences (of the order of, say, a nanosecond) and very minute difference of distance (down to, say, a nanometer), the phenomena of time dilation and space contraction can be precisely evaluated. Let us now understand in brief the concept of *time dilation* and *space contraction* of the special theory of relativity.

2.5.3 Time Dilation of Special Relativity

Let us consider two frames of references. One is M (which may be a train) that is moving with a velocity v (with a stationary passenger P_M inside the train) with respect to another frame of reference S (which may be the platform where we have a stationary passenger P_S).

A light source L (a bulb) is switched ON momentarily and then immediately switched OFF in the reference frame M. Suppose there is an arrangement in the reference frame M to get the light beam reflected from a reflecting surface R (may be a mirror) and get detected at a detector D, as shown in Fig. 2.6. Let this event be observed also from the reference frame S. For the passenger in the moving train (i.e., from reference frame M), the time interval between the two events (putting ON and then putting OFF the light source) would be measured as: Δt_M (*proper time*) $= 2H/c$, where H is the distance between the light source L and the reflector R and c is the velocity of light. But for the person standing on the platform (i.e., in reference frame S), the time interval between two events will take a longer time as

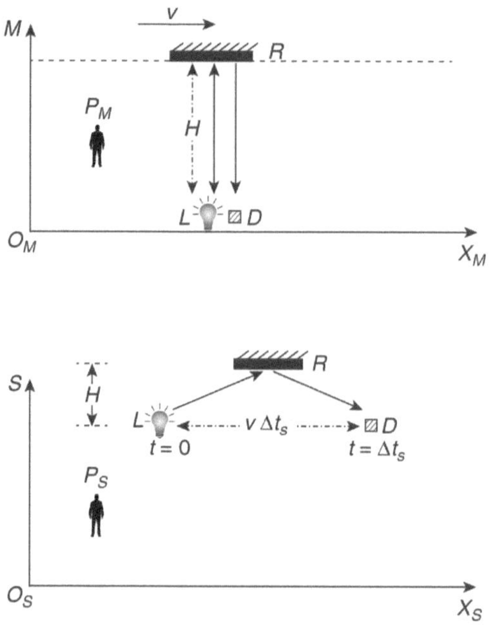

FIGURE 2.6
The concept of time dilation as per the special theory of relativity.

the train is moving with velocity v and the time interval between the two events measured by him will be Δt_S (*improper time*) = $(LR + LD)/c = (2/c)[H^2 + (v\Delta t_S/2)^2/2]^{1/2}$, where we have assumed that the speed of light observed by the passenger in the platform will be the same as that observed by the passenger inside the moving train (as per the special theory of relativity). In other words, we can write: $\Delta t_S = \gamma \Delta t_M$, where $\gamma = 1/[1-(v/c)^2/^{1/2}$ is called the *relativistic factor* or the *Lorentz factor*. Since γ is always greater than 1, $\Delta t_S > \Delta t_M$; that is, the *time is dilated* (i.e., expanded or advanced) *in the stationary frame of reference S* with respect to the moving frame of reference M (or this will be true for any two frames of reference that have a relative velocity with respect to each other). Let us make a simple calculation with $v = 320$ km/h, $H = 3.5$ m (common with the bullet train in Japan), and with $c = 3 \times 10^8$ m/s. The calculation shows that the improper time is $\Delta t_S = 2.330000133 \times 10^{-8}$ sec (since the proper time is Δt_M $(2H/c) = 2.33 \times 10^{-8}$ sec and $\gamma = 1.000000057$). Thus, the dilated time is: $(\Delta t_S - \Delta t_M) = 133$ nS. That is, the passenger in the stationary platform will find that his or her clock is advanced (or moving faster) by 133 nS with respect to the clock of the passenger in the moving train. In other words, the passenger in the moving train will find that his or her clock is slowed down by 133 nS with respect to the clock of the passenger in the platform that is stationary.

In 1941, Bruno Rossi and his colleague carried out an experiment on top of Mount Washington, which is about 1,920 m above sea level, to study the relativistic decay of muons (elementary particles that have exactly the same charge as electrons but have a mass about 207 times the mass of electrons) available in the atmosphere to establish the time dilation predicted in the special theory of relativity. Muons (also known as mu-meson) were used for the experiment because they travel with a velocity of 0.99*c* (*c* is the velocity of light). Their experiment conclusively proved for the first time the prediction of time dilation of the special theory of relativity. The mu-meson lives for only 2.22 microseconds as measured by their own time, but because of time dilation they live for 100 microseconds or more as measured by the physicist's time at rest in the laboratory.

2.5.4 Space Contraction of Special Relativity

In a similar way described above for time dilation, it may be established that if a rod of *proper length* 1 m is placed in the moving frame of reference M, then from the stationary frame of reference S the rod will appear to be of *improper length* ($1-x$) meter, where x is the amount the rod appears to be shortened along its length. For example, a moving spacecraft will appear to a stationary observer from Earth to be shortened in length if the spacecraft moves close to the velocity of light. It should be noted that the contraction only occurs in the dimension of the object's motion. That is, if the object is moving horizontally, then it is the horizontal dimension that is contracted; there would be no contraction of the height of the object. Just to have a feel of the space (length) contraction of a spacecraft of length, say 200 ft, would appear to the stationary observer at Earth to be 199 ft, if it is moving with a speed 10% of the speed of light; while it will appear to be only of 3 ft length, when moving with 99.99% of the speed of light.

2.5.5 Twin Paradox—A Thought Experiment to Demonstrate Time Dilation and Space Contraction

A simple thought experiment to demonstrate time dilation and space contraction is known as the *Twin Paradox* in the cosmic scale and may be understood as follows. One twin, ET, lives on Earth, and the other twin, ST, makes a journey to space to reach Proxima Centauri (the closest star to the Earth after the Sun), which is 4.2 light years away (one light year = one year × velocity of light = $365 \times 24 \times 60 \times 60 \times 3 \times 10^5$ km = 9.46 trillion km). Let the space vehicle move with a velocity very close to the velocity of light, $v = 0.99944c = 299{,}833$ km/s (this assumption is for the sake of discussion and for simplicity of calculation). With this data given, we have the value of $\gamma = 30$. According to ET (who is on the Earth), ST will take about 4.2 years to reach the star (as the velocity of the spacecraft in which he is traveling is assumed to be close to the velocity of light). Now ST will see a contracted distance of 4.2/30 = 0.14 lightyear = 1.316 trillion km. Thus, according to ST, he will reach the star in approximately 1 month and 20 days [i.e., 1/30th of the time recorded by ET]. Now, let ST return to Earth at the same speed. On his return journey, he finds that the Earth is moving toward him with a velocity 299,833 km/s and the contracted distance will be covered in 0.14 light years. When ST reaches back to Earth, his clock reads 1 year, while the clock of ET reads 8.4 years. Hence, the paradox is: ST will appear to be older than ET by 7.4 years. This is precisely due to time dilation and space contraction taking place in the cosmic domain.

Regarding thought experiments, a few words can be mentioned here about Einstein's liking for thought experiments—which in the German language is termed *Gedankenexperimente*—ideas that twirl around one's head rather than in a laboratory. At the age of 16, Einstein imagined chasing after a beam of light, and possibly this thought experiment played a memorable role in his development of the special theory of relativity at the age of 26.

2.5.6 Simultaneity Revisited in Relativity

The classical concept of *simultaneity*, incidences happening instantaneously at two or multiple places, demands a revisit to Einstein's relativity theory. In accordance with absolute time, classical mechanics postulated absolute and universal simultaneity. But according to relativity theory, the concept of simultaneity has no absolute significance like classical mechanics. Two events, viewed from a particular frame of reference, cannot be looked upon as simultaneous events when envisaged from a frame of reference that is in motion relative to the other frame of reference close to the velocity of light. Accordingly, the need arises for developing the theory of transformation of space coordinates and time from a system at rest toward a system in uniform rectilinear motion relative to the former (the special theory of relativity considers inertial systems). In developing this theory, Einstein arrived at the Lorentz transformations, but he arrived at these transformations in an original way, proceeding from his postulates, whereas Hendrik Lorentz introduced them a priori to obtain the invariance of Maxwell's equations for empty space. In Einstein's approach, the Lorentz transformations are organically linked with new properties of space and time—a four-dimensional space–time continuum.

In the framework of Einstein's relativity theory, two spatially separated events that are simultaneous in one frame of reference may be nonsimultaneous in the other reference frame—which follows from the invariance of the speed of light with respect to translations from one inertial reference frame to another. Let us take a simple example. Consider two inertial frames of reference x, y, z and x', y', z'. Let frame x', y', z' be traveling relative to frame x, y, z along x- and x'-axes with a uniform speed v (Fig. 2.7).

In frame x', y', z' there is a light source L and two light detectors, D_1 and D_2, that lie at equal distances from L along the x'-axis. Light source L sends out two light impulses simultaneously, one in the direction of D_1, the other in the direction of D_2. Since $LD_1 = LD_2$ and both signals travel at the same speed (the speed of light), the observer in reference frame x', y', z' will see detectors D_1 and D_2 operate *simultaneously*. Let us now turn to the observer in frame of reference x, y, z. In this frame of reference, the light signal that travels to the left will have to cover a smaller distance from production to registration than the signal that travels to the right. The speed of light in the x, y, z and x', y', z' frames of reference being the same, for the observer in frame of reference x, y, z, the detector D_1 will operate earlier than D_2. Hence, the simultaneity in the two frames of reference is not the same. In technical terms, we say that Einstein's relativity establishes that absolute simultaneity does not exist.

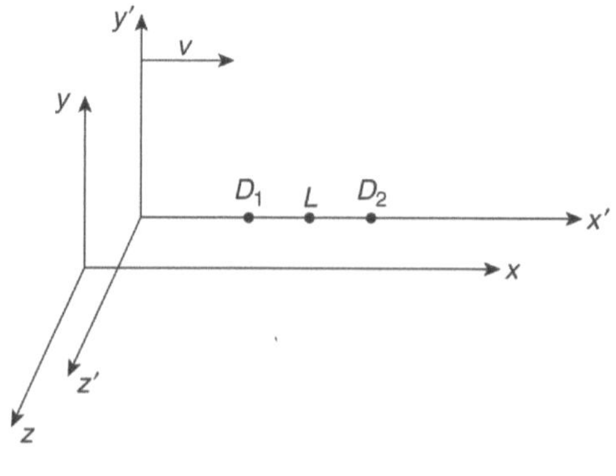

FIGURE 2.7
Two inertial frames of references, of which one is moving with respect to the other with a velocity v.

2.5.7 Extended Present in Space–Time Geometry

The lack of absolute simultaneity in Einstein's relativity puts a big question mark on the term *now* (which we understand to be instantaneous with our common sense and also with Newtonian mechanics) on Earth and that on the planet Mars. Suppose I hold a telephonic conversation (which takes place with radio signal, i.e., electromagnetic wave) with my friend who is staying on Mars, say. In response to my question, the reply I get from my friend on Mars takes almost a quarter of an hour; thus, the "now" for me and the "now" for him differs by this quarter hour. The key fact about nature that Einstein understood through his relativity and space–time continuum is that this quarter of an hour (for this particular case) is inevitable. In fact, when we talk on the telephone from, say, India to someone in the United States, we notice a finite time delay in our conversation may be very small ($13000/3 \times 10^5 = 43$ ms). This is inevitable; it is inseparably woven into the texture of the events of space and time. Also, the astronauts on the Moon experience a 1.5 second time delay in their communications with mission control on Earth due to the time it takes the radio signal (which is a form of light at lower frequency but travels with the velocity of light) to travel round-trip between Earth and the Moon. If any telephone conversation takes place between one person on Earth and the other person on our closest star (next to Sun), Proxima Centauri, then the "now" for the two persons will differ by 4.2 years. Thus, as per relativity and in the cosmic dimension of space and time, we need to think about "present" not being instantaneous, that is, not the same everywhere in the universe (as was the view prior to Einstein in the classical sense) but "present" is *extended over a time extent*, which increases with distance (see the space–time diagram, Fig. 2.8). It might sound strange, but we must get accustomed to the facts of nature when we are in the relativistic domain.

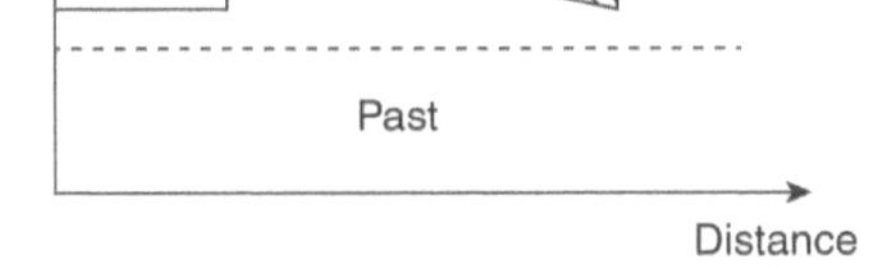

FIGURE 2.8
Extended present in space–time geometry.

2.5.8 Mass–Energy Relationship

Another landmark contribution of the special theory of relativity is the famous mass–energy relationship ($E = mc^2$), where E is the energy, m is the mass of the body, and c is the speed of light) envisioned by Einstein. In Einstein's own language: "It followed from the special theory of relativity that mass and energy are both but different manifestations of the same thing; a somewhat unfamiliar conception for the average mind" (the video: https://www.youtube.com/watch?v=44TQuHsZb_8). He also stated that "if a body emits energy L its mass decreases by L/V^2. . . the mass of a body is a measure of its energy content" (Einstein used the notation V instead of c for the velocity of light). According to Einstein's equation, the energy that an object has due to its motion will add to its mass. That is, it would be harder to increase its velocity at will. This effect is significant for objects moving at velocities close to the velocity of light. For example, at 25% of the velocity of light, an object's mass is only 3.2% more than its normal value (the rest mass), while at 90% of the velocity of light, it would be 2.29 times (i.e., 229%) its rest mass (since $m^* = m/[1 - (v/c)^2]$, where m^* is the relativistic mass, m is the rest mass, and v is the velocity of the body). As an object approaches the velocity of light, its mass rises even more quickly, so it takes more and more energy to speed it up further. It can never reach the velocity of light because by then its mass would become infinite, and by the equivalence of mass and energy, it would have taken an infinite amount of energy to get it there. For this reason, any normal object is restricted by the principle of relativity to move at velocities lower than the velocity of light. Only light or other electromagnetic waves like radio waves, microwaves that have no intrinsic mass (as they are constituted of photons and photons are known to be massless), can move at the velocity of light. Thus, the $E = mc^2$ relation of the special theory of relativity establishes that nothing in this universe can travel faster than the velocity of light.

Certain fundamental equations of physics and astrophysics help us to understand almost all physical phenomena in the universe, and one of them is of course Einstein's mass–energy relationship discussed above. Apart from this, other fundamental equations of physics include *Newton's laws of motion* and *Maxwell's electromagnetic equations*; discussed in Chapter 1. Also, in this chapter we come across some other fundamental equations of physics such as *Planck–Einstein's quantization equation, Einstein's field equation of general relativity, Heisenberg's uncertainty equation, Schrödinger's wave equation,* and so on. In Chapter 3, related to astrophysics, we come across *Hubble's equation* for an expanding universe and *Schwarzschild's radius equation* for the black hole. With regard to some more fundamental equations of physics and astrophysics, we may mention the *second law of thermodynamics:* $dS \geq 0$ (which is an inequality, and according to this the entropy (S) of the universe is increasing; in other words, the disorder of the universe is increasing); however, the details of this is not in the scope of our discussion in this book.

While mentioning the second law of thermodynamics, we need to say a few words about the tragedy in the life of Ludwig Boltzmann (being students of physics, we all have come across the Boltzmann Constant—$k_B = 1.38 \times 10^{-23}$ J/K.). Boltzmann is known primarily for his work in developing statistical mechanics and the statistical description of the second law of thermodynamics. He described the Second Law of Thermodynamics based on the atomic theory of matter in the early 1870s. This law states that the "entropy of the entire universe as an isolated system will always tend to increase." His famous quote is: "Thermodynamics, correctly interpreted, does not just allow Darwinian evolution; it favours it." Boltzmann was one of the first to recognize the significance of Maxwell's unification of electromagnetism. He was one of the most important advocates of atomic theory at a time when that scientific model was being highly argued. Unfortunately, Boltzmann spent most of his life defending his theories from his staunch opponents like Wilhelm Oswald and Ernst Mach, and gradually he became mentally unstable. On a holiday with his wife and daughter at Trieste, Italy, he committed suicide in 1906, at the age of 62. He hanged himself while his wife and daughter were swimming in the pool. Generally, his suicide has been attributed to the fact that stalwarts of his time did not accept his ideas. Thus, we all must remember that professional competition is of course a healthy process in every sphere of life, but that rivalry/enmity and zealousness are to be condemned. That hinders progress and ultimately leads to extreme consequences; Boltzmann is a burning example.

2.5.9 Experimental Proof and Application of the Mass–Energy Relation

In 1932, John Cockcroft and Ernest Walton first established the experimental proof of Einstein's mass–energy relationship. They demonstrated the first artificial splitting of the nucleus of lithium by bombarding the accelerated protons to produce two alpha particles. In such a nuclear reaction, a net decrease of mass $\Delta m = 0.033 \times 10^{-27}$ kg released an energy of 19 MeV, based on Einstein's formula: $E = mc^2$. Cockcroft and Walton won the Noble Prize in Physics in the year 1951 for "transmutation of atomic nuclei by artificially accelerated atomic particles."

The mass–energy equivalence relation mentioned above is the basis of nuclear energy generation with the fission process, which is humankind's boon and bane. In nuclear reactors, when uranium is bombarded with neutron, the uranium is fragmented into lighter elements like krypton and barium, and a small atomic mass difference of the parent element and the fission products manifests as huge nuclear energy. The mass of one gram of radioactive element (say, uranium) in a nuclear reactor can produce enough energy to illuminate a city and power the industries of a country for months. Conversely, this energy will be enough for a nuclear bomb and in just a second will be capable of destroying hundreds and even thousands of human beings, as happened in Hiroshima and Nagasaki, Japan, during World War II (August 6 and 9, 1945, respectively). J. Robert Oppenheimer, the "father of the atomic bomb" and wartime head of the Los Alamos Laboratory, that made the bomb, commented that the nuclear bomb explosion was "Brighter than a Thousand Suns." However, looking at the fury of devastation caused by the nuclear bomb dropped in Hiroshima and Nagasaki, Oppenheimer's soul cried out with a quote from *Srimad Bhagavad Gita*: "Now I have become Death, the destroyer of the world" (his own translation from the original Sanskrit language). After the war ended, Oppenheimer became chairman of the influential General Advisory Committee of the newly created United States Atomic Energy Commission. He used that position to lobby for international control of nuclear power to avert nuclear proliferation and a nuclear arms race with the Soviet Union.

Einstein was also instrumental in convincing President Roosevelt to initiate the Manhattan Project to make the nuclear bomb. However, for Einstein, "war is a disease," and he called for resistance to war. By signing the letter to Roosevelt, some argue he went against his pacifist principles. In 1954, a year before his death, Einstein said to his old friend, Linus C. Pauling, Nobel Laureate in chemistry and also Nobel Laureate in Peace: "I made one great mistake in my life—when I signed the letter to President Roosevelt recommending that atom bombs be made; but there was some justification—the danger that the Germans would make them." In fact, in the latter stage of his life, Einstein became an advocate of world peace and international cooperation for nuclear disarmament. Einstein was seriously hurt over the United States' use of the hydrogen bomb (based on his theory of $E = mc^2$) against Japan in World War II. He painfully commented that "if I had foreseen Hiroshima and Nagasaki, I would have torn up my formula in 1905." Einstein played only a minor role in the Manhattan Project, however. He also made a comment that we need to ponder: "I know not what weapons World War III will be fought, but World War IV will be fought with sticks and stones."

2.5.10 Space–Time Continuum

Relativity has thoroughly revolutionized our idea of space and time and has added a new term to the lexicon of modern physics dealing with the macro-world: The *space–time continuum*. This particular space–time concept, having been born from Einstein's relativity theory, radically modified the Newtonian image of the universe. The reader should be reminded that according to the Aristotelian school of thought, both space and time were considered to be absolute, while according to the Newtonian mechanics, space no longer remained absolute (which, however, Newton was skeptical to accept), but time was still considered to be absolute. The theory of relativity put an end to the idea of absolute time, in addition to absolute space with its unique postulate that the velocity of light is constant irrespective of the speed of motion of any one in the universe.

To make the point a little clearer, let us give an example. According to Newtonian mechanics, if a beam of light is sent from one point to another, different observers would agree on the time that the journey took by the light beam (since the time is absolute) but would not agree on how far the light has traveled (since space is not absolute). Because speed of light ($v = c$) in Newtonian mechanics has to be considered as just the distance (S) it has traveled divided by the time (t) it has taken (since $S = vt$), different observers would measure different speeds for light. According to relativity theory, on the other hand, all observers *must* agree on how fast light travels (as velocity of light is independent of the observer's state of motion). Thus, in addition to their disagreement on how far the light has traveled, all of them now *also* disagree about the time it has taken. This is because the time taken is the distance the light has traveled—which the observers do not agree on—divided by the speed of light—which they do agree on. In other words, *the theory of relativity put an end to the idea of absolute time* (i.e., identical clocks carried by different observers would measure different time depending on their state of motion and location in the space).

Prior to Einstein, space was supposed to be just a boring, static stage on which the cosmic drama unfolds. An invisible, rigid scaffolding—which Newton termed *God's sensorium* (no one has ever understood what he meant by that term; perhaps not even Newton understood it!) and time was an absolute entity unconnected with space. But Einstein visualized the organic link between space and time, which he considered a single four-dimensional continuum (space–time); everything is immersed in that gigantic and flexible cosmic mollusk (the metaphor is Einstein's), which can be squashed, stretched, and twisted. Einstein taught us that space and time are inseparable key actors in the cosmic drama. Space–time can curve into extreme extent close to black holes, it can ripple as gravitational waves, and it can stretch as an expanding universe. Soon after Einstein proposed his general relativity theory, each solution of its field equations had brilliant success in explaining or predicting various aspects of cosmology, which was unknown at that time. For example, the solution of Einstein's field equations of general relativity by Karl Schwarzschild gave us the current description of black hole; solution of the same equations by Howard Robertson and Arthur Walker gave us the description of the Big Bang and so forth.

The four-dimensional space–time coordinate system is the basis for understanding cosmic phenomena in terms of Einstein's relativity theory. The so-called space–time diagram and light-cone are used to elucidate the special theory of relativity and its consequences. It is very hard (if not impossible) to visualize, as it transcends our common-day experience of three-dimensional space, the four-dimensional space–time (the three coordinates x, y, z representing three-dimensional space), while time t is the fourth dimension (perpendicular to all three space coordinates). Thus, a space–time diagram is used in which the ordinate (y-axis) is taken as time, while the abscissa (x-axis) is taken as one of the spatial dimensions (representing distance). The other two spatial dimensions are ignored or, sometimes, one of them is indicated by perspective.

2.5.11 The Space–Time Diagram for the Light Cone of Special Relativity

The space–time diagram for a light cone of special relativity can be built by taking instantaneous snapshots of space (represented by a two-dimensional plane) at successive instants of time and stacking them vertically (i.e., in the time axis). In such a diagram, the history of motion of a stationary body will be represented by a vertical line, while an inclined line with respect to the vertical is for a body in motion, the angle of inclination being greater when the velocity of the body increases. The extreme inclination of the line is for a body that moves with the velocity of light—which is the bound of the space–time contour, *event horizon*. This is because, as per the special theory of relativity, nothing in this universe can move with a velocity more than the velocity of light (Fig. 2.9a). The prediction from Maxwell's electromagnetic equations is that the speed of light remains the same irrespective of the speed of the source. Thus, if a pulse of light is emitted at a particular time instant at a particular point in space (which may be termed as the present event, where the observer is located), then it will propagate out in space–time geometry as an expanding spherical shell in four-dimensional space–time (three-dimensional space and the fourth dimension being time). But in a two-dimensional space, which is easy to visualize, the light pulse will propagate with time as an expanding circle.

A similar picture may be found with the spreading out of ripples on the surface of water of a pond when a stone is thrown into it. If the snapshots of such circles with increasing diameter with the passage of time are stacked vertically (in the space–time geometry), we get the shape of a cone. Thus, the light spreading out from an event (representing the present, point *P*, where the observer is located) will form a three-dimensional cone, which we call the *future light cone* of the event. In the same way, we can draw another light cone, called the *past light cone*, which is the set of events from which a pulse of light is able to reach the point *P*, representing the present event location (Fig. 2.9b).

The past and future light cones originating from an event taking place at the present point *P* divide space–time geometry into two regions. The *knowledge space* for the present events is bounded by the past and future light cones, with the apex for both at point *P* where the observer at the present time is located. The future light cone is the region where all sets of events can be affected by what happens at *P*, with the signals traveling from *P* at or below the speed of light. Similarly, the past light cone is the region where all sets of events can affect what happens at *P*, with the signals traveling at or below the speed of light. But the *"hyperspace"* of the present events is the region of space–time that does not lie on the future and past light cone of *P*. Hence, the events in hyperspace cannot affect or be affected by events at *P*. For example, if the Sun ceases to shine at this very moment, it would not affect things at the present time (i.e., instantaneously) because the event when the Sun disappears falls in the hyperspace of the present. The event will be known; that is, it will touch the knowledge space only after 8 minutes (since the time taken by light to reach from the Sun to the Earth is approximately 8 minutes). Only then would events on Earth lie in the future light cone of the event at which the Sun went out (Fig. 2.9c). Thus, all the celestial bodies we see in the sky in our *present vision* actually relate to their respective individual past. Depending on their distance from the Earth, light takes approximately

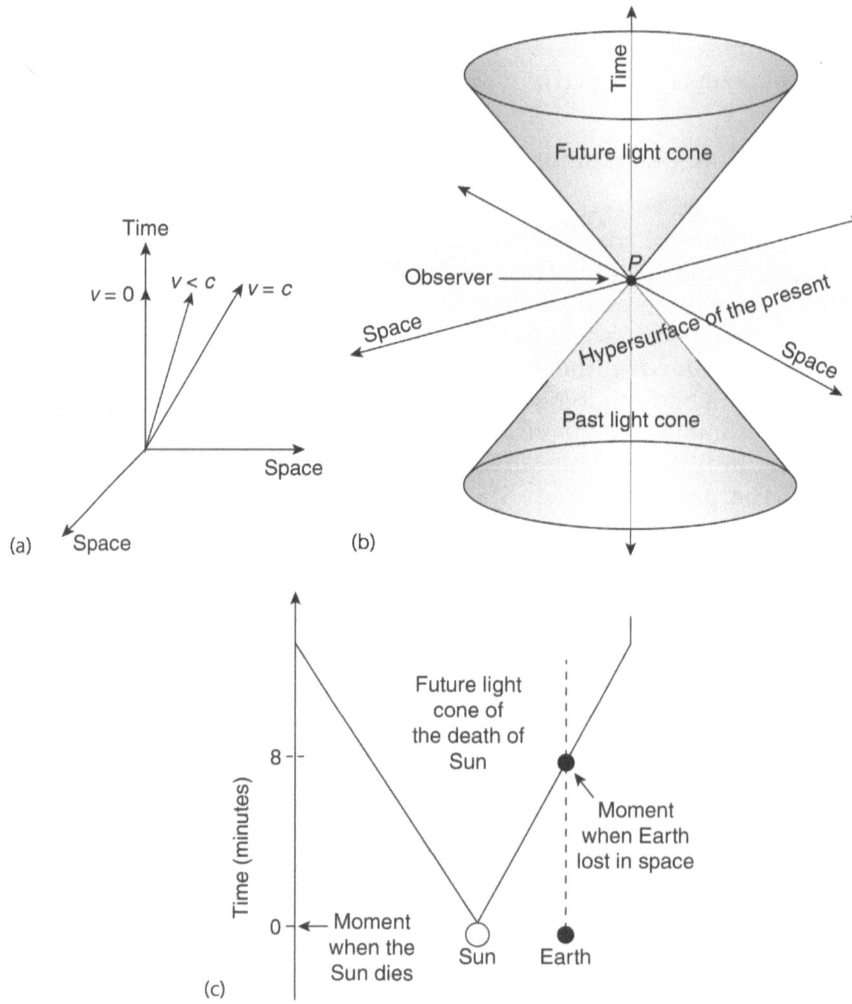

FIGURE 2.9
(a) Path of the motion of bodies with different velocity in space–time geometry; (b) the lightcone of the special theory of relativity; and (c) information about the disappearance of the Sun takes 8 minutes to reach the Earth.

8 minutes from the Sun, 4.2 years from our closest star, Proxima Centauri, and even a few million years from distant galaxies and so on. Thus, what we see about the Sun, Proxima Centauri, and distant galaxies *now* actually happened, respectively, 8 minutes, 4.2 years, and a few million years *earlier* in those celestial bodies.

2.6 General Theory of Relativity

2.6.1 Introduction

Einstein's special theory of relativity is concerned with the observation frames of reference that are in relative motion but with uniform velocity, that is, not being acted by any external force, including gravity. It was called the special theory of relativity because it specifically isolated from the effect of gravity any other form of external force. However, Einstein was dissatisfied because this theory did not accommodate the universal gravitation of Newtonian mechanics (or of any other from whatsoever): Acceleration is the missing issue in the theory. For nearly 10 years of agonizing search, missteps, and manic study, he finally succeeded in formulating a *gravitational field theory* based on relativity or the general theory of relativity (a generalization of the special theory of relativity in the sense that it now included the effect of gravity too) in November 1915. This theory became the basis of modern cosmology and in essence is also known as the relativity theory for gravitation. However, in his autobiography, Einstein expressed his disappointment that the idea of "general relativity" led him "merely" (in Einstein's word) to a theory of gravitation. This shows just how precious the idea of "general relativity" was to Einstein. He may have desired to take it a long way forward to address many other important issues related to knowledge of the universe. Einstein was correct in the sense that, in addition to the explanation of deflection of starlight by the Sun and the exact estimation of the perihelion shift of the planet Mercury, the general theory of relativity subsequently predicted the existence of black holes, gravitational waves, singularities of space–time in the heart of the black hole, and so forth, many of which have now been verified through experimentation.

Einstein began his research for a new law of gravity in 1907. Gravity was the biggest gap in his highly respected special theory of relativity. By early 1908, frustrated that he had made no real progress with the inclusion of gravity in relativity theory, Einstein temporarily gave up and turned his attention to the realm of the small—atoms, molecules, and radiation. The unsolved problems in that domain seemed to him at that moment more tractable and interesting. During 1908–1911 (when he moved from the patent office first to the University of Zurich as associate professor and then as a full professor in Prague), his research in the realm of the small moved forward impressively, producing insights that helped him to win the Nobel Prize in Physics in 1921—for his discovery of the wave/particle duality of light, photoelectricity (including the relativity theory), and the tentative quantum-mechanical laws of physics that were built around this discovery. In mid-1911, Einstein's fascination with the small waned and his interest returned to gravity, with which he would struggle almost fulltime until his triumphant formulation of general relativity in November 1915.

2.6.2 Chronogeometric Theory of Gravitation

Einstein's general theory of relativity is an integral fusion of his novel theory of space and time (which he introduced in his special theory of relativity) with the theory of gravitation. This unity is its primary characteristic. Thus, the Dutch physicist Adrian Fokker preferred to use the term *chronogeometric theory of gravitation* for Einstein's general theory of relativity (*kronos* is the Greek term meaning *time*). Interestingly, one of the most beautiful ideas in Einstein's theory of gravity is that geometry is not just mathematics: it is also physics. Specifically, Einstein's equations show that the more matter space contains, the more curved the space gets into. This curvature of space causes things to move not in straight lines, but in a motion that curves toward massive objects, thus explaining gravity as a manifestation of geometry around a body having mass.

The mass of the Sun warps (curves) space–time in such a way that, although in reality the Earth follows a straight path in four-dimensional space–time, it appears to move along a circular orbit in three-dimensional space around the Sun. All planets go around the Sun in elliptical orbit (i.e., an elongated circular path), with the Sun in one of the foci as predicted by Newton's law of gravitation and Kepler's law of planetary motion. Einstein recalculated planetary motion on the basis of space–time warping of general relativity and found that his result was in good agreement with Newton's and Kepler's calculations. However, the precession (change of orientation of the elliptic orbit in its orbital plane) of the perihelion (the closest distance from the Sun in elliptic orbit) of the planet Mercury is in better agreement than Newtonian calculation with the experimental observation of astronomers (made well before 1915): 1.38 arc-second (Fig. 1.7). This served as one of the first confirmations of Einstein's theory of general relativity. In recent years, even smaller deviations of the orbits of other planets have been measured by radar technology and have been found to agree with the predictions of Einstein's general theory of relativity.

Today, Einstein's general theory of relativity is arguably the most successful theory of all time for the macro-world. In this theory, Einstein revisited gravity and observed that it was not a mere "ghost force" of action between two masses (which was understood by Newton) but was produced by a curvature (a warpage) of space–time absolute (caused around a body due to its mass), the four-dimensional fabric. By mathematically manipulating his *field equation*, Einstein and other physicists not only explained the deflection of starlight by the Sun and the motion of the planets in their orbits, including the mysterious perihelion shift of the planet Mercury with exactitude, but also predicted the existence of the black hole, gravitational waves, singularities of space–time at the heart of the black hole, and perhaps the existence of the so-called wormhole—wholly and solely from that fabric. That is, each of them represents a specific type of space–time warpage. The predictions of general relativity have never once failed. This theory was first introduced in 1915 in an attempt to replace Newton's gravitation. Although it could reproduce the earlier Newtonian successes and explain the orbit of Mercury precisely (which Newton's law of gravitation could not), the most critical test to which it was put was to establish, with practical observations, its own prediction that light would bend when it passed close to a star. This follows from the concept of general relativity that space–time is curved by the mass of a body and that mass and energy are inseparably united by Einstein's famous equation: $E = mc^2$.

2.6.3 Recipes for the Development of General Theory of Relativity

Einstein was not a great mathematician; he actually had to struggle with mathematics. Many a time he took help from his former college classmate, Grossmann. (It was Grossmann's father who had helped Einstein to get the job at the Bern patent office.) With regard to Einstein's inadequacies in math, there is an interesting story about his response to a 9-year-old child with the name Barbara who wrote to Einstein about her difficulties with mathematics. Einstein's answer was: "Don't worry about experiencing difficulty with maths; I can assure you that my own problems are even more serious!" The story is told jokingly, but Einstein was actually not kidding.

When Einstein was formulating the relativity theory, he had at his disposal the mathematics of curved surface in four dimensions, which had already been formulated by Riemann, a German mathematician, and one of the greatest mathematicians of all time. Euclid (mid-4th century–3rd century B.C.) of Alexandria, Egypt (the center of Hellenic scholarship and science founded by Alexander the Great in 332 B.C.), the great mathematician and father of geometry, considered the idea of a "flat universe," even though it is a volume rather than a surface. However, according to Riemann, the universe is a four-dimensional analogue of a sphere, a hypersphere—a concept that is very difficult to comprehend with our three-dimensional commonsensical experience. On the basis of Riemann's four-dimensional mathematical model of the universe, Einstein could formulate his *chronogeometric theory of gravitation* or the so-called general theory of relativity.

Here special mention needs to be made of Hermann Minkowski, Einstein's teacher at ETH in Switzerland. It was Minkowski who labeled Einstein a "lazy dog" in Einstein's student days at ETH. (While at the ETH, Einstein was not at all interested in assignments based on the old Newtonian physics, and instead was seeking the "new physics," which ultimately developed into his relativity theory.) Minkowski left ETZ in 1902 to take up a more lucrative job at Göttingen, Germany. While at Göttingen, he read Einstein's article on special relativity and was impressed. Deep study of this theory led Minkowski to develop his absolute nature of four-dimensional space–time in 1908. In Minkowski's words: "Henceforth, space by itself, and time by itself are doomed to fade away into mare shadows, and only one kind of union of the two will preserve an independent identity." That is, Minkowski, building the mathematical background of Einstein's special theory of relativity, had now discovered that the universe was made up of a four-dimensional "space–time" fabric that was actually absolute and neither space nor time by itself was absolute.

This four-dimensional fabric is the same as that seen from all reference frames (if only one can learn how to "see" it): It exists independent of reference frames. But space and time taken individually have to be considered as relative. For example, my time differs from yours if I move relative to you (with a velocity comparable to the velocity of light), and likewise my space also differs from yours. My time is a mixture of your time and your space, and my space is a mixture of your space and your time. This is the crux of Einstein's special theory of relativity that demands space–time as a unit that is absolute and not space and time themselves in a segregated way. When Einstein learned of Minkowski's discovery, he was at first unimpressed. He even joked about the Göttingen mathematician's complicated mathematical approach, which he believed might obscure the beauty of the physical ideas that underlie his relativity theory. Though in 1908 Einstein dismissed Minkowski's unification of space and time mathematically as a unique space–time continuum; in 1912 he finally embraced it and proposed his general theory of relativity, which states that gravity warps or, in his own words, causes the curvature of space–time and that gravitation is no "ghost force," which Newton's law of gravitation indicated and confused Newton himself. Sadly, however, Murkowski did not live to learn of Einstein's acceptance of his work; he died of appendicitis in 1909, when he was only 45 years old. Thus, we must emphasize that the space–time continuum of Einstein's relativity theory was not Einstein's: The idea of time as the fourth dimension came from Hermann Minkowski.

Remarkably, Einstein was not the first to discover the correct form of the law of warpage either, the form that obeys his relativity principle. The first discovery possibly goes to David Hilbert, a great mathematician who also had a passionate interest in physics. He was a professor at Göttingen University, where Minkowski also worked. In autumn 1915, even as Einstein was struggling with the right law for general relativity, making mathematical mistake after mistake, Hilbert was mulling over what he had learned from Einstein's summer visit to Göttingen. While on an autumn vacation on the island of Rugen in the Baltic Sea, he derived a key idea, not using Einstein's arduous trial-and-error path, but by taking an elegant, succinct mathematical route. Hilbert presented his derivation and the resulting law at a meeting of the Royal Academy of Sciences in Göttingenon November 20, 2015, just 5 days before Einstein's presentation of the same law at the Prussian Academy meeting in Berlin. Einstein, through Gossmann's mathematical help and Minkowski's idea about the unification of space and time, arrived at the idea of warpage (or in his words the curvature) of space–time caused by gravitation. This became the cornerstone of his general theory of relativity. Though Hilbert carried out the last few mathematical steps of space–time warpage independently and almost simultaneously with Einstein (actually, 5 days before Einstein's result was reported), he described those remarkable equations as *Einstein's field equations* rather than appending his own name to the equations. This is because Hilbert recognized that 90% of the physical basis and mathematical understanding was developed by Einstein; general relativity, he fully acknowledged, was the brainchild of Einstein.

For his part, Einstein wielded his mathematical tools with some clumsiness. Thus it was that Hilbert sarcastically commented about the weakness of Einstein's mathematical capability: "Any youngster in the streets of Göttingen," he said, "understands more about four-dimensional geometry than Einstein." But Hilbert fittingly admitted that the full credit for the general theory of relativity to interpret gravitation as warpage of space–time must go to Einstein. It may be mentioned in passing that Einstein was not a bad mathematician, though in mathematical technique he was no towering figure, but his physical insight was prodigious. This is why it was Einstein and not a mathematician who formulated the general relativistic laws of gravity. Einstein finally formulated the general theory of relativity because he had an exceptional imagination and was able to form the physical picture of the universe in his visionary mind. For him, the equations came afterward; they were the analytical language with which to translate intuitive visions of objective reality.

2.6.4 Prediction of Gravitational Waves

The general theory of relativity predicts that space–time ripples like the surface of the sea, generating so-called *gravitational waves*—ripples in space–time created by the collision of massive objects. Such waves can arrive at the Earth from cataclysmic sources millions to billions of light years away. These ripples in the space–time curvature are waves generated by the so-called gravitational field similar to the electromagnetic field that makes our radio, television, and smartphone possible. The gravitational field is a very weak field. Compared to the electromagnetic field, it is trillions of times weaker; hence, the detection of gravitational wave is obviously very difficult. However, Rainer Weiss, Kip Thorne, Barry Barish, and the late Scottish physicist Ronald Drever spent decades building a hypersensitive instrument, LIGO (*L*aser *I*nterferometer *G*ravitational-Wave *O*bservatory), which recorded contractions and expansions in the fabric of space–time less than one-thousandth the width of an atomic nucleus and experimentally established the existence of gravitational waves in early 2016. Once more, the seemingly mad predictions of Einstein's relativity theory turned out to be precisely true, leaving the scientific world speechless. In 2017, Weiss, Thorne, and Barish were recognized with the Nobel Prize "for the detection of gravitational waves caused by the collisions of black holes." More details on gravitational waves and the black hole are presented in Chapter 3.

2.6.5 The Expanding Universe from the General Theory of Relativity

The general theory of relativity further predicted that the universe was expanding and that it emerged from a cosmic explosion—known as the Big Bang—some 14 billion years ago. But Einstein was a believer in a "static" universe, since that was the prevailing scientific idea of his time and was supported by the then prevailing astronomical data; he was actually disturbed by the expanding nature of the universe that follows from his own theory. To counter, he introduced the so-called *cosmological constant* into his equations, which was a factor representing "antigravity" whose effects occur only at extremely large distances or at a "cosmological" scale of distances. To force his equations—which theoretically predicted the expansion of the universe—to remain still, Einstein introduced the cosmological constant, Λ (lambda). He multiplied the metric tensor in his equation, $g_{\mu\nu}$, by the cosmological constant, leading to a term $\Lambda g_{\mu\nu}$, which adjusted his metric tensor acting on space–time. This mathematical trick assured him that his equations would yield a universe that was prevented from expanding or contracting. Thus, Einstein and most scientists until the 1930s held that the universe was "simply there," a static one with no beginning and no end. But it's interesting to note that creation myths across cultures tell the opposite story. The traditions of Indian, Chinese, pre-Columbian, and African cultures, as well as the biblical book of Genesis, all describe a distinct beginning to the universe—whether it's the creation in six days of Genesis or the Cosmic Egg of the ancient Indian text, the *Rig Veda*.

Alexander Friedman, a Russian physicist and mathematician, took general relativity at its face value and through his equations established the nonstatic, and hence expanding, nature of the universe in 1922, which was subsequently supported by the discovery of the red shift of far-off galaxies in the cosmos by Edwin Hubble in 1929. When Einstein heard about the results of the Hubble, in the early 1930s, he traveled to California and met with Hubble. At the Mount Wilson Observatory, he saw the massive data set on distant galaxies that had led to Hubble's law describing the expansion of the universe. Einstein, angry at himself, mused that had he not forced his equations to remain static with the cosmological constant he had introduced, he could have theoretically predicted Hubble's findings! Einstein exclaimed after his Mount Wilson visit: "If there is no quasi-static world, then away with the cosmological term!" Einstein abandoned the concept of the cosmological constant in 1931 after Hubble's discovery of the expanding universe.

Later in his life, Einstein said that changing his own equations of general relativity with the introduction of the cosmological constant was "the biggest blunder of [his] life." However, among the documents at the Einstein Archives at the Hebrew University in Jerusalem, in late 2013 a handwritten draft paper by Einstein was found, called *Zum kosmologischen Problem* ("About the Cosmological Problem"). In this paper, Einstein made a stubborn attempt to resurrect the cosmological constant he had vowed never to use again. In the 1930s, Einstein was still trying to return to his 1917 analysis of a universe with a cosmological constant—but now he tried to revise the model with the cosmological constant responsible for creating new matter as the universe expanded. Here the constant represents the *energy of empty space*—a powerful notion—and Einstein in this paper wanted, with time, to use this energy to create new particles. This possibly was a symbolic indication of the so-called dark energy (who knows?). Today we view the same energy of the vacuum as the reason for the acceleration of the universe's expansion, which is presumed to be due to the mysterious dark energy. The great mind has a great vision whose far-reaching possibilities may not be perceived by our average mind. Einstein's thirst for unfolding the "real" knowledge about the universe was in the genetic code of his inquisitive mind and being. This reminds us of his self-appraisal: "I am neither especially clever nor especially gifted. I am only very, very curious."

2.6.6 The Cosmological Constant—Not a Blunder but a Visionary's True Vision

The whole gamut of general relativity developed by Einstein, which established that space–time and the gravitational field are one and the same thing, has been expressed with a painter's eye and with a musician's master stroke in a mathematical expression known as Einstein's field equation:

$$R_{\mu v} - 0.5Rg_{\mu v} + \Lambda g_{\mu v} = 8\pi G T_{\mu v}/c^4$$

where $R_{\mu v}$ depends on Riemann's curvature and together with $0.5Rg_{\mu v}$ represents the curvature of space–time, $T_{\mu v}$ stands for the energy of matter, and G is the same universal gravitational constant introduced by Newton to represent the strength of the force of gravity.

The first two terms together (i.e., $R_{\mu v} - 0.5Rg_{\mu v} = G_{\mu v}$) are known as the Einstein tensor, $T_{\mu v}$ is the stress-energy tensor, and the term on the right of the equation (i.e., $8\pi GT_{\mu v}/c^4$) is known as the Einstein gravitational constant (= 2.077×10^{-43} N^{-1}). It may be mentioned that *Riemann curvature tensor* is a central mathematical tool in the formulation of Einstein's general theory of relativity. Riemann curvature tensor formula in terms of covariant derivatives is defined as $R(u,v) = \nabla^2_{u,v} - \nabla^2_{v,u}$; where u, v is a pair of tangent vectors and $R(u, v)$ is the linear transformation of the tangent space of manifold. The term $\Lambda g_{\mu v}$ was not present in the equation of general relativity that Einstein proposed in 1915. He introduced it in 1917 when he included the term *cosmological constant* or the so-called lambda factor in his equation to counter the expanding nature of the universe (indicated by the equation of 1915) in order to justify his model of the so-called static universe.

From the 1930s until the late 1990s, most physicists assumed the cosmological constant to be equal to zero. But when a genius such as Einstein makes a mistake, it tends to be a "good mistake": It can't simply go away, for too much thought has gone into it. So, like a phoenix, Einstein's cosmological constant quite unexpectedly made a remarkable comeback in 1998. That year, two groups of astronomers made an announcement that rocked the world of science. The Supernova Cosmology Project, based in California and headed by Saul Perlmutter, and the High-Z SN Search group at Harvard-Smithsonian and Australia, announced their results of the red-shifts of distant galaxies, leading to a conclusion that nobody had expected: The universe, rather than slowing its expansion since the Big Bang, was actually accelerating its expansion! The best theoretical way to explain the accelerating universe revived Einstein's discarded lambda factor—the cosmological constant. The possibility of a positive nonzero value for the cosmological constant was found to be the cause of the accelerating expansion of the universe and seems to be the best explanation for the mysterious "dark energy" seen to permeate space and to push the universe outward at an accelerating rate. To most physicists today, the lambda factor, cosmological constant, and dark energy may be closely synonymous. But unfortunately, Einstein died in 1955 and was not there to witness the reversal of his "greatest blunder." The value of this positive cosmological constant was measured at the end of the 1990s, bringing a Nobel Prize for Physics in 2011 for the astronomers Saul Perlmutter, Brain Paul Schmidt, and Adam Guy Reiss "for the discovery of the accelerating expansion of the universe through observations of distant supernovae."

2.6.7 Twin Paradox Due to Gravitational Dilation of Time and Correction of GPS Clocks

Another prediction of general relativity is that time should appear to run slower near a massive body like Earth, and upon being farther away from Earth's surface, time will be advanced correspondingly. Why this happens lies in the fact that energy is directly related to frequency (inverse of time), and the closer to a massive or energetic body, space–time is more distorted, slower will be the passage of time. Time is not universal and fixed; rather, it expands and shrinks according to the vicinity of masses (energy), mass and energy being synonymous according to Einstein's relativity theory ($E = mc^2$). An interesting twin paradox (Fig. 2.10) based on this prediction of general relativity is that two twins who have lived, respectively, at sea level and in the mountains will find that, when they meet up again, the one who lived on the mountain will have aged more than the other (due to the gravitational dilation of time); though by a very small amount. Thus, if I stayed at Shillong (the capital of Meghalaya, a northeastern state in India, which is a hill station 5,000 ft above sea level), then possibly I would have been older by a few microseconds compared to my present age (as I have been staying at Kolkata, India, since 1974).

FIGURE 2.10
Twin paradox due to gravitational dilation of time.

Today we have very precise atomic clocks in many laboratories, and we can measure this strange effect of gravitational time dilation even for a difference of altitude of just a few meters. In 1962, this prediction was tested using a pair of very accurate clocks mounted at the top and bottom of a water tower. The clock at the bottom of the tower that was nearer the Earth (i.e., at a lower gravitational potential with respect to the clock on top of the water tower) was found to run slower, in exact agreement with general relativity.

The difference in the speed of clocks at different heights above the Earth's surface is now of considerable practical importance with the advent of a very accurate satellite-based navigation system, the Global Positioning System (GPS). This is a multibillion-dollar growth industry with applications in airplane navigation, oil exploration, bridge construction, sailing, and innumerable other areas. GPS is based on 24 artificial satellites that contain atomic clocks to provide very precise time data that are ultimately converted to position data. It is a radio location technique for accurate determination of position which we all use even when driving a car using the Google map.

Because of the relativistic gravitational potential (as the satellites are in orbit 20,000 km above the Earth's surface), the atomic clocks in satellites of the GPS system get advanced by 38.2 μs every day with respect to the identical clocks in the ground station. Thus, the clocks in the satellites need relativistic correction, slowing down by 38.2 μs every day. If we ignore the relativistic correction of the atomic clocks in the satellites, the position determination by GPS will be inaccurate even by several kilometers! A representative diagram for such gravitational time dilation is shown in Fig. 2.11.

2.6.8 General Theory of Relativity and Gravitational Lensing

One of the primary predictions of Einstein's general relativity theory was the bending of light from stars by the Sun, which was experimentally verified in 1919. A similar effect is also expected by the more massive celestial objects in further reaches of the universe, which gave rise to the so-called *"gravitational lensing"* that was also predicted from the general theory of relativity. The bending of light from distant stars caused by a cluster of galaxies was discovered in 1979 by Dennis Walsh and his team using the Kitt Peak National Observatory 2.1 m optical telescope. They observed that QSO 0957+561, a twin quasar that lies 7.8 billion lightyears from Earth, is gravitationally lensed, so that two images of the quasar appear in the sky.

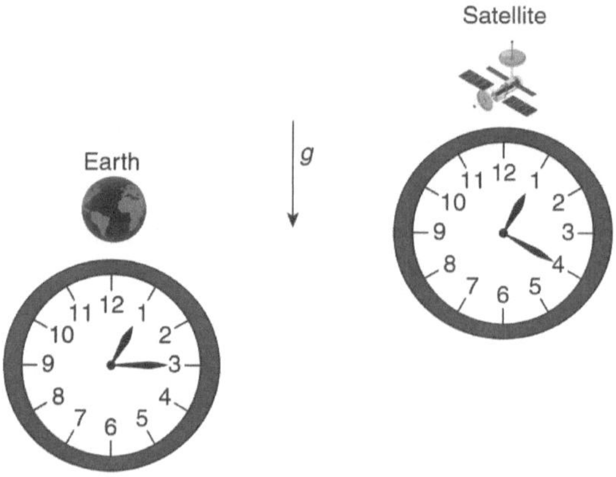

FIGURE 2.11
Representative diagram for gravitational time dilation showing clock at higher location from the ground (i.e., at higher gravitational potential) is advanced in time.

Gravitational lensing is different from our more familiar optical lensing where a convex lens focuses an object to a point and a concave lens does just the opposite. A gravitationally lensed object may get projected to the observer as two or more similar objects, or it might cause a bright ring around the object (see Fig. 2.12a), called an *Einstein ring*. Such phenomena have been observed with distant galaxies when light from them has been gravitationally lensed by the huge gravity of a black hole or that of dark matter. The Einstein ring was first discovered by astronomers in 1887 for a quasar known as MG 1131+0456. In 2020 its distance was estimated to be 10 billion light years.

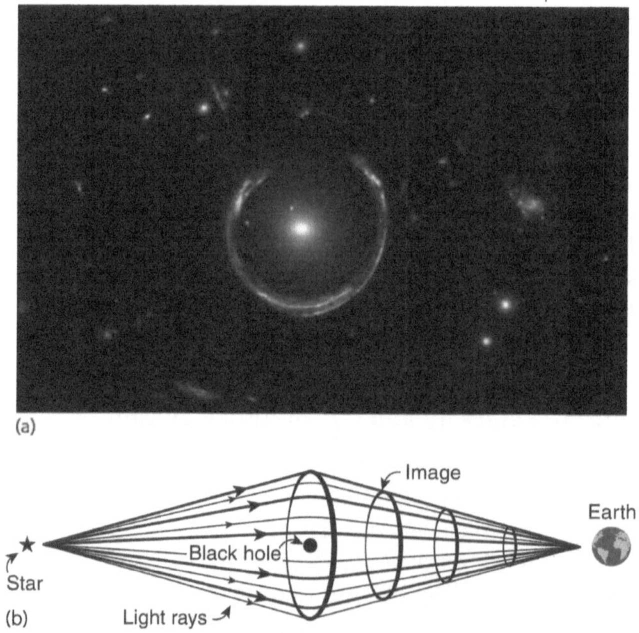

FIGURE 2.12
Gravitational lensing of star by the black hole: (a) photograph of a gravitationally lensed star, Einstein ring [3]; (b) principle.

The concept of gravitational lensing by a black hole can be understood as follows. The enormous gravity of a black hole causes bending of light from a distant star as seen from Earth if a black hole falls in the path between the star and the Earth (Fig. 2.12b). The result is *gravitational lensing*, as predicted from the general theory of relativity. The light from the star that reaches Earth can go over the top of the black hole, under the bottom, around the front, or around the back. All the light rays reaching Earth move outward from the star (due to the bending of the photon particles, of which the light is constituted, by the strong gravity of the black hole), forming a diverging cone. Thus, as the light rays from the star pass by the black hole, they get bent down toward the Earth and arrive on Earth on a converging cone. The resulting image of the star in the Earth's sky is a thin ring known as the Einstein ring (Fig. 2.12a), having a larger surface area, and hence with far larger total brightness (by a factor of 100 or more) than the star's image would have been if the black hole were absent.

2.6.9 General Relativity and Wormhole/Time Machine

H. G. Wells' novella, *The Time Machine* (1895), is generally credited with popularizing the concept of time travel by using a vehicle/device to travel purposely and selectively forward or backward through time. The term *time machine* was coined by Wells and is now almost universally used to refer to such a vehicle or device; a large number of feature films around the world have been based on the concept of Wells' time machine. The term has been synonymously used for "wormhole" (a hypothesized subway in space–time coined by Nobel Laureate Kip Thorne and predicted from the solution of the field equation of the general theory of relativity). This concept has fascinated scientists and common people alike because it states that through the wormhole we can even travel to distant galaxies (to our past?) through a shortcut. This has served as a hot topic for science fiction writers for a long time and continues to fascinate even today.

In science fiction, wormholes—tunnels through space and time—have long been the preferred means of travel across the universe. In the movie *Interstellar*, produced with scientific guidance from Kip Thorne, a venture has been shown, through the wormhole, to go from the solar system to another galaxy in order to explore possible replacement planets for a worn-out Earth in a billion years or so. Scientific investigation suggests that such a cosmic subway station may be found even at the center of our own Milky Way galaxy where the supermassive black hole Sagittarius A* resides. However, the wormhole is still a hypothesis, having no scientifically established proof of its existence; it just follows from Einstein's general theory of relativity, and in the scientific literature it is known as the Einstein–Rosen (ER) Bridge. The term *wormhole* was coined by the cosmology expert John Wheeler in 1957 for the ER Bridge. A question naturally arises: What this wormhole is and why it has such a peculiar name?

The scientific version of ER Bridge consists of a pair of black holes stuck back-to-back, each facing out into its own realm of the universe and connected by a "throat"—the wormhole. But for the common mind to understand it, let us present an analogical and demonstrative example. First, take a rectangular piece of paper and mark two points A and B, one on the top and another at the bottom at the middle line of the paper lengthwise. Now fold the paper along its third dimension—that is, in a direction perpendicular to the plane of the paper when it will look something like the one shown in Fig. 2.13, with point A now just above point B (of course without the back-to-back funnel-shaped net structures in the figure which represent two back-to-back black holes of ER bridge).

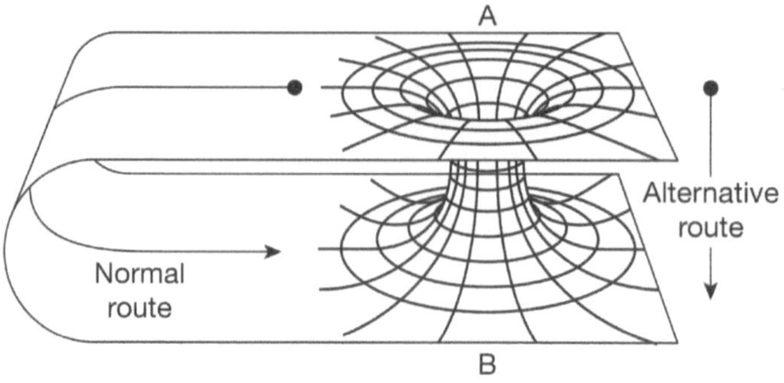

FIGURE 2.13
An artist's view of the wormhole or the Einstein–Rosen (ER) bridge.

Let us now imagine that a paper pipe is introduced via the small holes made at A and B. Observe the two ways to go from point A to point B for a worm (say an ant). If the paper was not folded, then the ant would have taken the "normal route" (a longer route), but now the ant can go from A to B via the hole of the pipe along the "alternative route" (a shorter route). The so-called alternative route is precisely the wormhole or the cosmic subway, which in the astrophysical terms is understood to be the possible ER Bridge—that is, a path between two back-to-back black holes as shown by the netted spider web. The journey through the wormhole, if it becomes possible at all, will help us to reach distant galaxies in a much shorter time. The vehicle that will take us through this alternative route (wormhole) via space that can be termed the *time machine*—as going to a faraway galaxy via the wormhole can be thought of as going to the past, perhaps a billion years ago.

To conclude this section on relativity let us not forget to mention herewith a famous limerick entitled "Relativity" that was published without attribution in an issue of the London humor magazine "Punch" in 1923 but subsequently in 1937 it has been substantiated that it was composed by A. H. Reginald Buller, a Botany professor. The limerick is given below:

There was a young lady named Bright
Whose speed was far faster than light;
She set out one day
In a relative way
And returned on the previous night.

2.7 The Game Changer of Atomic Physics—Quantum Mechanics

2.7.1 Introduction

The word *atom* means "indivisible" in Greek. Democritus, a preSocratic Greek philosopher of the 4–5th century B.C., believed that matter is inherently "grainy," that is, made up of a large number of atoms, and he was the first to propose a preliminary atomic theory of universe. Democritus combined a keen attention to nature, illuminated by a naturalistic clarity of ideas, and insight that allowed him to see matter as granular—composed of atoms that are indivisible as per his knowledge. Unfortunately, the world lost all the works of Democritus in the fire that destroyed the great library of Alexandria, Egypt, and so we know about his thought only through quotations and references made by Aristotle and other philosophers. Without the benefit of modern physics, Democritus had already concluded that everything is made up of indivisible particles. How did he come to this knowledge? His arguments were based on observation; for example, he imagined, correctly, that the wearing down of a wheel, or the drying of clothes, could produce the slow flight of particles, respectively, from wood and water (in fact, it is now known that evaporation is an atomic-level phenomenon). In ancient India, the concept of the atom was propounded by Kaṇāda of the 5–6th century B.C. in his atomistic approach to physics and philosophy in the Sanskrit text *Vaiśeṣika Sūtra*. A sublime and truly scientific thought about the atom by the modern mind may be found in Feynman's *Lecture on Physics*, where you read the answer to a thought-provoking question: "If, in some cataclysm, all of scientific knowledge were to be destroyed, and only one sentence passed on to the next generation of creatures, what statement would contain the most information in the fewest words?" The answer is "I believe that all things are made of atoms—the little particles that move around in perpetual motion, attracting each other when they are a little distance apart but repelling upon being squeezed into one another."

2.7.2 Rutherford's Atomic Model and Its Limitations

Joseph John Thomson discovered the electron (having a mass less than one thousandth of that of the lightest atom, hydrogen and having a negative charge) in 1897, and Ernest Rutherford's work in 1911 indicated that the atoms of matter do have an internal structure. According to Rutherford's atomic model, atoms are made up of an extremely tiny, positively charged nucleus, around which the electron orbits in a circular orbit—resembling the planetary organization in our solar system. Rutherford came to this conclusion by analyzing the way in which α-particles (which are positively charged particles given off by radioactive materials) are deflected when they collide with atoms of a gold foil. Rutherford's so-called gold foil experiment is the basis of his atomic model. A series of experiments were performed between 1908 and 1913 by Hans Geiger and Ernest Marsden under the guidance of Rutherford at the Physical Laboratories of the University of Manchester. After studying how an alpha particle beam is scattered when it strikes a thin metal foil made of gold, scientists learned that atoms are almost empty except for a tiny volume where its mass is concentrated and that part of the atom also has positive charge as the positively charged alpha particles are deflected by that volume at the atomic center.

In the solar system-like structure of the atom, proposed by Rutherford, the attraction between the positively charged protons in the nucleus and the negatively charged electron in the orbit around the nucleus keeps an electron in its orbit in the same way that the gravitational attraction between the Sun and the planets keeps the planets in their orbit around the Sun. But in such a classical atomic model, we need to include one more effect: The electron that is accelerating (changing its speed or direction of motion in order to be in the circular orbit according to Newtonian mechanics) will radiate away electromagnetic energy, just as an antenna/aerial of our radio or the mobile set does. Since energy is conserved, this energy must come from somewhere. In the radio or phone, this comes from the energy source (the battery), but in the atom it gets taken from the motion energy of the electron, eventually causing it to spiral inward until it collides with the nucleus. Thus, the stability of the

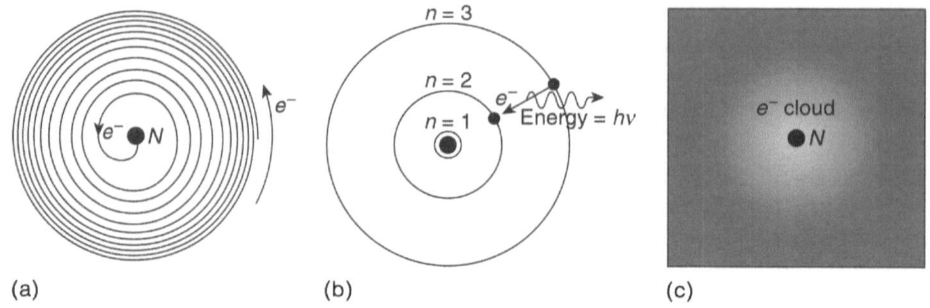

FIGURE 2.14
Atomic model: (a) Rutherford's solar system-like model; (b) Bohr's quantum model; (c) Schrödinger's model in which a single electron has the probability of being at once in different places, in an "electron cloud" whose shape is given by the so-called wave function Ψ.

atom and the matter as a whole are jeopardized. As per the classical approach, in the solar system-like Rutherford's atomic model, the electron orbit is not a circle, but rather a death spiral (Fig. 2.14a); after about 100,000 orbits, the electron would crash into the nucleus and the atom would collapse—at the ripe old age of about 0.02 nanoseconds! Further, the spiraling path of the electron approaching the nucleus would make the emitted radiation have constantly increasing frequency, which contradicts the experimentally observed atomic spectra at specific frequencies. The solution to the problems of Rutherford's atomic model was henceforth provided by Niels Bohr, who sometimes worked as a postdoctoral fellow in Rutherford's laboratory in Manchester, England.

2.7.3 Bohr's Atomic Model—A Quantum cum Classical Approach

In 1913, Niels Bohr presented a partial solution to the problem of Rutherford's atomic model of spiraling electrons to the nucleus by resorting to a quantum-cum-classical physics approach for the case of the simplest of the atoms, the hydrogen atom. Bohr's atomic model also accounted for the experimentally observed line spectra of the hydrogen atom. According to Bohr's theory, there exist certain energy states of electrons, known as stationary states, in which the atoms do not radiate. The atom emits/absorbs energy (radiation) during transition of the electron from one stationary state to another, the emitted/absorbed quanta of energy being the difference between the energies of the corresponding stationary states. Thus, for a transition from a state with higher-energy E_U (outer orbit) to a state with lower-energy E_L (inner orbit), a quantum of radiation with energy $h\nu = (E_U - E_L)$ is emitted, where h is known as Planck's Constant and ν is the frequency of radiation emitted. Thus, a line with frequency: $\nu = (E_U - E_L)/h$ appears in the spectrum; this is the well-known *Bohr's frequency condition.* In Bohr's theory, the nth stationary state of hydrogen atom corresponds to a circular orbit of radius r_n around which the electron revolves around the nucleus. In order to compute r_n, Bohr suggested to equate the centripetal force caused by the Coulomb's force between electrical charges (negative charge of electron in the orbit and positive charge of proton in the nucleus), based on Newtonian classical mechanics, with the condition of quantization (based on Planck's quantum theory) of the angular momentum of the electron as: $mv_nr_n = nh/2\pi$. From these he calculated, with $n = 1$, the Bohr radius as $r_1 = 5.29 \times 10^{-11}$ m.

The velocity of the electron in this orbit is given by $v_1 = 2.18 \times 10^6$ m/s. Thus, the velocity of electrons in an atomic orbit is much less than that of light (3×10^8 m/s), and so the atomic phenomena, in general, are nonrelativistic. De Broglie's calculations also showed that the outer orbits have greater energy than the inner ones. Thus, whenever an electron jumps into a lower orbit, the extra energy must be emitted from the atom in the form of a photon (Fig. 2.14b), and in order to jump back up to a higher orbit, the electron must absorb energy from an incoming photon. Since there is only a discrete set of orbit energies, this means that the atom can only emit or absorb photons with specific energy, that is, at specific frequencies given by Bohr's frequency condition: $\nu = (E_U - E_L)/h$.

2.7.4 Fallout of Bohr's Atomic Model and Its Shortcomings

Bohr's atomic model could account for the long-standing mystery in the rainbow of sunlight—the presence of dark lines at certain mysterious frequencies (certain colors were missing). Further, by studying the hot, glowing gases in the laboratory, it was observed that each type of atom had its unique fingerprint in the form of the frequencies of light that it could emit or absorb. Bohr's atomic model explained not only the existence of emission and absorption spectral lines due to photon emission and absorption, but also their exact frequencies corresponding to hydrogen atom.

The value of the Rydberg Constant calculated from Bohr's atomic model matches excellently well with Balmer's empirical formula (1885), thereby providing direct, striking confirmation of Bohr's model. The result of Bohr's atomic model suggested the presence of other series spectra for the hydrogen atom—those corresponding to transition resulting from $n_1 = 1$ and $n_2 = 2, 3, \ldots$; $n_1 = 3$ and $n_2 = 4, 5, \ldots$, and so on. Such series (line spectra) were identified in the course of spectroscopic investigations and are known as the Lyman, Balmer, Paschen, Brackett, and Pfund series (see Fig. 2.15). The explanation of the hydrogen atom spectrum provided by Bohr's model was a brilliant achievement, which greatly simulated progress toward developing a new quantum theory in subsequent years. In 1922, Bohr was awarded the Nobel Prize in Physics "for his services in the investigation of the structure of atoms and of the radiation emanating from them."

R. A. Millikan's oil-drop experiment in 1911, considered to be one of the most beautiful tests in the history of physics, showed that the electron was a fundamental and discrete particle and convincingly provided the first proof of Bohr's quantum theory of atom. Millikan received the 1923 Nobel Prize in Physics "for his work on elementary charge of electricity and on the photoelectric effect."

Bohr's model did not work for atoms with more than one electron and was excellent for the hydrogen atom with only one electron in a single orbit around the nucleus. In addition, his model was unable to explain the relative intensities of the frequencies observed in the spectrum. Further, Bohr's theory (or the *old quantum theory*, as it is now called) suffered from internal contradictions. In order to determine the radius of the atomic orbit, Bohr used the classical relation of Newtonian mechanics and the quantum relation given by Planck. In this connection, a letter from Rutherford to Bohr, written in 1913, is illuminating: "Your ideas as to the mode of origin of the spectrum of hydrogen are very ingenious and seem to work out well; but the mixture of Planck's ideas with the old mechanics (of Newton) makes it very difficult to form a physical idea of what is the basis of it. There appears to me a grave difficulty in your hypothesis which I have no doubt you fully realize namely, how does an electron decide what frequency it is going to vibrate at when it passes from one stationary state to the other? It seems to me that you would have to assume that the electron knows beforehand where it is going to stop." Thus, the need of the hour was a bold new idea that could establish a noncontradictory picture of Bohr's atomic model, including the idea of discretization, that is, the quantum character of energy. Here the new

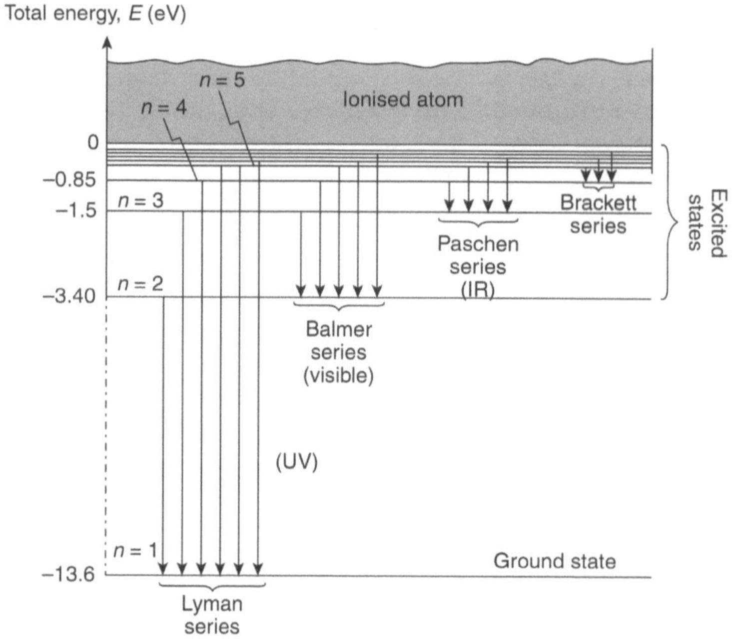

FIGURE 2.15
Different line spectra originating from electron transition between energy levels in the atom.

quantum theory or quantum mechanics was born, which could be attributed to the collected efforts of de Broglie, Heisenberg, Schrödinger, Dirac, and others.

2.7.5 The Wave–Particle Duality of de Broglie in the Quantum World

Until 1923, the internal contradiction of Bohr's theory of quantum transition of the electron from one stationary orbit to another emitting/absorbing radiation was not resolved. However, it was Louis de Broglie, who intended a career in the humanities, receiving his first degree in history, and who afterward turned his attention to mathematics and physics and finally resolved the above issue. In 1924, he wrote that just as light (electromagnetic radiation) exhibited wave-like as well as corpuscular (photonic) behavior, so micro-particles (electrons, protons, etc.) of matter must also show wave-like behavior in addition to their particle-like (corpuscular) behavior. In this respect, note the following remark by de Broglie: "For a century, the corpuscular method of analysis in optics was too much neglected in comparison with the wave method. Hasn't the converse been the case in the theory of matter? Haven't we thought too much of the particle picture and neglected the wave aspect far too long?" His argument was that since light could be seen to behave under some conditions as particles (the photoelectric effect) and at other times as waves (diffraction), we should consider that matter has the same ambiguity of possessing both particle and wave properties. De Broglie proposed associating with every micro-particle corpuscular characteristics (energy E and momentum $p = mv$; where m is the mass of the particle and v is the velocity with which it moves) on the one hand and wave characteristics (frequency ν or wavelength λ) on the other hand. This wave–particle duality of micro-particles is the fundamental basis of the new quantum theory—quantum mechanics. He argued that the relation $p = h/\lambda$, which is applicable for photons (as per photoelectric effect of Einstein, 1905), is equally applicable for micro-particles. To establish the wave–particle duality, de Broglie assumed that a moving particle of matter always has a wave associated with it (he called it a "pilot wave") and that the particle is controlled by the wave in a manner similar to how a photon is controlled by a light wave.

According to this view, each electron circling around an atomic nucleus must be considered as being accompanied by a *standing wave* that runs around the electronic orbit with a wavelength that depends on its velocity. A standing wave is a wave that we find in connection with the sound wave of a vibrating string or the circular waves in a water tank and also in a microwave/optical cavity. When the forward and backward waves in a vibrating string or two circular waves in a water tank pass through each other, at any instant of time, their effects simply add together depending on the time phase of each wave. In some places, we see the peaks of two waves adding up to an even higher peak (the so-called constructive interference); in others, we see a peak from one wave canceling the trough from the other to leave the string/water completely undisturbed (so-called destructive interference).

In 1924, de Broglie in his PhD thesis applied the same reasoning for the so-called electron wave going around the nucleus in the hydrogen atom and obtained exactly the same frequencies and energies that Bohr's atomic model predicted. According to de Broglie's concept, the only orbits (Bohr's allowed quantum orbit) that would be possible are those whose wavelengths are integral multiples of the wavelength of the corresponding de Broglie wave (Fig. 2.16). This is because for such an orbit the length of the orbit would correspond to a whole number (as opposed to a fractional number) of wavelength of the electron wave. For these orbits the wave crest would be in the same position each time, so that the waves will add up. However, for orbits whose lengths are not a whole number of wavelengths, each wave crest would eventually be canceled by a trough as the electrons went round; these would be the forbidden orbits. With this physical concept of the electron wave, de Broglie derived the relation $\lambda = h/mv$ (*de Broglie wavelength*)—a relation that combines the corpuscular or particle nature (represented by the momentum $p = mv$ of electron) with the undulatory or wave-like nature (represented by the wavelength λ of the electron wave). In this connection, it is worthwhile to quote from de Broglie's own writing: "I was convinced that the wave-particle duality discovered by Einstein in his theory of light quanta was absolutely general and extended to all of the physical world, and it

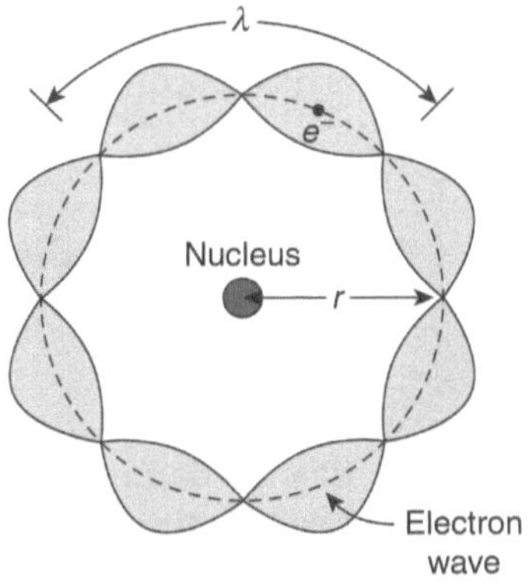

FIGURE 2.16
Louis de Broglie wave or electron wave associated with electron motion in the orbit around the nucleus of an atom (four de Broglie wavelengths shown to fit to the circumfurence of the orbit).

seemed certain to me, therefore, that the propagation of a wave is associated with the motion of a particle of any sort—photon, electron, proton or any other."

The idea of duality was first applied to electromagnetic radiation (light). As early as 1905, Einstein suggested that the quanta of radiation, introduced by Planck, should be considered as *particles* possessing not only a definite energy: $E = h\nu$ but also a definite momentum: $p = h\nu/c$. The corpuscular (particle) properties of radiation were clearly demonstrated in the Compton Effect in 1923 by Arthur Compton. He showed that when a beam of X-rays is scattered by atoms of matter, the wavelength of scattered wave is greater than the initial wavelength of the rays (though according to the classical concept, the scattered rays should have the same wavelength as the incident rays). Moreover, the difference between the wavelengths depends on the angle of scattering. The wave–particle understanding of the Compton Effect establishes that the X-ray (a type of electromagnetic radiation) also behaves as a photon (a packet of electromagnetic particles) that undergoes elastic collisions with the electrons of the atoms, in conformity with the laws of conservation of energy and momentum for colliding particles.

It should be noted here, however, that the waves associated with electrons do not subscribe to the deterministic wave of the classical physics. It would be clear subsequently that by "electron wave" we mean to indicate a so-called electron cloud arising out of the probability of the presence of electrons over a zone around the nucleus (given by Schrödinger's wave function) inside the atom (Fig. 2.14c).

In this context, it is necessary to make a clear distinction between the classical and quantum approach for two types of motion: *corpuscular* and *wave* motion. According to the classical approach, the corpuscular motion is characterized by localization of the object in space and the existence of a definite trajectory of its motion. On the contrary, the wave motion is characterized by delocalization in space. No localized object corresponds to the motion of a wave; it is the motion of a medium. In the world of micro-phenomena (dealt with in classical physics), corpuscular and wave motions are clearly distinguished. The motion of a stone thrown upward is entirely different from the motion of a wave breaking a beach. These usual concepts of classical mechanics cannot, however, be transferred to the quantum domain. In the world of micro-particles, the above-mentioned strict demarcation between the two types of motion is considerably obliterated. The motion of a micro-particle is characterized simultaneously by wave and corpuscular properties. They are not "purely" (in the classical sense) corpuscular, and at the same time they are not "purely" wave-like. In other words, it may be said that a micro-particle to some extent is akin to a corpuscle, and in some respects, it is like a wave. Further, the extent depends, in particular, on the conditions in which the micro-particle is considered. This is the essence of *wave–particle duality* in quantum mechanics.

2.7.6 Experimental Verification of Wave–Particle Duality

De Broglie's bold idea received experimental confirmation in 1927 with the discovery of *electron diffraction* through the Davisson and Germer experiment. While studying the passage of electrons through a single crystal of nickel (during 1923–1927), Clinton Davisson and Lester Germer of Western Electric (later renamed Bell Labs) observed characteristic diffraction rings on the detector screen. Their experimental result resembles the X-ray diffraction from crystals and so led them to believe that a beam of electrons is diffracted from crystals just like X-rays (as in the Compton Effect) are, which thus established the wave nature of electrons. For electron waves, the crystal lattice served as a diffraction grating. The diffraction pattern observed through experiment could be explained if the waves associated with electrons were assumed to have a wavelength given by the de Broglie relation.

Just as Davisson and Germer performed an experiment to show the diffraction of electrons by a periodic structure of matter, proving that electrons can behave as waves, Peter Kapitsa and P.A.M Dirac in 1933 went further and predicted that electrons should also be diffracted by light waves, known as the Kapitsa–Dirac effect. It is analogous to the diffraction of light by a grating but with the roles of the wave and matter reversed. (Kapitsa, incidentally, was awarded the Nobel Prize in Physics in 1978 "for basic inventions and discoveries in the area of low-temperature physics.") Shortly after the work of Davisson and Germer was published, in 1928, George Paget Thomson carried out electron diffraction experiments by passing electrons through thin polycrystalline metal targets. He observed that the wavelength associated with the electron beam was in agreement with the de Broglie relation. In 1937, C. Davisson and J. P. Thomson shared the Nobel Prize in Physics "for their experimental discovery of the diffraction of electrons by crystals," while de Broglie was awarded the 1929 Nobel Prize in Physics "for his discovery of the wave nature of electrons."

G. P. Thomson was the son of J. J. Thomson, who discovered that the electron was a constituent particle of the atom. For his part, G. P. Thomson established that even being a particle, the electron also possesses wave behavior. This represented an interesting family legacy: Both were Nobel laureates, with both being awarded in physics—J. J. Thomson received it in 1906 and G. P. Thomson in 1937. Yet another Nobel winning family legacy involves Niels Bohr. He received the Nobel Prize in Physics in 1922 for his contribution to the structure of the atom, whereas his son, Aage Bohr, received the same prize in 1975 for his work on the structure of the atomic nucleus.

Yet another family legacy is that of the Curie family. Marie Salomea Skłodowska Curie, popularly known as Madam Curie, discovered radium, while her daughter, Irène Joliot-Curie, discovered artificial radioactivity, and both of them received the Nobel Prize in Chemistry; Marie Curie in 1911 and Joliot-Curie in 1935. Marie Curie received an earlier Noble Prize in Physics, in the year 1903, along with her husband Pierre Curie, while their daughter, Irène Joliot-Curie, received the Nobel Prize jointly with her husband Frédéric Joliot-Curie in the year 1935. Thus, the Curie family has the unique legacy of five Nobel Prizes in the same family.

As is well known, Marie Curie was the first woman to receive the Nobel Prize and the only scientist to receive the prize in two different disciplines of science. Less well known is that she was also a great human being and gave her selfless service to soldiers and the general public during World War I, temporarily shelving her scientific research so that she could render service to others. She developed mobile radiology units that were used at the front. Together with her then 17-year-old daughter Irène, she visited the Belgian front hospitals in Furnes, Hoogstade, Adinkerke, De Panne, Beveren, and Roesbrugge. There she examined patients, both soldiers and civilians. She also installed X-ray equipment and offered medical advice. The three mobile X-ray cars she used at the Yser Front were nicknamed petites Curies or Little Curies.

2.7.7 The Double-slit Experiment with Electrons to Establish Its Dual Nature

An oft-repeated iconic experiment, the *double-slit experiment*, was first performed by Thomas Young in 1801 to establish the wave nature of light with the formation of interference fringes of white and black stripes on the screen. This experiment can be interpreted

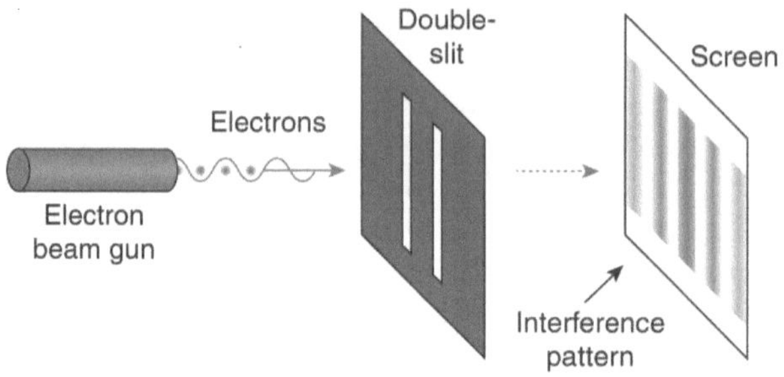

FIGURE 2.17
Double-slit experiment with electron.

in terms of the wave/particle duality of de Broglie to explain the apparently confounding result obtained with photons or electrons passing through the slits (Fig. 2.17).

When a beam of light shines through two parallel slits that are cut into a barrier and falls on a strip of photographic film (screen) beyond the barrier, photons (the particle content of light) are expected to pass through one slit or the other en route to the film. When the experiment is set up with a photon detector beyond the barrier with slits, at each slit photons are found to hurtle randomly through either the first slit or the second, which results in two separate clumps of dots forming on the film. However, slight adjustment of the experiment by removing the photon detectors, keeping only the film, profoundly alters the result. The pattern created on the film/screen now changes completely, and alternating light and dark bands appear across the film (called interference pattern) instead of those clusters of dots (Fig. 2.17). Such an interference pattern could form only if each individual photon somehow spread out like a wave and went through both slits simultaneously. Bright bands develop on the film where two wave crests coincide; overlapping crests and troughs create the dark bands. The inference drawn from this double-slit experiment is that photons are found to behave like particles with detectors present and like waves without detectors.

Further, when the electron beam is shot from an electron gun through the slits instead of shining the slits with a light beam, quantum weirdness becomes still more prominent. If instead of light shining through the slits, an electron beam is shot through the slits and again an interference pattern is seen on the screen. But if electrons are sent one at a time, one would expect each to pass through one slit or the other, and so behave just as if the slit it passed through were the only one there—giving a uniform distribution on the screen. In reality, however, even when the electrons are sent one at a time, the interference fringes still appear. Each electron, therefore, must be passing through *both* slits at the same time!

This apparently strange result of a two-slit experiment in terms of the photon/electron may be understood on the basis of the wave/particle duality of quantum mechanics propounded by de Broglie, who first proposed that particles could behave like waves. He developed an interpretation of quantum mechanics called *pilot wave theory*, where waves and particles are equally probable and eventually real. A particle always goes through one slit or the other, at the same time its "pilot wave" travels through both slits simultaneously. But this does not result in any wave–particle paradox because the experimental apparatus and the wave-surfing particle all form one interdependent system described by a pilot wave. Adding or removing a detector from the experiment just changes the system's pilot wave and hence the pattern on the screen.

2.7.8 The Uncertainty Principle: An Inescapable Property of the Material World

In Bohr's institute in Copenhagen, the most brilliant young minds of the 20th century gathered, seeking to find the golden key that would unlock the door of the true mystery of the quantum world. One of these seekers was a German, Werner Heisenberg, who was just 24 years old when he wrote the quantum mechanics equations. He was nearly the same age as Einstein when the latter wrote his three seminal papers in 1905. In many ways, the destinies of Heisenberg and Einstein were strangely interwoven, although they created theories that are a world apart. Both were revolutionary iconoclasts who challenged the established wisdom of their predecessors; both were of German origin; and both did their best work at an astonishingly young age. However, Heisenberg won the Nobel Prize when he was just 32 years old, whereas Einstein received it when he was 42. Not only were the personal destinies of these two men linked, but their scientific creations were also intricately related, though in two different worlds—the macro-world for Einstein and the micro-world for Heisenberg.

There are three distinct chords in quantum mechanics—*granularity*, *relational*, and *indeterminism*. We have already discussed the first one, *granularity* (via Planck, Einstein, and the Bohr quantum theory). As for the *relational* aspect of quantum mechanics, which is due to Heisenberg's *uncertainty principle*, it is a consequence of the wave–particle duality of de Broglie. The uncertainty principle states that if the x-component of the position of a particle is known to an accuracy of Δx, then the x-component of the momentum cannot be determined to an accuracy better than $\Delta p \approx \hbar/\Delta x$, where $\hbar = h/2\pi$ (quantum-mechanical Planck's Constant or reduced Planck's Constant). In other words, $\Delta x\, \Delta p \geq \hbar$. We do not feel the effect of this inequality in our everyday experience because of the microscopical smallness of the value of Planck's Constant ($h = 6.63 \times 10^{-34}$ J-s). For example, for a tiny particle of mass 10^{-6} gram, if the position is determined with an accuracy of 10^{-6} cm, then according to the uncertainty principle, its velocity cannot be determined within an accuracy better than $\Delta v \approx 6 \times 10^{-16}$ cm/sec. This value is much smaller than the accuracies with which one can determine the velocity of the particle. For a particle of greater mass, Δv will be even smaller. Indeed, had the value of Planck's Constant been much larger, then the uncertainty principle would possibly have been perceivable by our senses and the world around us would have been totally different. Possibly we would have been in the Wonderland-like *Tompkins* as fantasized by George Gamow, who propounded the Big Bang theory of the origin and expansion of the universe.

The relational aspect of quantum mechanics based on the uncertainty principle may be understood as follows. In order to predict the future position and velocity of a particle, one has to measure its present position and velocity accurately. In the subatomic regime (i.e., to find the position and velocity of, say, the electron), we need to do this by irradiating the atom with electromagnetic radiation (light). Some of the waves of the light will be scattered by the particle with which the position of the particle can be determined. However, the precision of this position determination depends on the wavelength of the electromagnetic radiation used; the smaller is the wavelength (or the higher is the frequency) of the radiation, the better is the precision of position determination. But as per Planck's quantum hypothesis ($E = h\nu$), higher frequency will demand a higher amount of energy of the irradiation signal even for a single photon of the radiation. This large quantum of energy is sure to disturb the particle and hence would change its velocity unpredictively. In other words, the more accurately one tries to measure the position of a particle, the less accurate becomes its velocity measurement, and vice versa. Heisenberg showed that the uncertainty in the position of the particle times the uncertainty in its velocity times the mass (i.e., momentum) of the particle can never be smaller than the reduced Planck's Constant. However, one could argue, with a better measuring apparatus: could not the velocity and the position of the particle (electron or any other particle) be measured without altering it? According to Heisenberg, the answer is *NO*—the uncertainty principle is a fundamental, inescapable property of the material world. Heisenberg importantly stated: "Not only [is] the universe stranger than we think, it is stranger than we CAN think."

With Heisenberg's uncertainty principle of quantum mechanics, the so-called deterministic approach of the classical mechanics of the Newtonian era, further elaborated by the French mathematician Pierre Simon Laplace, received a great blow at least in the micro-world. According to Heisenberg, the deterministic concept of Laplace, that whether we wind up in heaven or hell can be determined ahead of time, is all nonsense. Our fate is not sealed in a classical or quantum heaven or hell. The uncertainty principle makes it impossible to predict the precise behavior of individual atoms, let alone the universe. Moreover, according to this theory, *in the subatomic realm, only probability of occurrences can be calculated;* since, for example, it is impossible to know the exact position and velocity of an electron, and it is impossible to predict much about the electron's individual behavior. We can, however, predict with amazing accuracy the probability that a large quantity of electrons will behave in a certain way. These days we all know that using probability and statistics, we can predict with uncanny accuracy the average performance of the entire class in an examination (the bell-shaped curve, in fact, changes very little year to year). However, we cannot make any prediction about the performance of individual students in a class.

2.7.9 Application of the Uncertainty Principle in the LIGO Design

Although the uncertainty principle was discovered in connection with the phenomena related to the micro-world such as for electrons, atoms, and molecules, it also needs to be considered in regard to making extremely accurate measurements even in the macro-world. A typical example is its use in the design of the LIGO instrument used for measuring feeble gravitational waves where as small as a 10^{-19} m length difference, that is, $1/10{,}000^{th}$ the width of a proton, needs to be measured. It follows from Heisenberg's uncertainty principle that if one desires to make a highly accurate position/displacement measurement of an object, then in the process of this measurement a "kick" or "shake" of the object will be experienced, as the perturbation of the object's velocity will take place in a random, unpredictable way. The more accurate the position measurement desired, the more strongly and unpredictably the object's velocity will be perturbed.

No matter how clever one is in designing the measurement, one cannot circumvent this innate natural phenomenon of uncertainty. However, because a large object has large inertia, a measurement's kick or shake will be negligible as its velocity perturbation will be imperceptible (since velocity perturbation is inversely proportional to the object's mass). While designing the LIGO instrument based on sensing an extremely small length difference (i.e., position determination) for the detection of gravitational waves, in 2016 the high sensitivity of measurement demanded a very long (4 km) length of the two arms of the interferometer, with 40 kg weight mirrors at the end of each arm of the interferometer in order to have negligible kick or shake on the arm to be caused by velocity perturbation (as per uncertainty principle). This shows the need for the applicability of the uncertainty principle even in the design of the LIGO instrument in the macro-world. In addition to this, however, other arrangements were made in LIGO design to guard against any earthquake or traffic in nearby roads in order to detect the telltale flicker of light caused *only* by the gravitational waves.

2.7.10 Atomic Stability on the Basis of Quantum Mechanics

The stability of an atom is ensured on the basis of the quantum-mechanical concept. All our practical experience shows that atoms of matter are stable and do not collapse as per the limitations of Rutherford's atomic model. However, Bohr tried to account for the stability of the atom by assuming that electrons have only some stationary states (orbit) around the nucleus. But his atomic model gives no clue as to why the electrons should have only such selected stationary orbits prohibiting its existence in any other orbit around the nucleus. The wave nature of electron motion, propounded by de Broglie, and the uncertainty principle, due to Heisenberg, ultimately give an intuitive picture of why atoms do not collapse with electron spiraling to the nucleus on radiating electromagnetic energy while in motion around the nucleus of the atom (as was the case in terms of classical physics and the failure of Rutherford's atomic model). In quantum-mechanical terms, when we try to confine the electron wave to a very small region of space, then it will have lots of random momentum (according to Heisenberg's principle), which causes it to spread out and become less confined. This means that if a hydrogen atom tries to collapse, as shown in Fig. 2.14a, by sucking the electron into the proton, then the increasingly confined electron will get enough momentum and speed to come flying back (as if a repulsive force were acting) out to a higher allowed orbit (the so-called stationary states of Bohr's atomic model) as determined by the de Broglie wavelength. This gives the stability of an atom with electron(s) going around the nucleus in Bohr's stationary state orbit, which is possible with de Broglie's wave–particle duality concept and Heisenberg's uncertainty principle.

2.8 Quantum Mechanics in the Hands of Schrödinger, Dirac, and Others

2.8.1 Introduction

Indeterminism is the cornerstone of quantum mechanics and is inseparably woven into the heart of this theory, which started with Heisenberg's uncertainty principle. Indeterminism ultimately formed the true basis of quantum mechanics—*probability is in the genetic code of the atomic world*—that was formalized by the principal experts of *new quantum mechanics*, chiefly by Erwin Schrödinger, Paul Dirac, and others such as Richard Feynman and Max Born. However, it is to be noted that the term *probability* has different meanings in classical physics and quantum physics. In the classical world, it means a quantitative and deterministic chance, whereas in the quantum world it means "lack of determinism." That is, the particle may be present here, or it may even be present simultaneously here and there. The atomism of ancient times had also anticipated this aspect of modern physics, but in a different way: The appearance of the laws of probability at the deeper level of matter. Epicurus (2nd century B.C.) stated that atoms can on occasion deviate from their course of motion by chance, whereas Lucretius (1st century B.C.) explained this principle in beautiful language: This deviation, he said, occurs *incerto tempore. . . incertisque loci* (meaning "at an uncertain place and uncertain time"). The same randomness, the same appearance of probability at the elementary level, was visualized/anticipated by those ancient thinkers, too, in a philosophical way rather than in a systematic and mathematical way as in modern physics.

2.8.2 Schrödinger's Wave Mechanics

Erwin Schrödinger, the Austrian physicist, developed a number of fundamental results in the field of quantum theory that formed the basis of *wave mechanics* of quantum physics. The term *wave mechanics* was introduced by Schrödinger himself. The following obvious question arises in quantum mechanics: Is the electron (or proton or an alpha particle) a wave or a particle? The answer may be given by quoting Richard Feynman: "It is neither a wave nor a particle." According to quantum theory, it is described by the wavefunction Ψ, which depends on the position and contains all information that is known about the system. In the nonrelativistic domain, the wave function satisfies what is known as the *Schrödinger equation*:

$$i\hbar\frac{\partial\Psi(\vec{r},t)}{\partial t} = -\frac{\hbar^2}{2m}\nabla^2\Psi(\vec{r},t) + V(\vec{r})\Psi(\vec{r},t)$$

where $i = \sqrt{-1}$, $\hbar = h/2\pi$ (quantum-mechanical Planck's Constant or reduced Planck's Constant or Dirac's Constant—which represent the quantum of angular momentum and are considered by many to be more fundamental in quantum mechanics than h since 1930), Ψ is the wave function, m is the particle mass, V is the potential, and ∇^2 is the Laplacian in Cartesian coordinate.

In November 1925, in a seminar held in Zurich, Schrödinger spoke about the wave nature of the electron going around the orbit in an atom, as proposed and established by de Broglie. When he had just finished, Peter Debye, Nobel Laureate in Chemistry in 1936, asked him: "You speak about waves (associated with electron orbit), but where is the wave equation (Just as we have the wave equation for electromagnetic wave hypothesized by J. C. Maxwell)?". Schrödinger ultimately produced his famous wave equation, the master key of quantum mechanics, which is the basis for modern physics. But what were these waves that Schrödinger's equation described? This central puzzle of quantum mechanics remains a potent, controversial issue to this day. However, in the language of quantum mechanics given by Schrödinger's wave function, the wave nature of particles implies that they can be in several places at one time in a so-called *superposition*, unlike the classical view that they're either here or there. Why superposition is a must property for particles in the micro-world (dealing with quantum mechanics) but not in the macro-world (dealing with classical mechanics) may be understood as follows. An electron takes only a quadrillionth (10^{-15}) second to orbit the nucleus in an atom. Thus, it can demonstrate the funky quantum-superposition behavior (to be in two or even all sides of the atom at once) when we consider it for the next collision if the intervening period is even a very small fraction of a second. But if we consider a bowling ball outdoor observed by me with the photons being bounced off from it, then its quantum superposition will get ruined even before I have a chance to become consciously aware of it. Thus, quantum observation and the concept of quantum superposition are relevant where the transfer time of information is much smaller than the process taking place. This is a phenomenon of serious concern in the micro-world.

Let us again quote Feynman: "Where did we get that (the Schrödinger equation) from? Nowhere. It is not possible to derive it from anything you know. It came out of the mind of Schrödinger, … invented in his struggle to find an understanding of the experimental observations of the real world." Of course, Schrödinger had some reasoning to get to this equation. In spite of the fact that it is not possible to have a rigorous derivation of the Schrödinger equation, it readily got accepted because its solution agreed extremely well with the experimental data.

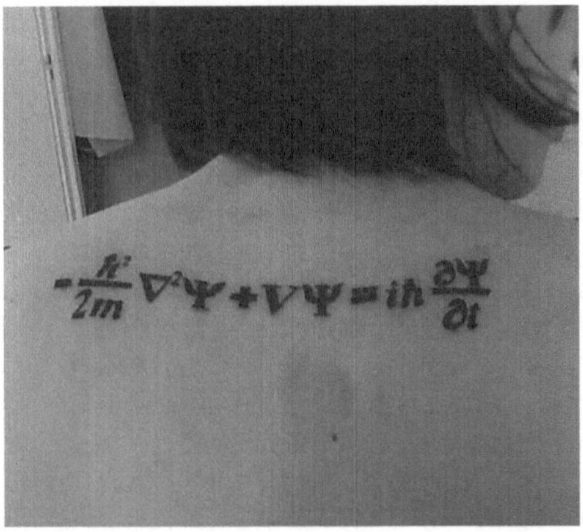

FIGURE 2.18
Schrödinger's equation for the wave function of a particle tattooed on the back of a lady [4].

Schrödinger equation is so very popular that people love to tattoo the equation even in their body (Fig. 2.18).

With regard to Schrödinger's wave equation, two topics are of special importance: *Copenhagen interpretation*, proposed by Bohr and Heisenberg in the 1920s, and a thought experiment popularly known as *Schrödinger's cat*, devised by Schrödinger himself in 1935 to take care of quantum weirdness which says that things can be at different places at once—that is, the so-called *superposition* in quantum mechanics—that left people tearing their hair out in utter confusion.

The essence of *Copenhagen interpretation* is that if something is not being observed (in physics, observation and measurement are considered to be synonymous), then its wave function changes according to Schrödinger's equation, but if it is being observed, then its wave function collapses so that one finds the object in only one place (as in the classical concept). This collapse process is both abrupt and fundamentally random, and the probability that one finds the particle at any particular position is given by the square of the wave function. There are other elements (in which there is disagreement among quantum physicists) in the Copenhagen interpretation as well, but the part above is what is most agreed on. Although with the Copenhagen interpretation, business as usual continued in the world of physics, not everyone was thrilled or satisfied. If wave function collapse really happened, then this would mean that a fundamental randomness was built into the laws of nature in the micro-world. Quantum weirdness on a microscopic scale possibly declares a counterintuitive claim: Reality itself may be subjective in respect to the observation/experiment. The bedrock principle of the objectivity of age-old view of science at every level before the advent of quantum mechanics possibly becomes incompatible at the micro-level. Einstein was deeply unhappy about this interpretation and expressed his preference for a deterministic universe with the oft-quoted remark "God does not play dice." After all, the very essence of physics had been to predict the future from the present, and now it was supposedly impossible not just in practice, but even in principle.

Schrödinger himself was perplexed with the Copenhagen interpretation, and in terms of a thought experiment popularly known as Schrödinger's Cat, tried to illustrate the essence of the Copenhagen interpretation of quantum mechanics applied to macro-objects. In simple terms, the thought experiment stated that if you placed a cat and something that could kill the cat in a box and sealed it, you would not know if the cat was dead or alive until you opened the box, so that until the box was opened, the cat was (in a sense) *both* "dead and alive." Schrodinger constructed his imaginary experiment with the cat to demonstrate that simple misinterpretations of quantum theory can lead to absurd results that do not match the real world. Unfortunately, many popularizers of science sometimes embrace the absurdity of Schrodinger's Cat and claim that this is how the world really works. In quantum theory, quantum particles can exist in a superposition of states at the same time and collapse down to a single state upon interaction with other particles. Some scientists, at the time that quantum theory was being developed (1930s), drifted from science into the realm of philosophy and stated that quantum particles only collapse to a single state when viewed by a conscious observer. Schrodinger found this concept ludicrous and devised his thought experiment to clarify the absurd, yet logical, outcome of such claims. "Schrodinger's Cat" was just a teaching tool invented to try to make the fact of quantum randomness more of an obvious fact by reducing the observer-driven notion to absurdity.

In addition to many Western philosophers like Voltaire, Schopenhauer, and Max Müller, the Western scientists were also impressed by the sublime philosophical and metaphysical thoughts of ancient Indian texts like the *Bhagavad Gita* and *Upaniṣhads* (a collection of Sanskrit texts transmitted orally from teacher to student over thousands of years and concerned with the nature of reality, mind, and the self). An ardent student of the *Upaniṣhads*, Schopenhauer had declared, "In the whole world there is no study so beneficial and as elevating as that of the Upaniṣhads. It has been the solace of my life. It will be the solace of my death." We already know that Robert Oppenheimer, who led the Manhattan Project which developed the world's first nuclear weapons and witnessed the first atom bomb explosion in Japan during World War II; uttered from the *Bhagavad Gita*: "Now I am become death, the destroyer of worlds." These were the words Lord Krishna spoke to Arjuna while showing his multitude of divine forms. Schrödinger, who was first exposed to Indian philosophy around 1918, through the writings of the German philosopher Schopenhauer, also liked some of the sublime metaphysical thoughts in the Upaniṣhads. Some overenthusiasts who believed in Indian supremacy in shaping world thought are prejudiced with the absurd claims that Schrödinger and other scientists possibly baked the lessons of the Upaniṣhads into quantum theory. They believed the crazy idea of quantum mechanics: "the reality we see exists only when we are there to observe it" is somehow supported by the implications of subjective reality in the metaphysical thought of the Upaniṣhads. Such thinking is utterly misleading as Schrödinger was primarily a physicist deeply entrenched in the methods of modern science. Indian philosophy soothed his soul, but it is unlikely that it helped him frame mathematical equations of quantum mechanics even in a remote way.

Schrödinger was often critical of many ancient Indian ideas, pointing out that they were prone to superstition. He believed that modern science represented the zenith of human thought. He sought Indian philosophy, not to replace the methods of science but as an inspiration to all of us from a philosophical point of view. He was aware that mixing two systems of thought separated by thousands of years was not easy. He believed Western thought needed to borrow ideas from Indian philosophy, though with great care. As he wrote, "I do believe that this is precisely the point where our present way of thinking does need to be amended, perhaps by a bit of blood-transfusion from Eastern thought. That will not be easy, we must beware of blunders— blood-transfusion always needs great precaution to prevent clotting. We do not wish to lose the logical precision that our scientific thought has reached, and that is unparalleled anywhere at any epoch."

The metaphysics of ancient thought, be it Indian or any other oriental or occidental school, tried to understand the truth about universal phenomena using a different language of knowledge compared to the modern science. A healthy ambience can indeed be created if we respect each other without claiming the supremacy of one to the other.

2.8.3 Paul Dirac's Contribution to Quantum Mechanics

The English theoretical physicist Paul Dirac contributed significantly to the development of new quantum mechanics and quantum electrodynamics. Dirac constructed the entire mathematical formalism of Heisenberg's theory based on the abstract mathematics of Hilbert space. Many consider Dirac to be the greatest physicist of the 20th century after Einstein, though he is much less well-known than Einstein. This may be due partly to the rarefied abstraction of his science and partly to his reserved and silent character. His physics has the pristine clarity of music that comes from within. Dirac made a thoughtful comment in his comparison of science and poetry (in a discussion with Robert Oppenheimer): "In science one tries to tell people, in such a way as to be understood by everyone, something that no one ever knew before. But in poetry, it's the exact opposite." Here we may also quote William Butler Yeats, the Nobel Prize winner in Literature in 1923, who said: "What can be explained is not poetry." In fact, in my opinion, poetry is the language of the soul and thus need not be explained (only to be realized in the heart and soul of a reader). For Dirac the world was not made of things but of an abstract mathematical structure. Dirac's famous comment was: "God used beautiful mathematics in creating the world." Dirac can be considered to be the consolidator of the complete theoretical foundation of quantum mechanics. Einstein made this significant remark about Dirac: "Dirac to whom in my opinion we owe the most logically perfect presentation of quantum mechanics."

Dirac once said: "A great deal of my work is just playing with equations and seeing what they give." In fact, we intend to give here the flavor of his contribution in quantum mechanics only in terms of some logical views rather than entering into the mathematical intricacies. Schrödinger analyzed quantum mechanics in terms of wave mechanics based on his famous equation making quantum mechanics quite "crazy" to common people and experts alike. But Dirac consolidated the whole probabilistic craziness of quantum mechanics with a logical vision of the whole issue on the basis of his intricate mathematics in terms of the so-called eigenvalues of the operator (*eigen* is a German term meaning characteristic or one's own) associated with the physical variables in question. Unlike Heisenberg, who considered the position and momentum of a particle to be undefined simultaneously with the same precision, Dirac considered that every object in the microworld is defined in an abstract space (the Hilbert space) and has no property in itself, apart from those that are unchanging, such as mass. Its position and velocity, its angular momentum, its electrical potential, and so on, acquire reality only when their "interaction" with another object takes place. Thus, the *relational* aspect of quantum theory now becomes universal in Dirac's visualization.

Dirac's quantum mechanics provides the general recipe for calculating the spectra (the *granular* nature) of the variables (i.e., the set of the particular values a variable can assume) and also provides information on which value of the variable in the spectrum will manifest itself in the next interaction—but only in the form of probabilities. We do not know with certainty where the electron will appear, but we can compute the probability that it will appear here or there. The absence of determinism or *indeterminacy* is intrinsic to small-scale physics and hence of quantum mechanics. The probability of finding an electron or any other micro-particle at one point or another may be imagined as a diffuse cloud, denser where the probability of finding the particle is stronger. These electron clouds are termed *orbitals*—a term that we all come across in our school/college textbooks on atomic physics. The orbital is a three-dimensional region in space around the nucleus of the atom, where there is high probability of finding the electron. In contrast, the orbit is a definite circular path around the nucleus in which electrons are considered to revolve around the nucleus as per Rutherford's and Bohr's atomic model.

The efficacy of Dirac's consolidation of quantum mechanics in a useful manner for practical use was soon realized and proved to be extraordinary. Dirac's quantum mechanics is the unavoidable mathematical tool used today by engineers, chemists, and molecular biologists. It is this quantum mechanics that helps us to build modern computers and to advance in molecular chemistry and biology, including the advancement that is taking place in laser and semiconductor electronics.

In our evaluation of Dirac's contribution to quantum mechanics, we need to talk about the *Dirac equation*. In the development of quantum theory/quantum mechanics, there are five pivotal/fundamental equations: Planck's radiation law, Louis de Broglie's wave–particle duality relation, Heisenberg's uncertainty principle, Schrödinger's wave equation, and finally the Dirac equation. We have already talked about the first four, and now we will say a few words about the Dirac equation. The Dirac equation brings together the two cornerstones of modern physics: quantum mechanics and relativity (the special theory of relativity and not the general theory). It helps to describe how particles like electrons behave when they travel close to the speed of light. It was the first step toward the so-called quantum field theory, which has given us the standard model of particle physics and the Higgs boson. This equation is very powerful, mainly because of what it signifies and the role it played in the history of 20th-century physics. The Dirac equation predicted the existence of antimatter—the mirror image of all known particles. Antimatter was subsequently found to be possible to generate in particle accelerators, which was first experimentally observed by Carl Anderson in 1932 while studying cosmic rays. The Dirac equation in the form originally proposed by P.A.M. Dirac in 1928 is as follows:

$$\left(\beta mc^2 + c\sum_{n=1}^{3}\alpha_n p_n\right)\psi(x,t) = i\hbar\frac{\partial\psi(x,t)}{\partial t}$$

where $\psi(x, t)$ is the wave function for the electron of rest mass m and space–time coordinates x and t. The p_1, p_2, p_3 are the components of the momentum, understood to be the momentum operator in the Schrödinger equation. Also, c is the speed of light, and $\hbar$ is the reduced Planck Constant. These fundamental physical constants reflect the presence of both special relativity and quantum mechanics in the Dirac equation.

So far we have briefly discussed the scientific contribution of Dirac. However, on passing, it may be interesting to compare the personality of Dirac with that of Newton; the former being the stalwart of quantum mechanics while the latter was that of classical mechanics, and both of them were Britons.

Newton was undoubtedly one of the most brilliant minds in the history of science. No other scientist has had such an important, wide-ranging, and enduring influence on the entire world as Newton did. But he was an eccentric, an egoist, and a troublemaker who tolerated no criticism; was uncompromising, vengeful, and conniving. When two other scientists, Robert Hooke and Gottfried Leibniz, offered criticism or competed with Newton for claim over the revolutionary ideas of gravity and calculus, Newton pursued personal vendettas against them. These grudges persisted even after Hooke and Leibniz were in their graves, with Newton trashing the reputations and discoveries of both Leibniz and Hooke while elevating his own.

On the other hand, Dirac even being a great figure in the development of quantum mechanics was a soft spoken and timid person and had no arrogance with any person whatsoever. He was so reserved within himself in personal relationship and absorbed in his scientific thinking that his colleagues in Cambridge jokingly defined a unit called "dirac," which was one word per hour. Dirac was also noted for his personal modesty. He called the equation for the time evolution of a quantum-mechanical operator, which he was the first to write down, the "Heisenberg equation of motion." Most physicists speak of Fermi–Dirac statistics for half-integer-spin particles and Bose–Einstein statistics for integer-spin particles. While lecturing later in life, Dirac always insisted on calling the former "Fermi statistics" and referred to the latter as "Bose statistics." In fact he coined the terms "fermions" and "bosons" for the fundamental particles following the Fermi–Dirac statistics and the Bose–Einstein statistics, respectively. The venerable Bohr said about him, "Of all physicists Dirac has the purest soul."

2.8.4 Quantum Mechanics and the Periodic Table of Elements

The exclusion principle propounded by the Austrian physicist Wolfgang Pauli gives the electron distribution in different orbitals of atoms and is the basis of the modern periodic table of elements. In 1926, Pauli published a paper stating his exclusion principle. According to this principle, no two electrons in the same atom can have identical values for all four of their quantum numbers. In other words, (1) no two electrons can occupy the same orbital and (2) two electrons in the same orbital must have opposite spins. Pauli received the Nobel Prize in Physics in 1945 for the "discovery of the Exclusion Principle, also called the Pauli Principle."

The modern periodic table was developed by Henry Moseley in 1913. According to this periodic table of elements, *properties of elements are a periodic function of their atomic number.* The atomic number gives the number of protons in the nucleus of an atom, and this number increases by one in going from one element to the next. In the modern periodic table, there are 18 vertical columns known as *groups* and 7 horizontal rows known as *periods*, as per the International Union of Pure and Applied Chemistry (IUPAC; Fig. 2.19). The group number denotes the number of valence electrons, while the periodic number indicates the number of shells (orbits). The periodic table (that is probably pinned to and adorns every chemistry classroom in the world) was first developed in the late 1860s by the Russian chemist Dmitri Mendeleev in which elements are arranged in terms of atomic mass (the mass of an atom, that is, approximately the sum of the mass of protons and neutrons in the nucleus of an atom as mass of electrons is practically negligible compared to that of proton or neutron).

Chemists of the two preceding centuries fully understood that all the substances in this universe, whether in Earth, Mars, or the Galaxies, were just the combination of a relatively small number (less than a hundred) of simple elements: hydrogen, helium, oxygen, and so on to uranium. Mendeleev put these elements in order of the atomic mass in the periodic table. However, the question arises: Why these specific elements? Why does one element have certain properties that others do not? Why, for instance, do some elements combine easily, whereas others do not? What is the secret of the curious structure of the Mendeleev's periodic table of elements? The properties of elements, with everything else, follow from the solution of the equation of quantum mechanics that determines the form of the orbitals of electrons. This equation has a certain number of solutions that correspond exactly to hydrogen, helium, oxygen, and the other elements. Thus, Paul Dirac was correct when he said that "quantum mechanics could reduce all of chemistry to a set of

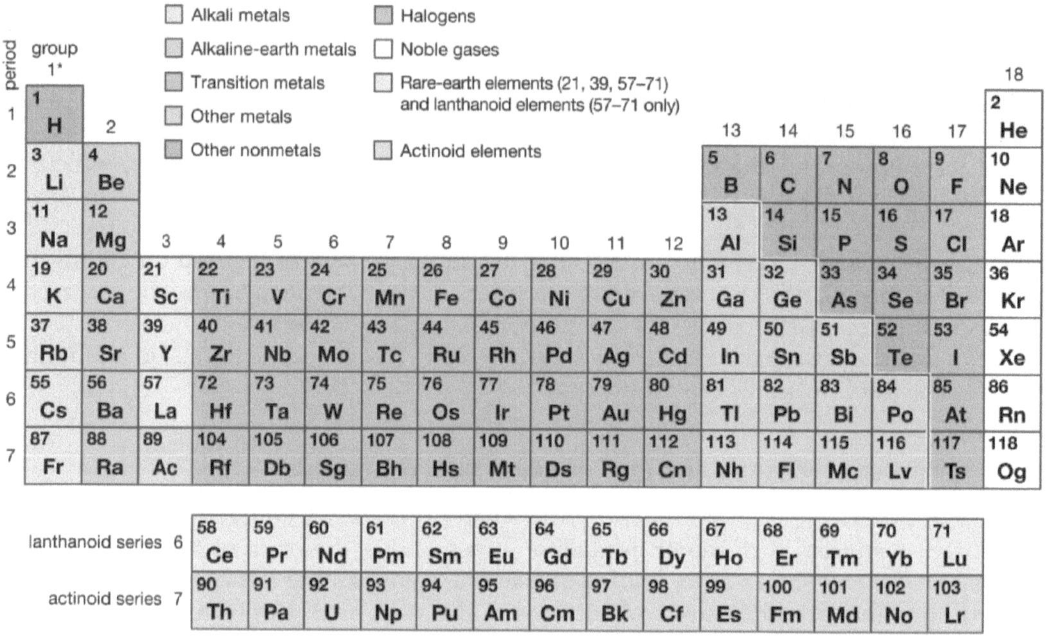

FIGURE 2.19
Modern periodic table of elements [5] (*group:* elements with similar properties, *period:* begins when a new principal energy level begins filling with electrons (following Pauli's exclusion principle).

mathematical equations." Accordingly, quantum mechanics deciphers perfectly the secret of the structure of the periodic table of elements—the modern form of which is shown in Fig. 2.19. In addition, quantum mechanics allows us to calculate the properties of chemicals that we have yet to see in nature.

2.8.5 Feynman and the Feynman Diagram

The Feynman Lectures on Physics is perhaps the most popular physics book ever written. Feynman presented these lectures before undergraduate students at the California Institute of Technology (Caltech) during 1961–1963. His primary contribution is the path integral formulation of quantum mechanics, which is an elegant way to visualize the wave–particle duality. He was awarded the Nobel Prize in Physics, along with Sin-Itiro Tomonaga and Julian Schwinger, in 1965 "for their fundamental work in quantum electrodynamics, with deep-ploughing consequences for the physics of elementary particles." In Feynman's approach, a particle is not supposed to have a single history or path in space–time, as would be the case in the classical or nonquantum approach. Here the particle is supposed to go from S to T by every possible path (Fig. 2.20a). Each path is associated with a couple of numbers: One representing the size of a wave and the other the position in the cycle (i.e., whether it is a crest or a trough of the wave). The probability of going from S to T is obtained by adding up the waves for all the paths. When one compares different sets of neighboring paths, some sets are seen to cancel out (as the phase or position in the cycle for different paths here differ significantly), while the other set in which the phase of neighboring paths do not practically differ will not cancel out; these later paths correspond to Bohr's

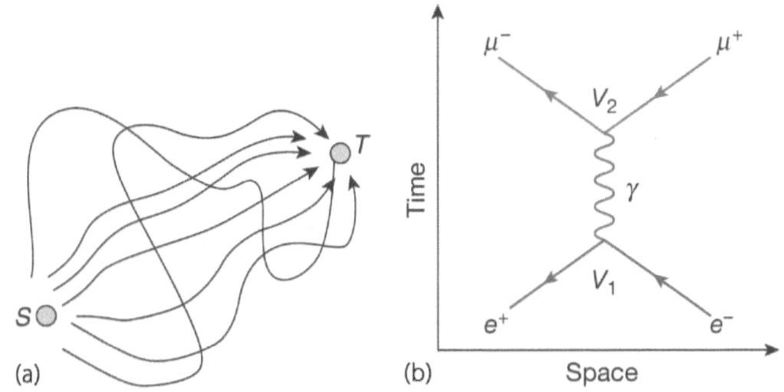

FIGURE 2.20
Feynman diagram: (a) schematic showing the electron's possible paths of motion from S to T; (b) a typical Feynman diagram showing the annihilation of the electron (e^-) by a positron (e^+) leading to the formation of a muon (μ^-) and an antimuon (μ^+).

allowed orbits. This technique of computing the probability of a quantum event is known as Feynman's sum over paths, or *Feynman's integral.*

Feynman diagrams are used in quantum mechanics and show what happens when elementary particles collide. Feynman introduced these diagrams in 1948 to serve only as a bookkeeping device for simplifying lengthy calculations in the area of quantum electrodynamics. These diagrams are a simple, yet elegant, visualization of the interaction of subatomic particles that can otherwise be very complex and difficult to understand. We know that the fermions, referred to as matter particles, are the elementary particles with half-integral spin; electrons and protons are examples of fermions. In contrast, the bosons with integral spin are termed the force carriers or the field particles. Both of these particles are the basis for all the interactions that take place in the universe. Whenever two particles interact, some other particles get exchanged between them, and the Feynman diagram represents a handy two-dimensional picture of these interactions happening in space–time.

While drawing the Feynman diagram, certain rules are followed. Here different types of lines are assigned to particles belonging to different categories: For fermions (like electrons), use straight line, while for bosons (such as photons) use wavy/curly lines. The point where any three lines meet is termed the vertex (V)—the point where interaction actually occurs and a particle is either absorbed or emitted. The incoming particle is shown with an arrowhead pointing toward V, whereas the arrowhead for the outgoing particle points away from V. These rules for particles are also applicable for antiparticles, but the latter move backward in time (i.e., the direction of the arrowhead is reversed). An incoming antiparticle has its arrowhead pointed away from V, while the outgoing one has it pointed toward V. Figure 2.20b shows a Feynman diagram, including both particles and antiparticles. It shows the annihilation of an electron (e^-) by a positron (e^+). This annihilation of the particle-antiparticle pair leads to forming a muon (μ^-) and an antimuon (μ^+) via a photon indicated by the wavy line γ. Both antiparticles are shown moving backward in time, while the particles are moving forward.

These diagrams are one of the fundamental tools used to make precise calculations for the probability of occurrence of any process by physicists. Different diagrams can represent a single interaction process, and the contribution from each diagram is taken into consideration while calculating this probability. David Kaiser, an American physicist and historian of science, has commented, "Since the middle of the 20th century, theoretical physicists have increasingly turned to this tool to help them undertake critical calculations. Feynman diagrams have revolutionized nearly every aspect of theoretical physics." Undoubtedly, these diagrams are one of Feynman's finest contributions to the physics fraternity.

2.9 Quantum Weirdness and Quantum Entanglement

2.9.1 Quantum Mechanics—A Weird Subject

Nearly 100 years since quantum mechanics—the physics of atoms, photons, and other elementary particles—was introduced, it remains as baffling and weird a subject as ever. Experiments have repeatedly confirmed the theory's strange predictions with phenomenal accuracy. Technologies derived from it drive the world's economy: The electronics industry and the newly developing photonics as we know it today wouldn't have existed without quantum mechanics. Most of the things that people are doing on almost every floor of every physics department in the world now are based on quantum mechanics in one way or another. And yet, despite the theory's unquestioned dominance and practical significance, physicists still don't agree on what it means or what it says about the nature of reality. Physicists are practically stumped with quantum mechanics.

The confusion of quantum weirdness dates back to the early days of quantum mechanics, in the 1920s, when Bohr clashed with Einstein. Bohr, an almost oracular figure in 20th-century physics, argued that when studying the atomic world, physicists must give up the notion of a reality that exists independently of their own measurements. The message of quantum mechanics is inescapable, he said, and exceedingly strange: Atoms and all other particles do not possess definite positions, energies, or any properties until they are measured in an experiment. For Bohr, the objects we consider elementary particles don't have a definite existence until they are observed in a measurement. On the very smallest scales, reality is blurry, not sharply defined; at least when no one is looking; a single, definite state crystallizes only upon measurement. To be clear, it's not just that physicists don't know what the properties are; the properties literally only come into being at the time of the measurement. However, Einstein categorically rejected Bohr's view. While strolling around the grounds of the Institute for Advanced Study in Princeton University one moonlit night, Einstein famously asked a colleague, "Do you really believe that the Moon is not there when you are not looking at it?"

At least a dozen interpretations of quantum mechanics vie for physicists' hearts and minds, each with a radically different take on reality. The central issue is that physicists don't know what the most basic equation of quantum theory (unlike electromagnetism which is governed by Maxwell's Equations and relativity which is governed by Einstein's field equations). A question that has been asked ever since its inception is the following: "Does Schrödinger's wave equation describe a fundamental feature of the physical world, or is it instead just a handy way to predict experimental results?" Whether it is the "Copenhagen interpretation" of Bohr and Heisenberg or Schrödinger's famous thought experiment, "Schrödinger's Cat," nothing settles the uncanny issues of quantum mechanics.

A question that naturally arises is that if everything is ultimately understood to be constituted of those blurry particle-waves (governed by wave mechanics), why don't we see quantum effects in our everyday lives? Why aren't people, trees, and everything else as wavy and indistinct as the atoms they're made of? The short answer is: "No one really knows"—hence the crazy cornucopia of quantum interpretations. However, one thing is for sure: The micro-world may be "weird" as per quantum mechanics, but the macro-world is possibly not! The apparent determinism of the macroscopic world is due to the fact that the microscopic randomness cancels out on average, leaving only fluctuations that are too minute for us to perceive in our everyday life. Anyway, among all controversies in physics, a few are so great that they tower over the rest and last for generations. The controversy about how to interpret quantum mechanics may clearly be one of the most important of them. As Roger Penrose, recipient of the 2020 Nobel Prize in Physics, states: "There are probably more different attitudes to quantum mechanics than there are quantum physicists."

2.9.2 Einstein's Thought on Quantum Mechanics and Quantum Entanglement

It is sometimes alleged that Einstein rejected quantum mechanics, just as some scientists rejected the theory of relativity. In actual fact, Einstein never rejected quantum mechanics as a scientific theory, although, as regards the philosophical aspects of the discussion of those times, he held a negative attitude to the indeterminism inbuilt in quantum mechanics. Einstein remained convinced until his death that quantum mechanics was only a steppingstone toward a deeper and more comprehensive theory that would make sense of the uncanny phenomena of the quantum world. The notion of a completely objective reality is the bedrock principle of science (of course, that is for nonquantum concepts), which is the main reason Einstein was so uncomfortable with Bohr's "nothing exists without observation" take on quantum theory. In a letter to Max Born on December 4, 1926, Einstein made his famous remark regarding quantum mechanics: "Quantum mechanics is certainly imposing. But an inner voice tells me that it is not yet the real thing. The theory says a lot, but does not really bring us any closer to the secret of the 'old one'. I, at any rate, am convinced that He is not playing at dice." This quotation is often paraphrased as "God does not play dice." Niels Bohr reportedly replied to Einstein's sentimental comment by advising him to "stop telling God what to do."

Interestingly, Einstein nominated Heisenberg and Dirac, the quantum mechanics stalwarts, for the Nobel Prize, recognizing that they had understood something fundamental about the mysteries of nature. However, he was seriously disturbed by the fact that quantum mechanics was dependent on indeterminacy. The young minds of the Copenhagen school of quantum mechanics were dismayed that their spiritual father, the man who had the courage to think the unthinkable, was now pulling back and apparently feared to accept this master leap into the unknown—the very leap that he himself had triggered. They felt extremely depressed that the same Einstein who had boldly taught that time was not universal and was now afraid to accept the fact that Mother Nature could be this strange in the micro-world, which was profoundly supported by experimental facts.

The trouble with quantum physics is that it seems to defy the common-sense/classical notions of causality, locality, and realism. For example, we know that the Moon exists even when we are not looking at it; that's realism. Causality tells us that if I or you flick a light switch "here", the bulb will illuminate "there" instantaneously, with the electromagnetic signal traveling with the velocity of light (3×10^8 m/s). But due to this hard limit on the speed of light, if I/you flick a switch now, the related effect cannot take place instantly a million lightyears away according to locality. However, these principles break down in the quantum realm. Perhaps the most famous example is *quantum entanglement*, which says that particles that are even on opposite sides of the universe can be intrinsically linked so that they can share information instantly—an idea that made Einstein scoff.

Einstein endeavored to formulate an interpretation of quantum mechanics that was free from noncausality, nonlocality, and nonrealism but failed to solve the problem. In 1935, Schrödinger used the term *entanglement* to indicate the nonlocal nature of quantum mechanics. According to Schrödinger's concept, two particles can get entangled with a binary, yes-or-no-like property or state, such as spin or phase or polarization. But that state remains fuzzy (i.e., in "superposition") until one particle is observed or measured. Then, at the exact moment of observation/measurement, even if the particles are separated by lightyears of space, the other particle takes on the opposite state of its twin. To Einstein such nonlocal teleportation-like effect was so absurd that he described it as "spooky action at a distance." How and why small particles can get entangled makes no sense in the context of our everyday experience of classical physics. However, at tiny scales, as per quantum mechanics, the universe appears to play by different rules, many of which appear to be paradoxical and defy reason. But Einstein strongly believed in "locality"; that is, measuring the state of one particle will not affect the other one light years away as demanded by quantum entanglement. However, if nonlocality has to be true, then there must be some "hidden variables" that result in the so-called entanglement, which Einstein dubbed as "spooky action at a distance."

To justify his view, Albert Einstein, Boris Podolsky, and Nathan Rosen published a paper in 1935 in which they argued that if Schrödinger's entanglement concept was to be verified, then two particles (which are millions of kilometers apart) can be *entangled* in the sense that by determining a property of one of the particles the property of the second particle can be *instantaneously* changed. But the special theory of relativity forbids the transmission of any signal faster than the velocity of light and hence appears to violate the causality. This theory of quantum entanglement came to be known as the EPR paradox (so named based on the first letters of their last names). Einstein used to believe that because of the EPR paradox, quantum mechanics could not be considered a complete theory of nature. Bohr vehemently disagreed with this view of Einstein and defended the far stricter Copenhagen interpretation of quantum mechanics. The two men often argued passionately about the subject, especially at the Solvay Conferences of 1927 and 1930, neither ever conceded defeat. Today, most physicists regard the EPR paradox more as an illustration of how quantum mechanics violates classical physics than as evidence that quantum theory itself is fundamentally flawed, as Einstein had originally intended.

In fact in 1964, John Stewart Bell, an Irish physicist, based on a statistical inequality theorem showed that quantum mechanics actually permits instantaneous connections between far-apart locations without any so-called hidden variables, thus proving "spooky action at a distance" to be real as demanded by the weird quantum mechanics. Bell's theorem proves that quantum physics is incompatible with local hidden-variable theory. This constraint later came to be known as the Bell inequality. But Einstein was a believer in the locality, meaning that no event in one part of the universe can instantly affect what happens far away and must be limited by the velocity of light for information transmission. This is the bedrock principle of classical physics, especially that of relativity. Thus, Einstein, along with his two colleagues, when dealing with the quantum entanglement of quantum mechanics, ended up with the so called EPR paradox. But since the 1970s, physicists have made increasingly precise experimental tests of Bell's theorem, and the predictions of quantum mechanics of nonlocality or quantum entanglement have been confirmed with great precision. Thus, as far as the weirdness of quantum mechanics is concerned, Einstein was proved incorrect at least in this occasion: The micro-world is thus proved to be as "crazy" as predicted by quantum mechanics.

Quantum entanglement is a quantum-mechanical phenomenon in which the quantum states of two or more objects can be correlated with each other in terms of some physical properties of the objects (spin, phase, polarization, and so forth), even though the individual objects may be spatially separated far apart. For example, it is possible to set two particles in a single quantum state, such that when one is observed to be spin-up, the other one will always be observed to be spin-down and vice versa—this despite the fact that it is impossible to predict, according to quantum mechanics, which set of measurements will be observed. As a result, measurements performed on one system seem to instantaneously influence other systems entangled with it.

Today, quantum entanglement forms the basis of several cutting-edge technologies. Quantum entanglement has the advantages of wireless capacity, unlimited transmission speed, and absolute security, and has been widely used in various fields of quantum communication. Quantum entanglement has applications in the emerging technologies of quantum computing and quantum cryptography, and has been used to realize quantum teleportation experimentally. Quantum computers will be able to do amazing things, solving complex problems infinitely faster than classical computers. This has significant implications for a wide range of fields from weather prediction to medical research and the development of artificial intelligence, not to mention the already known impact on cybersecurity and the ability to break current encryption methods. In quantum cryptography, entangled particles are used to transmit signals that cannot be intercepted by an eavesdropper without leaving a trace. The first viable quantum cryptography systems are already being used by several banks. As regards quantum teleportation, quantum dense communication coding and quantum spatial secret information sharing are the main applications of quantum teleportation.

Are quantum communications faster than the speed of light? Even though entangled quantum particles seem to interact with each other instantaneously, regardless of the distance, the prevailing position among scientists appears to be that breaking the speed of light is not yet possible in quantum communication. In other words, quantum entanglement does not enable the transmission of so-called classical information faster than the speed of light. In order to "communicate" in the classical sense, we need to send data/messages, but as of today it is impossible to send data/messages using quantum entanglement. In quantum entanglement, the entangled particles/systems can share information in terms of correlation with each other instantaneously, even though they may be light years away. But no real information (data/messages) is passed when the entangled particles affect each other. That is, they only teleport the quantum states between the systems that bring about revolutionary applications in quantum computing, quantum cryptography, quantum teleportation, and so forth. This is the crux of quantum entanglement as of today.

3

Miscellaneous Developments: In the Realm of and Beyond Relativity and Quantum Mechanics

Incessant search of knowledge is to look beyond

3.1 Introduction

After the basic foundation/formulation of relativity and quantum mechanics, further the developments in physics and astrophysics took place both in the realm of and beyond the boundaries of relativity and quantum mechanics. On the one hand, miscellaneous developments took place primarily in the realm of relativity—that includes the expanding universe, Big Bang theory of the origin of the universe, the black hole, and gravitational waves, all of which in the course of time have been established with experimental support. Although the concept of the expanding universe followed from the solution of Einstein's relativity theory, as he believed in the static nature of the universe, like his predecessors, he introduced the so-called cosmological constant in 1917 to counter the expanding nature of the universe given by his field equations of gravitation, originally developed by him in 1915.

It was Georges Lemaître who tried to convince Einstein that the expanding nature of the universe was correct on the basis of the then available astronomical data, but Einstein did not agree with Lemaître. Subsequently, Alexander Friedman, who opposed the conventional wisdom of the time—the static universe—worked out the complete theoretical basis of the *expanding universe* based on Einstein's relativity but received a cool response from the astronomical community. In 1929, when Edwin Hubble and his team experimentally established the expanding nature of the universe, Einstein and the rest of the physicists/astrophysicists finally accepted the concept of the expanding universe. The *Big Bang theory*, an important contribution by George Gamow in the 1940s regarding the origin of the universe, was substantiated by Arno Penzias and Robert Wilson's experimental observation of cosmic microwave background (CMB) radiation in 1965. The possibility of the *black hole* was first predicted by Karl Schwarzschild in 1916 as a solution emerging out of Einstein's relativity and was finally established by astronomical observations in the 1970s. In addition, the first black hole image was obtained in 2019 with Event Horizon Telescope (EHT). Finally, Einstein's prediction of the existence of *gravitational waves* as the ripples in space–time curvature when two extremely large masses make vigorous movement was also verified a century later. The gravitational waves generated by the collision/coalescence of two massive black holes were detected in 2015 with a supersensitive instrument known as Laser Interferometer Gravitational-wave Observatory (LIGO).

On the other hand, a unification effort also took place beyond the fundamental developments in relativity and quantum mechanics leading to Einstein's unified field theory, quantum field theory (QFT), and the theory of loop quantum gravity. The history of science

DOI: 10.1201/9781003215721-3

has seen many efforts to unify two apparently dissimilar entities in nature beginning with the time of Sir Isaac Newton. The unification of the terrestrial and celestial physics was made by Newton himself, followed by the unification of electricity and magnetism in a single domain of electromagnetism by James Clerk Maxwell. Einstein in his relativity theory linked seemingly dissimilar entities like space and time in one cosmic unity, the space–time; also, Einstein showed that mass and energy were manifestations of each other. De Broglie observed the dual behavior—the corpuscular and wave nature—of the particles in the micro-world (like electron, proton, etc.), which was the seed for the development of quantum mechanics. In this perspective, Einstein sought to unite his chronogeometric theory of gravitation of general relativity with Maxwell's electromagnetic theory (light); however, he neglected to include quantum mechanics in his so-called *unified field theory*. Another unification effort, primarily by Dirac and Feynman, was to unify classical field theory (electrodynamics), special theory of relativity, and quantum mechanics leading to the QFT, which in turn led to the formulation of the standard model of physics. A number of scientists also tried to unite the gravity of general relativity with quantum mechanics and proposed the so-called *quantum gravity* and, consequently, the *theory of loop quantum gravity*.

New concepts such as *string theory/superstring theory*, or the so-called *theory of everything* (TOE), sought to unite the four diverse forces/fields found in nature (electromagnetism, the weak nuclear force, the strong nuclear force, and gravity) into one comprehensive theory. This theory assumes that the ultimate building blocks of nature consist of tiny vibrating strings and that the different particles of matter are but a manifestation of these vibrating strings in different resonant modes of vibration. However, string theory and its competitor, theory of loop quantum gravity, have not come as far as relativity and quantum mechanics in terms of establishing themselves as testable scientific theories because they are yet to obtain experimental support.

In the realm and beyond the relativity and quantum mechanics concepts, in this chapter we will discuss *antimatter, dark energy*, and *dark matter*, of which the presence of antimatter has already been established in particle accelerators, while dark energy and dark matter are still debatable issues. In 2020, a huge machine was set in motion to investigate the presence and properties of dark energy and dark matter. In 2021, the DESI (Dark Energy Spectroscopic Instrument) team obtained an initial map of dark matter. Last but not the least, the chapter will conclude with some comments regarding the present scenario of science and technology research after the developments of relativity and quantum mechanics.

3.2 Miscellaneous Developments in the Realm of Relativity Theory

3.2.1 The Expanding Universe

3.2.1.1 Introduction

In 1917, two years after he formulated his celebrated general theory of relativity, Einstein was disturbed by an observation: Every time he solved his own equations, the universe was found to change with time, that is, it was heading into a version of so-called Bentley's paradox. In that era, it was a common concept in every sphere of knowledge that the universe was eternal and static. Thus, Einstein found the results derived from his own equations to be alien, and he was forced to conclude that his equations demanded corrections. Einstein then added a "fudge" factor, which he named the *cosmological constant* or lambda factor (applicable in cosmic scale), which implied the existence of a repulsive force pervading space, counteracting the gravitational attraction (an implosive force) holding matter together. Even Einstein, the great revolutionary who overturned 300 years of the so-called unchallengeable Newtonian physics, could not believe in his own equations and he became perplexed that he had made, in his own words, "the biggest blunder."

3.2.1.2 Work of Lemaître and Friedman on the Expanding Universe

In 1927, a Belgian priest and astronomer, Georges Lemaître, then an MIT graduate student, studied Einstein's equations and found—just as Einstein had—that they predicted the universe must expand. Unlike Einstein, he looked for whatever scarce astronomical data for galaxies were available at that time to test the prediction of the theory. At that time, the galaxies were known as nebulae (the Latin word for clouds or fog) because, seen through the then available telescope, they looked like small, opalescent clouds among the stars. It was not yet known that they are distant, immense islands of stars like our own Milky Way galaxy. The young Belgian priest-cum-scientist could obtain a rough idea from the available astronomical data of the time that all the galaxies were moving away from each other with great speed, the distant ones flying away with higher speed compared to the near ones to the Earth—as if the universe was swelling like a balloon. Pictorially, it may be visualized as follows. Imagine that spots are marked on the surface of a balloon from its mouth to the end lengthwise. As the balloon is inflated, the spots (stars/galaxies) move away from one another. If our Milky Way galaxy is on the surface of the balloon, then all stars/galaxies would appear to move away from us—the farther the star/galaxy from us, the faster they are flying away.

However, in 1922, Russian physicist Alexander Friedman found perhaps the simplest theoretical solution to Einstein's equations, which gives us the most elegant description of the concept of the expanding universe. His theory can be interpreted in terms of two models of the expanding universe. According to the first model (which Friedman found), the universe is expanding with sufficient slowness that the gravitational attraction between the different galaxies causes the expansion to slow down and eventually stop. The galaxies then start to move toward each other and the universe contracts, leading to a Big Bang–Big Crunch model. Figure 3.1a shows how the distance between two neighboring galaxies changes as time increases. It starts at zero, increases to a maximum, and then decreases to zero again. The second model suggests that the universe is expanding so rapidly that the gravitational attraction can never stop it, though it does slow it down a bit with time. This model suggests an *accelerating universe*, as shown in Fig. 3.1b.

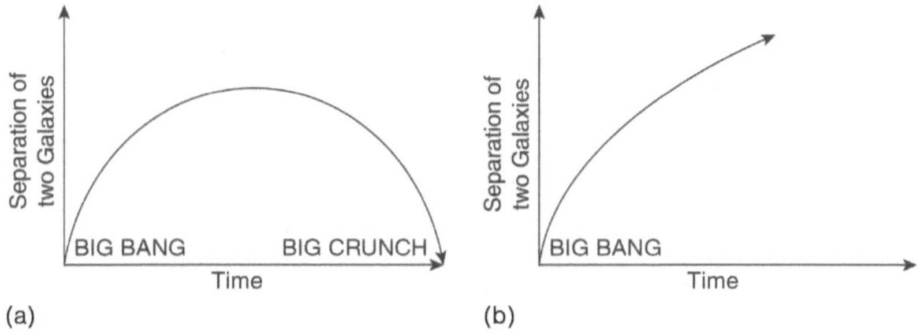

FIGURE 3.1
Big Bang models by Alexander Friedman: (a) Big Bang–Big Crunch model of the universe; (b) the accelerating model of the universe.

But which Friedman model describes our universe? To answer this question, we need to know the present rate of expansion of the universe and its present average density. If the density is less than a certain critical value, determined by the rate of expansion, the gravitational attraction will be too weak to halt the expansion. On the other hand, if the density is greater than the critical value, gravity will stop the expansion at some time in the future and cause the universe to be collapsed or crunched. The present evidence, however, suggests that the universe will probably expand forever (an accelerating universe). But all we can really be sure of is that even if the universe is going to be collapsed/crunched, it won't do that until a few billion years from now. Both of these models of the universe derived from Friedman's solution of Einstein's equation of general relativity have one common feature that at some time in the past (10–20 billion years ago) the distance between neighboring galaxies must have been zero—which is known as the Big Bang. We will discuss the details of Big Bang in the next section.

Like Einstein's solution (contained in the general theory of relativity), no one took Friedman's idea of an expanding universe seriously when he proposed it because it ran counter to the conventional wisdom of the time—the static universe—a cosmological model in which the universe is both spatially and temporally infinite, and space is neither expanding nor contracting. In fact, if you think ahead of your time, you will be misunderstood. In my opinion, Friedman is one of the great unsung heroes of cosmology. Ignoring great insights is a venerable tradition in cosmology (and indeed in science more generally).

In this context, it is interesting to note that Lemaître met Einstein and tried to dissuade him from his prejudicial view of the static universe (and to accept the concept of the expanding universe that followed from his own theory of general relativity). But Einstein resisted going so far as to answer Lemaître: "*Correct calculations, abominable physics*," he declared. Later, Einstein came to recognize that Lemaître was correct when the experimental support provided by Hubble's experiment in 1929 in favor of an expanding universe emerged. All of which is to show that even the greatest among us may make mistakes and are prey to the prejudice of preconceived ideas.

In this context, it is worthwhile to mention the theme encapsulated in the motto *Nullius in verba*, a motto of Great Britain's Royal Society (the world's oldest independent scientific academy), dedicated to promoting excellence in science. All of Einstein's fame, his influence on the scientific world, his immense authority counted for nothing in respect of his doubt of the expanding universe and his support of a static universe. The experimental observations proved Einstein to be wrong and the obscure Belgian priest to be right. For this reason, scientific thinking has a distinctive power compared to art and literature.

3.2.1.3 Experimental Support in Favor of an Expanding Universe

As we have noted, good news in support of an expanding universe came in 1929 with Hubble's experimentation. Using the 100-inch Mount Wilson telescope of the Palomar Observatory, after years of dedicated effort, Hubble and his colleagues collected precise data and announced their dramatic findings confirming that the galaxies were moving away at a speed proportional to their distance from Earth. In this effort, Hubble and his team used the period-luminosity relation of Cepheid variable stars (known as the Leavitt Law) discovered by Henrietta Leavitt in 1908, which are used as "standard candles" for measuring distance to faraway stars and galaxies. To measure the enormous velocities of the galaxies, the Doppler Effect was used. According to the Doppler Effect, light or sound waves from an object coming toward the observer get its frequency increased but if the object moves away from the observer its frequency gets decreased. We experience this effect in our everyday life. For example, when we listen to the whistle of a train/ambulance passing by us, for the approaching train/ambulance, we hear the whistle with a higher pitch (corresponding to an increased frequency of the sound wave), and for the receding train/ambulance, we hear the whistle at a lower pitch (with decreased frequency). Therefore, in the case of light, this means that stars moving away from us will have their spectra shifted toward the red end of the electromagnetic spectrum (red-shifted), and those moving toward us (if at all that happens!) will have their spectra blue-shifted. As in VIBGYOR (Table 5.1), the red color has a lower frequency compared to blue. Hubble verified that this Doppler Effect was happening to the light from distant stars and galaxies, creating a "red-shift" of the light being received from the stars and galaxies. If the stars/galaxies were coming toward the Earth, there would have been a "blue shift," which has never been observed. The discovery that the universe is expanding was one of the great intellectual revolutions of the 20th century. In the words of Stephen Hawking, it was "probably the most remarkable discovery of modern cosmology."

Georges Lemaître was the first to derive what is now known as Hubble's law, or to be more precise, the Hubble-Lemaître law, and made the first estimation of what is now called the Hubble Constant, which he published in 1927, two years before Hubble's article based on experimental observation. Hubble's law/Hubble-Lemaître law is given by $V = HD$, where V is the velocity of recession of a galaxy in miles per second and D is its distance from Earth in millions of light years; the proportionality constant H is popularly known as the *Hubble's Constant*. If we have good reliable figures for both D and V of a single distant galaxy or group of galaxies, then we can solve for H, and that value of H will hold for all galaxies. Lemaître also proposed what we now know as the theory of the Big Bang. Originally, he called it the "hypothesis of the primeval atom" or the "Cosmic Egg." Today, the confirming evidence of the Big Bang theory of the origin of the universe is overwhelming. It is now known, for example, that in the far distant past, this observable universe was extremely hot and extremely compact, and it has expanded since. We can reconstruct in detail the history of the evolution of the present universe, starting from its initial super-hot and super-compressed state. We now know exactly how atoms, elements, galaxies, and stars formed and how the universe we see today developed. Recent extended observations of the CMB radiation that fills the universe, carried out mainly by the Planck Satellite, once again confirmed the theory of Big Bang. We now know, with a reasonable degree of accuracy, what happened to our universe in the last 14 billion years (the age of the universe with latest data is 13.8 billion years, to be precise), from the time when it was a super-dense and super-hot ball of fire.

In 1931, Einstein dropped his "fudge" factor of "the cosmological constant" and reintroduced his old theory of the expanding universe, which he had abandoned 14 years earlier. But here again, Lemaître tried to persuade him to change his mind—that is, to accept the expanding universe concept together with the cosmological constant. And once again Lemaître was right and Einstein was wrong as the cosmological constant has nothing to do with a static universe, and this small quantity is not against the theory of the expanding universe. Rather, it might be reinterpreted to account for an acceleration of the expansion of the universe; and this acceleration has recently been measured by Saul Perlmutter, Brian Paul Schmidt, and Adam Guy Reiss. They received the Nobel Prize in Physics in 2011 "for the discovery of the accelerating expansion of the universe through observations of distant supernovae." The value of cosmological constant or lambda factor they found for an accelerating universe was 10^{-52} m^{-2}. It is a testament to Einstein's genius that even his blunders proved to be significant. *Lambda factor* implied the existence of a repulsive form of gravity, and such a thing appears to be the driving force behind cosmic acceleration.

If one moves with the velocity of light, one will take approximately one second to reach the Moon, 8 minutes to reach the Sun, and 200,000 years to move out of our own Milky Way galaxy. But one cannot move away out of the universe even if one travels at the velocity of light (the maximum speed at which one can move in this universe, as per relativity theory), as the universe keeps on expanding more and more with time!

3.2.2 The Big Bang theory of the Origin of the Universe

3.2.2.1 Introduction

In ancient times, most people accepted their religion's explanation of how the universe began. With the progress of science during the last three centuries, however, the new discoveries revealing the mysteries surrounding the birth of the universe shifted people's views from religion to mostly science. In the 1920s, the idea of Big Bang theory was proposed as a possible scientific explanation for the creation of the universe. As noted earlier, Friedman and also Lemaître gave their own version of the concept of Big Bang on scientific basis, and Edwin Hubble and his team provided experimental support to the theory of an expanding universe and hence the possibility of a Big Bang as the beginning of the universe. An artist's rendition of the Big Bang is given in Fig. 3.2.

According to Christian belief, God created the universe, but this belief is found in almost all the religions in the world with different stories and symbolism. Two stories of how God created the world are found at the beginning of the book of Genesis in the Bible; most Christians accept Genesis 1 as authoritative. It states that on the first day of the creation, God commanded, "Let there be light"—and light appeared. It took six days for God to create this universe, but modern science says that the present age of the universe is approximately 14 billion years. Why is there this discrepancy? Possibly because science and theology speak in two different languages, however, some theologists and some scientists misunderstand each other.

Genesis isn't prepared to give a scientific account of the world's beginnings. Genesis was written to be understood by ancient people who had no knowledge of modern science. However, when the idea that the universe had emerged from a Big Bang began to be accepted, Pope Pius II declared in a public meeting on November 22, 1951, that the theory confirmed the account of Creation given in Genesis. Again, here it was Lemaître who was able to persuade the Pope to refrain from making a reference to links between divine

FIGURE 3.2
Artist's rendition of the Big Bang [6].

creation and the scientific theory of a Big Bang. Lemaître believed that it was foolish to mix (or to set into rivalry) science and religion in this way; he emphasized that the Bible knows nothing about physics and that physics knows nothing about God. Myth is myth and science is science—a clear dividing line separates the two. To mix up theology with science—and to drown genuine science in the process, thereby making mockery of theology—was not merely irresponsible but also blasphemous. Pope Pius II allowed himself to be persuaded, and the Catholic Church never again made a public allusion to the subject.

In this respect, Carl Sagan made a pertinent comment, "The Hindu dharma (religion) is the only one of the world's great faiths dedicated to the idea that the Cosmos itself undergoes an immense, indeed an infinite, number of deaths and rebirths. It is the only dharma in which time scales (at least the order—billion years) correspond to those of modern scientific cosmology. Its cycles run from our ordinary day and night to a day and night of Brahma (the Creator), 8.64 billion years long, longer than the age of the Earth (4.543 billion years) or the Sun (4.6 billion years) and about half the time since the big bang." The *Upaniṣhad* says: Truth is one we reach through different paths of knowledge. Thus, both of these paths of knowledge, science and religion, need to respect each other's viewpoint from a respectful platform without letting down other or claiming superiority over the other. I think this respectfulness for each other will be good for a healthy coexistence of both science and religion in their respective domain and would also be good for humankind.

3.2.2.2 Experimental Evidence Supporting the Big Bang Theory

The Big Bang theory posits that around 14 billion years ago, all the matter and energy in the universe were at a point of infinite density and temperature. It then expanded rapidly, and eventually stars, galaxies, and planets formed, producing our present vast universe. The beginning of time is also reckoned from the moment of the so-called Big Bang. The theory is supported by evidence that space is expanding, including the red-shift of light from distant galaxies and the existence of CMB radiation that is found in all directions, turning around 360° in the azimuth (Fig. 3.3). CMB is the oldest radiation we can observe; it was emitted just 380,000 years after the Big Bang and provides a treasure trove for cosmologists studying the origin of the universe. The age of the universe (13.77 billion years) can be calculated by taking the inverse of the Hubble's Constant ($H = D/V = 71$ km/sec/mega parsec; D is the distance of a particular galaxy from the Earth, and V is the velocity with which it is receding from us; 1 parsec = 30.86 trillion km. See the footnote in Table 5.4), just as one can estimate how long ago (t) a speeding car left the bank after a bank robbery

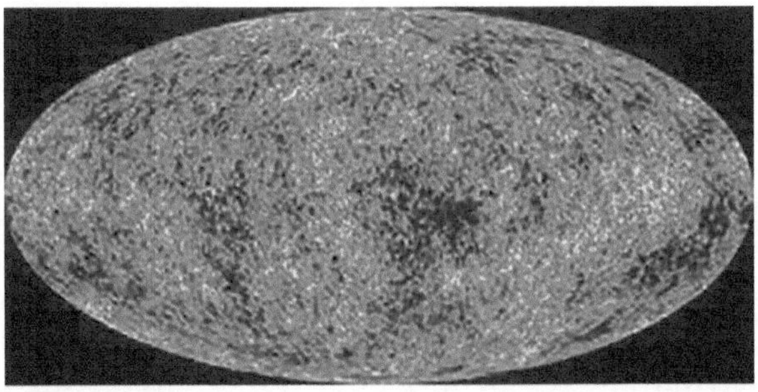

FIGURE 3.3
Cosmic microwave background (CMB) radiation left over from the Big Bang; the oldest electromagnetic radiation of the universe [7].

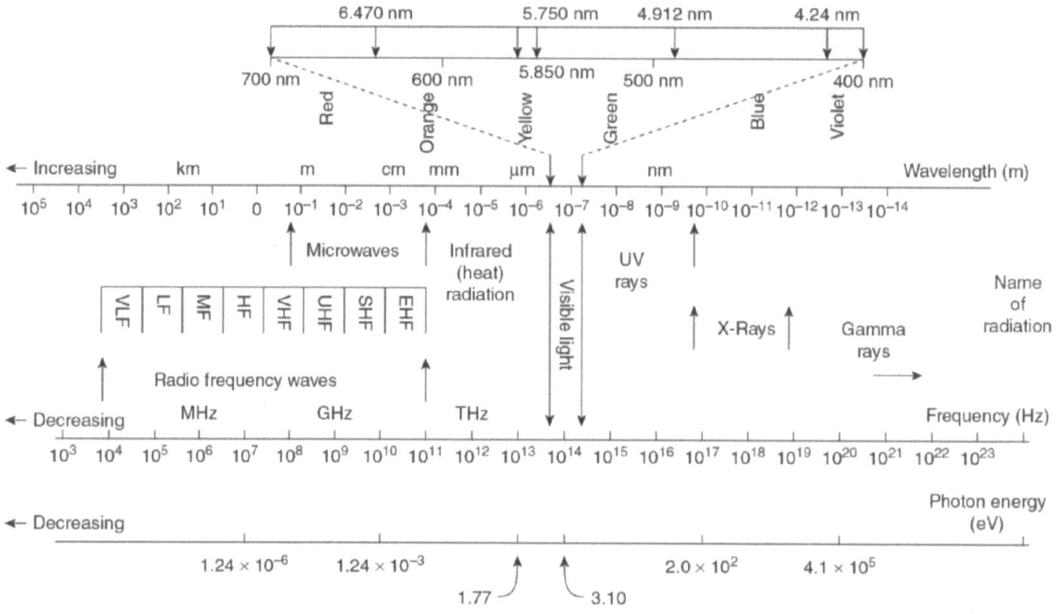

FIGURE 3.4
The electromagnetic spectrum indicating the microwave and other frequency and wavelength regions with photon energy.

by dividing the distance (s) of the car by its speed (v), that is, from $s = vt$. The present age of the universe is found to be 13.799 ± 0.021 billion (10^9) years, which is based on the measurements by Planck's satellite, Wilkinson Microwave Anisotropy Probe (WMAP), and other space probes including the cooling time of the universe from the time of the Big Bang. It may be noted that Hubble's Constant H obtained from different astronomical measurenets gives varying data: 67 (from CMB radiation), 73 (from a supernova study), 70 (from red giant star studies), and 77 (from gravitational lensing study). However, in the standard model of cosmology, astronomers have now settled to a value of H as 71 km/sec/mega parsec.

In this connection, note that microwaves are electromagnetic radiation with wavelengths from 10 cm down to 1 mm (a frequency of 300 MHz–300 GHz; Fig. 3.4). Microwaves are used for satellite communication, in the last mile of cell phone connectivity, in microwave ovens for cooking, and so on. For cancer treatment, microwave hyperthermia and microwave ablation too are used. However, the CMB radiation left behind by the Big Bang is an extremely weak signal and would heat our pizza only to –271.3°C—which would not be much good for even defrosting the pizza, let alone cooking it! Of course, the small percentage of snow that we see on our television screen when we are tuned to an empty channel is due to this leftover background microwave radiation of the Big Bang explosion that happened some 14 billion years back.

In 1940, the Russian physicist George Gamow predicted that there might be a way to verify experimentally that the Big Bang actually took place. He maintained that the original radiation left over from the Big Bang should still be circulating around the universe, although its temperature would be quite low after 14 billion years. He predicted that this "echo" from the Big Bang would be distributed evenly around the universe, so it would appear to be the same no matter where we looked. His Big Bang model of the beginning of the universe predicted that when we gaze farther into space we should encounter old galaxies nearby, then young galaxies beyond them, further to that transparent hydrogen gas, and then a wall of glowing hydrogen plasma. Plasma is a soup of free electrons and proton (the fourth state of matter beyond solid, liquid, and gas phases); the fifth state of matter incidentally is known as the Bose–Einstein condensate, which is produced when the matter is cooled to temperatures very close to the absolute zero—0 Kelvin or –273.15°C.

In 1948, Gamow's collaborators, Ralph Alpher and Robert Herman, even calculated the temperature to which the cosmic fireball (the hydrogen plasma) had cooled down to by now: 5 Kelvin (–268.15°C). It is known from the characteristics of black body radiation that as the temperature of a body decreases, the peak of the radiation it emits shifts toward the longer wavelengths or lower frequency (Fig. 2.5). For example, for visible radiation when an object is less hot, it emits red light (λ = 700 nm), while when it is heated more, it emits blue light (λ = 400 nm), and so on. Thus, on traveling 14 billion years through space to reach us, the original radiation of the Big Bang has cooled enough, and its frequency has shifted from higher-frequency electromagnetic radiation to correspondingly lower frequency—down to microwaves. Thus, we can expect this background radiation to be CMB radiation. Unfortunately, however, they failed to convince any astronomers to search for this CMB in the sky, and their work was largely forgotten, just as Friedman's expanding universe discovery was.

However, in 1965, there was a spectacular verification of the Gamow–Alpher–Herman prediction of this "echo" or background radiation left by the original Big Bang. Arno Penzias and Robert Wilson of the Bell Telephone Laboratories in Holmdel, New Jersey, while picking up signal between the Earth and a communication satellite via a microwave telescope, picked up some unwanted (noise-like) faint background radiation in the microwave range in all directions (though they expected to detect signals only when they pointed the telescope at particular objects in the sky, such as the Sun or a satellite transmitting microwaves). Initially, they took all precautions to eliminate all recognizable interference from their receiver, removing the effects of any nearby radar and radio broadcasting and suppressing interference from the heart of the receiver itself by cooling it with liquid helium. But the problem was not solved, and they even thought that it might be due to the droppings from pigeons roosting in the horn-shaped antenna. Thus, they contrived a pigeon trap to oust the birds and spent hours removing pigeon dung from the contraption. Finding no other way out, they finally decided to put the instruments and huge radio antenna on jet planes and balloons to get rid of interference from Earth.

But to their surprise, they observed that the strange unwanted signal became even stronger. When the scientists plotted the relationship between the intensity of the radiation and the frequency, they found that it resembled the curve Gamow has predicted many years before. The measured temperature of 3 Kelvin of the present radiation was remarkably close to the original prediction of the temperature of the comic fire ball or the plasma sphere predicted in Gamow's Big Bang model.

This 3 Kelvin radiation is still the most conclusive evidence that the universe started with a cataclysmic explosion, the Big Bang. This brilliant piece of cosmic detective work was a stunning confirmation of the Big Bang theory. Penzias and Wilson earned the Nobel Prize in Physics in 1978 "for their discovery of cosmic microwave background radiation." Now almost everyone agrees that the universe started with a Big Bang singularity (a point of infinite density and space–time curvature, where time has the beginning). However, success did not come easy: Gamow's hypothesis—that the universe at its beginning was a fireball, hotter than the core of the Sun, and expanded so fast that it doubled its size in a very small fraction of a second—received a cool reception. Indeed, the name "Big Bang" was coined by one of its detractors, the respected scientist Fred Hoyle, as a means of ridiculing it. Hoyle first used the term "Big-Bang" on BBC radio's *Third Programme* broadcast on March 28, 1949 only to turn down satirically the great theory proposed by George Gamow.

3.2.3 The Black Hole

3.2.3.1 Introduction

The notion of a black hole seems more at home in science fiction and ancient myth than in the real universe; nonetheless, well-tested laws of physics predict firmly that black holes do exist. In 1972, astronomers first discovered the physical existence of black hole in Cygnus X1. It is predicted (on the basis of Einstein's general theory of relativity) as a "hole" in space with a definite edge (the event horizon) into which anything can fall (star, planet, and even a whole solar system), but nothing can escape; it is a monstrous hole, where the gravity's grip is inexorable, and nothing can ever escape from there—not even light (hence the name black hole in the optical sense)—a hole that curves space and warps time. These gravitational goliaths pack so much matter into such a small volume that they form a class of objects unlike any other in the universe.

However, a question of the inquisitive mind—why light cannot escape the grip of a black hole—may be understood in simple terms, as follows. The inexorable and stupendous gravitational pull of the black hole close to its event horizon makes the escape velocity of any object more than the velocity of light (3×10^8 m/s or 186,000 miles per second). Hence, light going close to the event horizon of the black hole gets swallowed by the black hole and is thus lost to the rest of the universe forever. Using Newton's law of gravitation, we come across the term *escape velocity* ($v_e = [2GM/r]^{1/2}$, where G is the universal gravitational constant, M is the mass of the body to be escaped from, and r is the distance of the center of mass of the body to the object) in connection with space vehicles going out from Earth to the distant space (the value of v_e is nearly 25,000 miles per hour for such applications). It is this velocity that the thrust of the rocket of a space vehicle must provide to the vehicle so that the space vehicle can get over the gravitational attraction of the Earth and "escape" to the deep space.

Another question naturally arises in the mind of general readers—*as to why the gravitational attraction of black hole becomes stupendously high close to its event horizon?* As is known, the event horizon (determined in terms of the Schwarzschild radius), which when crossed from our universe into the black hole by anything (even light) gets lost in the black hole by its monstrous gravitational pull and cannot emerge back to the universe around it. We all know that a body that possesses mass produce gravitational effect on another body (calculated either with Newton's formula: $F = G(m_1 m_2/r^2)$ or by Einstein's space–time warpage of general relativity)—and the more massive a body the more its gravitational pull on another body. But if a body with a given mass is squeezed to a very small volume making it denser, its gravitational pull at the surface will eventually be very intense—this is precisely the case with a black hole which is formed out of a dead star when nuclear fuel of the star gets exhausted. If the Sun would have become a black hole, it will be just a stupendously dense mass of only 3 km radius. The gravitational force experienced by a body of unit mass (1 kg) on the surface of Sun (whose mass is 1.99×10^{30} kg) can be calculated (for simplicity with Newton's formula) as 274 million Newton (where we use radius of Sun = 6,96,340 km and $G = 6.67428 \times 10^{-11}$ N m^2/kg^2). But if the Sun becomes a black hole of radius 3 km, the gravitational force to be experienced by the same 1 kg body on the surface of Sun will be 14.7 trillion million Newton. Thus, the gravitational force due to Sun, if it becomes a black hole, on a body at its surface gets monstrously increased by a trillion times! Hence, anything coming close to a black hole will be swallowed by its stupendously high gravitational attraction; though the attraction on a body decreases as square power of distance as per Newton's law of universal gravitation, however, with Einstein's concept of gravitation (based on general relativity) the whole issue is more complicated and avoided here for simplicity.

When the possibility of a black hole as an outrageously bizarre object that swallows everything that falls into it emerged, Einstein and most physicists of his time immediately rejected its possible existence and believed that such an object should not be allowed to be present in the real universe. Somehow, the laws of physics must protect the universe from such beasts. However, it was Karl Schwarzschild in 1916 who, first solving Einstein's field equations of general relativity, observed that such an object might exist with a singularity at the heart of it known as "Schwarzschild singularity." Schwarzschild even derived a mathematical relationship that provides the radius, known as the *Schwarzschild radius*, of the black hole to which a massive star would exist in the cosmos after its so-called death (with the exhaustion of the nuclear fuel) given by $R_{Sch} = 2GM/c^2$, where R_{Sch} is the Schwarzschild radius, G is the universal gravitational constant, M is the mass of the object, and c is the speed of light. Even today Schwarzschild's formulation is the key foundation for black hole researchers.

It is to be noted that in 1783, John Michell, a British natural philosopher, talked of something called a "dark star" that he believed did not allow light (thought to be corpuscular according to the Newtonian concept at that time) to escape from it if the star had a critical circumference. However, with the abandonment of the Newtonian corpuscular theory of light, the discovery of the interference of light by Thomas Young (1801), and the subsequent development and success of the wave theory of light proposed by Christiaan Huygens in 1678, Michell's dark star concept began to lose ground and was soon totally forgotten.

In the 1930s, it was unacceptable even to the Don of astronomy, Arthur Eddington, an expert in the area of general relativity, that a star could collapse or shrink ideally to zero size (at singularity) with infinite density (as is the case with a black hole) when a massive star (that crosses much beyond the so-called Chandrasekhar Limit, which is discussed in Chapter 5) dies. The problem of understanding the fate of such a massive star after death was first solved, following Chandrasekhar, by a young American, Robert Oppenheimer in 1939. However, with the telescopes of the day, it was impossible to verify the consequence. Even Oppenheimer himself was taken aback at the astonishing conclusion he came by using general relativity to the ultimate fate of a massive star after its death—the existence of the black hole! Then World War II intervened and Oppenheimer himself became closely involved in the atom bomb project. After the war, the problem of the gravitational collapse of a massive star was largely forgotten, as most scientists became caught up in what happens in the atomic scale and its nucleus, that is, in nuclear physics. In the 1960s, however, interest in the large-scale problems of astronomy and cosmology was revived thanks to the great increase in the number and range of astronomical observations made possible by applying modern technology to radio astronomy. Oppenheimer's work was then rediscovered and extended by a number of scientists.

In 1964, at the beginning of the golden age of theoretical research on black holes, these bizarre objects were just thought to be what their name suggests—holes in space, down which things can fall but out of which nothing can emerge. Also prior to 1971, all theoretical studies of black holes had been based on Einstein's general relativistic laws, and those studies were unequivocal: A black hole cannot radiate. But from 1972 onward, one calculation after another, by more than 100 physicists (including John Wheeler, Yakov Zel'dovich, Stephen Hawking, Rozer Penrose, and others), using Einstein's relativity equations and also quantum mechanics, changed that picture. Black holes were ultimately regarded not as mere quiescent holes in space, but rather as dynamical objects. Black holes are now known to spin, and as they spin, they create a tornado-like swirling motion in the curved space–time around it when they radiate—not only gravitational waves, but also electromagnetic waves (photons) at different frequencies, neutrinos, and all other forms of radiation that can exist in nature. Stephen Hawking went a step further, showing that with radiation when the spin energy is exhausted, a black hole still radiates and now the energy comes from inside the hole—known as *Hawking radiation*. Thus, Hawking commented that "black holes ain't so black." Further, it was theorized that "a black hole has no hair"—a pithy phrase coined by John Wheeler to indicate a proved conjecture that there is no trace left in the event horizon of a black hole to know any details of the star (whether it was of this shape or that, whether it had magnetic field or not) from which a particular black hole was formed.

The black holes in the active galactic nuclei and supernovae explosion of stars are believed to be the source of cosmic rays that routinely come toward the Earth and other planets with energy a million times higher than our latest accelerators can deliver. Just as the ozone layer in Earth's atmosphere shields us from the Sun's harmful ultraviolet radiation, the Earth's magnetic field shelters us from the lethal cosmic rays. These rays, upon entering the Earth's atmosphere, collide with atoms and molecules, mainly oxygen and nitrogen present in the atmosphere, producing a cascade of lighter particles called a cosmic ray shower. Typical particles produced in such collisions are neutrons and charged mesons such as positive or negative pions and kaons. Some of these subsequently decay into muons and neutrinos, which are able to reach the surface of the Earth. Some high-energy muons even penetrate for some distance into shallow mines, and most neutrinos traverse through the Earth without further interaction.

3.2.3.2 What Are Black Holes—A Brief Description

The ultimate fate of a massive star after its death (when its nuclear fuel gets exhausted) is supernova explosion—which eventually become a neutron star or a black hole depending on the massiveness of the died star beyond the so-called Chandrasekhar (mass) Limit (see Section 5.4.3). If the death-time/final mass of the star is more than three times the mass of the Sun, then we expect its ultimate fate to be a black hole (details of white dwarf, neutron star, and black hole formation are presented in Chapter 5). The term *black hole* was given to this eventual state of a star (after its death) by the relativity expert John Wheeler in 1967 (initially, he thought its name should be frozen star/collapsed star, and earlier it was also known as the Schwarzschild singularity-type object) because the gravitational pull of a black hole is unimaginably so strong that even light cannot escape its gravitational pull. Anything and everything get swallowed up by this monstrous cosmic object that comes into its reach. It reminds us of the Minotaur of Greek mythology, which was believed to be a monster with the head of a bull and the body of a man that lives in the center of a great spiral maze known as the Labyrinth and devours anyone foolish enough (or lost) to come near to it. At the center of our Milky Way galaxy (which is a spiral galaxy)—and quite possibly in the center of all spiral galaxies—there is a black hole (Sagittarius A*) that swallows anything and everything (gaseous cloud, dust, stars, planets, even solar system) that comes close to its reach.

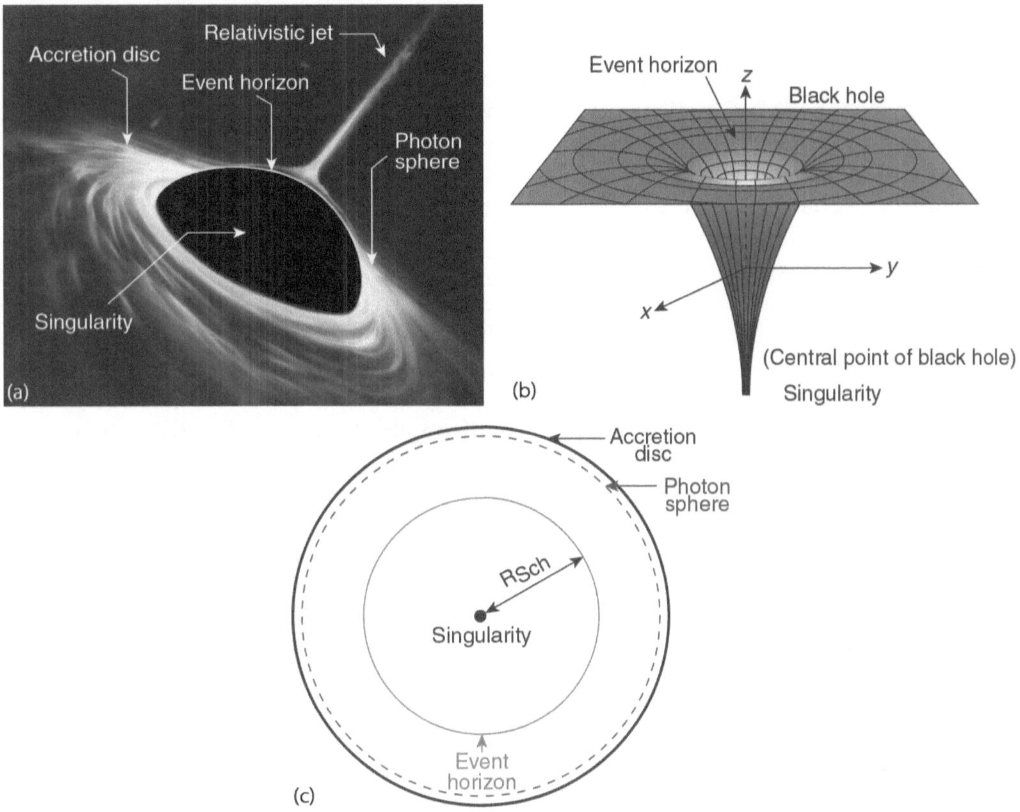

FIGURE 3.5
(a) An artist's impression of a rapidly spinning supermassive black hole surrounded by an accretion disc [8]. (b) Black hole regions showing singularity and event horizon, and (c) plan view, R_{Sch} is the Schwarzschild radius.

The black hole is like a cosmic sinkhole, swallowing all the surrounding material (with exceptionally high gravitational field that it possesses) into the monstrous cooking-hose at its center. As stars and planets or any other matter that happens to be in the neighborhood are drawn into the vortex, they begin to spin around it, forming what is called an *accretion disc* (see Fig. 3.5a). Scientists can determine the mass of the black hole by analyzing the motion of the material in the accretion disc using the light emanating from it. The stupendously high gravitational field of a black hole results in an extreme dent in the space–time curvature leading to the so-called *singularity* (the single point in space–time where the mass of the black hole is understood to be concentrated) at the heart (central part) of the black hole (Fig. 3.5b). The singularity in scientific terms means that all physical laws fail in this extreme situation, even the general relativity theory whose solution led to the possibility of black hole. Possibly for this reason, initially Einstein himself was loath to accept the existence of such a monstrous celestial object. In Fig. 3.5b and Fig. 3.5c, we have also marked the event horizon of the black hole past which nothing (including light) can escape, that is, once a particle crosses the event horizon into the black hole, it cannot leave. Gravity is constant across the event horizon.

Chapter 5 discusses how a star (our nearest star being the Sun) is born. A star exists as a stable object emitting light (as we see in our night sky) and other electromagnetic radiation like radio waves and X-rays, with a delicate balance between an implosive force (caused by gravitation) that wants to crush the star down to a point and an explosive force (due to the nuclear fire generated by a thermonuclear fusion process that converts hydrogen to helium within the star), which tends to blow apart the star. This delicate balance is destroyed when the star's nuclear fuel (basically hydrogen, helium, and the lighter elements) is used up over billions of years. Once the nuclear fuel is exhausted, the gravitational force takes over and the star collapses. After the collapse or the so-called death of the star, it ultimately becomes either a white-dwarf or a neutron star and in extreme cases, a black hole depending on the massiveness of the star. A *white-dwarf* is formed if the death-time mass of the star is less than or equal to 1.4 times the mass of the Sun when the cold star exists, with the balance between the implosive force of gravity and the explosive force of the so-called electron degeneracy pressure. If the said mass of the star is more than 1.4–3 solar mass, the gravitational force of the star will crush the atoms in it to a dense ball of neutrons, making the dead star a *neutron star.* In a neutron star, the quantum-mechanical repulsion of neutrons in its confined state in the nucleus gives the implosive force, known as neutron degeneracy pressure, to balance the enormous gravitational crushing force. However, if the dead star is a massive one (i.e., its death-time mass is at least more than three times the solar mass), the neutron star itself will be unstable. The gravitational force will be so great that all the neutrons will be pushed into one another, finally crushing the neutron star down to an extremely small size (practical size is calculated on the basis of Schwarzschild radius)—which is the so-called *black hole.*

3.2.3.3 Types of Black Hole

There are four different types of black holes, and each type is a mysterious beast. *Stellar black hole* is formed out of stars when it dies with death-time mass/final mass of the dead star more than three times the solar mass $M_{\odot}$. Such black holes while packing more than three times the mass of the Sun into the diameter of a city become extensively dense and possess a crazy amount of gravitational force pulling on objects around and thus consume the dust and gas from their surrounding which keeps them growing in size. Astronomers estimate that the Milky Way has anywhere from 10 million to 1 billion stellar black holes, with death-time masses roughly three times that of the Sun. *Supermassive black holes* are those that are million or billion times as massive as the Sun but having diameter comparable to or even larger than that of the Sun itself. Such black holes are found to lie at the center of most of the galaxies. The black hole Sagittarius A^* that lies at the center of our Milky Way galaxy is a supermassive black hole with diameter 44 million kilometer. Scientist believes that such black holes are formed either due to the merger of thousands of stellar black holes or due to the collapse of a stellar cluster. Once these monstrous giants have formed, they gather mass from the dust and gas around them, material that is plentiful in the center of galaxies, allowing them to grow even more massive with time. Astronomers once thought that black holes came in only small and large sizes, but recent research has revealed the possibility that midsize, or intermediate, black holes (IMBHs) could exist. IMBHs fall between the stellar-mass black holes and the supermassive black holes. This type of black holes are thought to form when multiple stellar-mass black holes undergo a series of mergers with one another, however, they are rare. These mergers are believed to happen in crowded areas of galaxies. In 2014, astronomers found what appeared to be an IMBH in the arm of a spiral galaxy while in 2021 astronomers took advantage of an ancient gamma-ray burst to detect one. Lastly, we'd be remiss if we didn't briefly discuss a hypothetical type of black hole called a *primordial black hole.* As their name suggests, primordial black holes were born when the universe was still young—within a mere second of the Big Bang. This was a time long before stars, galaxies, and other black holes existed. But primordial black holes wouldn't have started out as a star anyway. They would have popped into existence when the newly created universe was not yet homogenous and evenly distributed. At this point, some scientists think that certain parts of the universe were unbelievably rich in energy. It's these tiny, insanely energetic points in space that could have theoretically collapsed directly into primordial black holes. And depending on just how soon after the Big Bang these first black holes formed, they could range from about 0.00001 times the mass of a paperclip to about 100,000 times the mass of the Sun. No practical primordial black hole has yet been discovered by astronomers. Anyway, some astronomers believe that such primordial black holes may be a candidate to account for dark matter called MACHOs, which stands for massive compact halo objects, because astronomers think they are found in the halos, or outskirts, of galaxies.

3.2.3.4 Black Hole Identification and Experimental Evidence of Its Existence

The black holes were catalogued at the initial stage of their experimental studies only indirectly through their gravitational tugs on stars in their neighborhood and fireworks left behind by the superheated matter being swallowed into it by its monstrous gravitational pull. Like the Cheshire cat in Alice's Wonderland, the black hole disappears from view (in the optical sense), leaving only its "smile"—the signature of the distortion of space–time caused by the super intense gravitational field.

What is the simplest way to identify the presence of a black hole in the cosmic canvas? We all know that X-rays cannot penetrate Earth's atmosphere, so we are saved from exposure to the X-rays emanating from different celestial objects, including the Sun, which would cause cancer and undesired mutation in our body. Thus, to study X-rays from celestial bodies, telescopes are installed in space stations (Chandra X-ray Observatory) located in space far outside the Earth's atmosphere. However, for black hole identification, we need an X-ray telescope in addition to an optical telescope. Of the several black hole candidates discovered by X-ray detectors and X-ray telescopes, Cygnus X-1 (Cyg X-1 in short) is the first, its companion star being HDE 226868 (the two together forming a binary star). To search for black holes, one could use a combination of optical telescopes and X-ray telescopes. The black hole candidates are mostly binaries in which one object is optically bright but X-ray dark (the non-black hole star of the binary), and the other is an optically dark but X-ray bright object (the black hole one). Since a neutron star could also capture gas from a companion star (just like black hole), heat it in shock fronts, and produce X-rays, a note of caution is there for final inference of the astronomical observation. To be confident that the optically dark but X-ray bright object is definitely a black hole, and not a neutron star, one has to be sure that the identified object has a mass that is at least more than three times the solar mass.

In 1972, radio astronomers identified the first black hole in distant space: Cygnus X-1, 6,000 light years from Earth, by studying the X-rays that pour in very close to the hole's vicinity. It may be noted that, very far from the hole's event horizon, the gas atoms are cool (just a few degrees above absolute zero); hence, they vibrate slowly, producing slowly oscillating electromagnetic waves, say radio waves. Nearer to the hole, where gravity has pulled the gas atoms into a faster stream, they collide with each other strongly and heat up to several thousand degrees, thereby generating microwave/millimeter-wave or even much higher frequencies. Being too close to the hole, where the gravity becomes stupendously strong and the stream of atoms moves violently, collisions heat up the atoms to several million degrees, and they vibrate very fast, producing X-rays or even γ-rays. In our Milky Way galaxy alone, there are approximately 10 million black holes (at the center is the supermassive black hole Sagittarius A*), but their dimness hides them from view. However, in 2019 the EHT, an Earth-size virtual aperture radio telescope, was able to take a breathtaking photograph of black hole M87* in the constellation Virgo some 55 million light years away from Earth.

In 1994, the Hubble telescope first discovered that the galaxy M87 (55 million light years from the Earth) contained a black hole. In January 1995, a second black hole was identified in the galaxy NGC 4258 (21 million light years away) using an array of radio telescopes. Numerous black holes have also been discovered at the center of our own Milky Way galaxy and other galaxies, too, with the advent of the latest generation radio telescopes installed in space like the Hubble telescope, Planck telescope, Chandra X-ray telescope, and so on.

3.2.3.5 What Is Inside a Black Hole?

A big question that has haunted physicists since the 1950s, especially beginning in the 1970s, is, what's inside a black hole? Oppenheimer insisted that the inside of a black hole was hidden from view by the event horizon of a black hole (where one encounters nothing except space–time curvature). According to general relativity, nothing can back away from the "singularity" (either Schwarzschild type or Oppenheimer–Snyder type or Belinsky–Khalatnikov–Lifshitz [BKL] type) that is at the heart of a black hole; everything that hits it is instantaneously destroyed. A singularity is a region where—according to the laws of general relativity—the curvature of space–time becomes infinitely large, and space–time ceases to exist (i.e., all physical laws, including general relativity fails), producing a deadly death chamber for everything conceivable in this universe.

The laws of physics also break, as of today, above the velocity of light, below the Planck length/Planck time and before the Big Bang. Anyway, John Wheeler believed that understanding the hole's core was a holy grail worth pursuing; it might help us to discover the full marriage of quantum mechanics with gravity (the partial marriage of quantum mechanics with Maxwell's laws of electromagnetism, carried out by P.A.M. Dirac, has given birth to quantum electrodynamics [QED]). And perhaps the nature of the black hole's core holds the key to other mysteries of the universe that are still to be unfolded. Wheeler believed that the marrying of tidal gravity (which becomes infinite at the core of a black hole and results in singularity, leading general relativity to fail there), a concept related to general relativity introduced by Einstein, with quantum mechanics leading to quantum gravity will unfold new physical phenomena inside the black hole, phenomena unlike any we have ever met.

Various ideas were proposed in 1980 and 1990s regarding what happens inside a black hole when something falls in it. For a young hole, everything that falls in it gets torn apart by tidal gravity in a violent, chaotic way before quantum gravity becomes important. However, things that fall into a comparatively older hole might survive unscathed until they come face to face with the laws of quantum gravity. This is possible because everything is tamed with age: Stars consume their fuel and die; Earth gradually loses its atmosphere by evaporation into space and will ultimately become an airless, dead planet; and humans grow wrinkled and wise with aging.

Two questions arise: When does quantum gravity take over, and what happens to the space–time curvature? The understanding is that when the oscillating tidal gravity (space–time curvature) becomes so large, it completely deforms all objects in about 10^{-43} seconds

(Planck–Wheeler time) or less. The quantum gravity then radically changes the character of space–time: It ruptures their bonding and unglues space and time from each other, and then it destroys time as a concept and also demolishes the definiteness of space. In the language of quantum gravity, "time does not exist." In other words, we cannot say that "things happen before or after that" because without time functionality there is no concept of before or after. Space that was once the soulmate of time and was a unified entity called space–time as per the general theory of relativity now becomes a random, probabilistic froth, like soapsuds (i.e., quantum foam). The geometry and topology of space do not remain definite; instead, they become probabilistic. Honestly speaking, the fate of the black hole and what goes on when something falls inside it are still based on some speculative theoretical possibilities. Perhaps our existing theories will get the correct answer someday, or it might be so that new theories are needed to understand that holistically. Only time will tell what treasure-trove is still hidden in the Pandora-box of Mother Nature.

3.2.3.6 First Experimental Imaging of Black Hole with EHT

In April 2019, at the heart of the Messier 87 or M87, discovered by the French astronomer Charles Messier in 1781 and catalogued it as a nebula, (a supergiant elliptical galaxy in the constellation Virgo), that lies about 55 million light years away, the image of a supermassive black hole M87* (which is six billion times more massive than our Sun) was taken by the sharpest astronomical instrument ever assembled, known as the EHT. In the coming days, EHT planned to image the supermassive black hole Sagittarius A* at the center of our own Milky Way galaxy. This was not done at this time because imaging Sagittarius A* is difficult due to its dimness and its smaller size compared to M87*. In addition to its size, M87* interested scientists because, unlike Sagittarius A*, it is situated in the active galaxy Messier 87, with matter falling into it and spewing out in the form of jets of particles (relativistic jets; Fig. 3.5a) that are accelerated to velocities near the speed of light. But its distance made it even more of a challenge to capture than the relatively local black hole, the Sagittarius A*.

The EHT used a technique known as very-long-baseline interferometry (VLBI). The long-baseline interferometry (LBI) was first proposed by R. Hanbury Brown and was in use for radio astronomical studies since the 1950s. But EHT uses a much-advanced version of LBI and a collection of eight radio telescopes, operating at millimeter-wave frequency: 230 GHz, that is, 1.3 mm wavelength. (Use of submillimeter-wave frequency, 345 GHz, will improve the resolution of the image; planned especially for Sagittarius A*.) EHT spans the Antarctic, Europe, Africa, North America, and South America (Fig. 3.6a). The constituent telescopes are used to make simultaneous observations and time-stamp the data using atomic clocks. The information received in all eight constituent telescopes is then combined with a proper algorithm in a supercomputer. A supergiant telescope array of this magnitude has resolution sharp enough to read the lettering on a 10p coin in London from 3,500 km away in Cairo or even to see a golf ball placed on the Moon very clearly from Earth.

In the biggest experiment of its kind, a network of eight linked telescopes forms the EHT that enormously increases the effective aperture of the antenna of the radio telescope array artificially to image an object that is more than 50 million light years away from Earth. They all turn on simultaneously, combining to make a single and an extremely powerful virtual telescope. Each constituent telescope is located high up at a variety of exotic sites, including volcanoes in Hawaii and Mexico, mountains in Arizona and the Spanish Sierra Nevada, the Atacama Desert of Chile, and Antarctica, spanning almost the whole surface of the planet Earth (Fig. 3.6a). Professor Sheperd Doeleman of the Harvard-Smithsonian Centre for

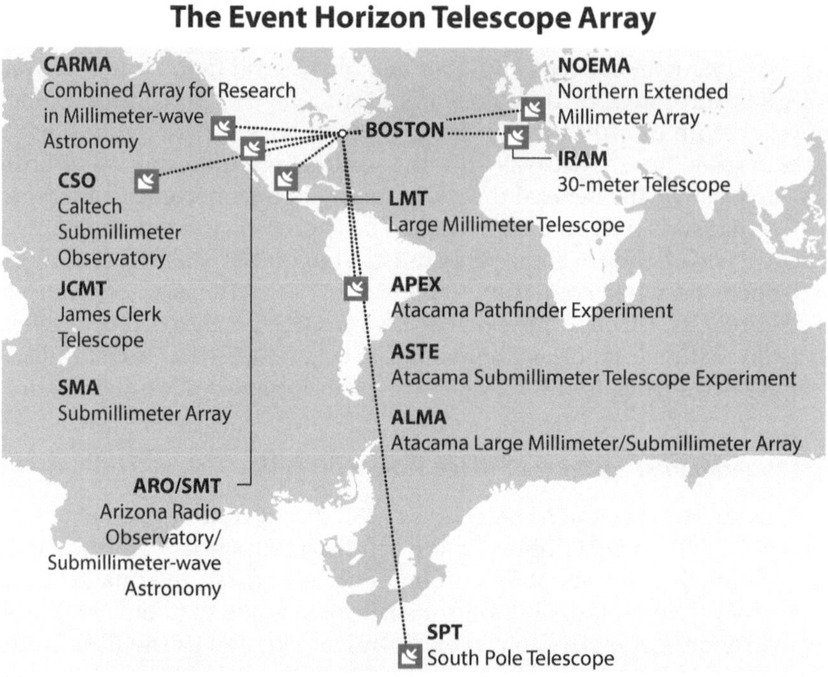

(a)

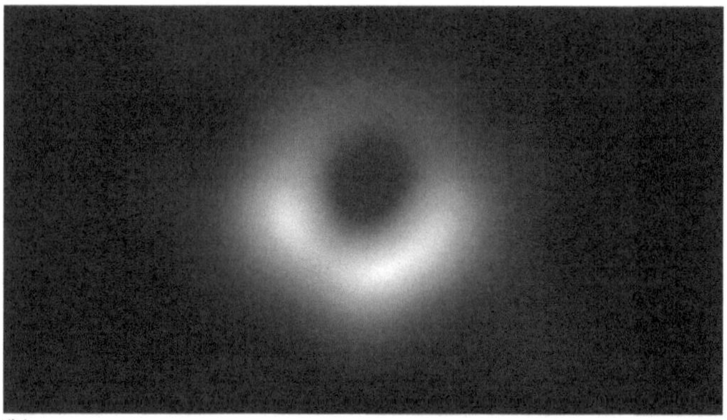

(b)

FIGURE 3.6
(a) EHT radio telescopic array arrangement; (b) first ever image of black hole M87* pictured by EHT [9].

Astrophysics led the project, and a team of 200 scientists pointed the networked telescopes toward M87* and scanned its heart over a period of 10 days to obtain a clear image of the black hole close to its event horizon that is surrounded by the photon sphere (Fig. 3.5).

Why can a single radio telescope not be used for the purpose instead of an array of radio telescopes located at such a large distance from one another? The answer may be as follows. A radio telescope gathers information from a distant object whose resolution (i.e., to see the details of the object/to distinguish two given points on the object clearly with its radio-eye) is dependent on the antenna aperture size/capture area (which is directly proportional to πr^2, where r is the radius of the open face of the dish-type antenna used much as the roof-top dish antennas are for receiving our DTH TV signal) and inversely proportional to the distance of the object. Thus, to image the object (black hole M87*) which is 55 million light years away from Earth with high resolution, we need an enormously large antenna aperture size/capture area, which is not possible using a single antenna that demands that the antenna face diameter be unimaginably large and technically impossible to design. The antenna technology of array antenna can virtually/synthetically/artificially increase the overall antenna aperture of EHT using an array of radio telescopes (each having its respective antenna) at far distances from each other and using a proper computer algorithm to combine the signals received from them (being synchronized with atomic clocks) in a supercomputer. Through EHT, those eight radio telescopes placed at different locations around the Earth made it possible to realize an Earth-sized virtual antenna capture area that facilitated clear imaging of black hole M87*.

It is commonsense that capturing an image of a black hole is impossible because an image of something from which no light can escape would appear completely black. The image of the supermassive black hole seen at the heart of M87 by the EHT (Fig. 3.6b) shows an intensely bright ring of fire surrounding a perfectly circular dark hole. The observed image with EHT may be understood as follows. When any matter comes close to the event horizon of the black hole, chunks of matter grind against each other due to the stupendous gravity of the black hole, creating massive friction and releasing huge amounts of energy. Thus, the matter close to the event horizon glows brightly with heat equivalent to several hundred stars. It is so bright that it becomes observable from Earth with a supergiant radio telescope array like the EHT. The glowing bright ring is basically a ring of photons emitted by a swirling mass of ultra-hot plasma—not in a visible spectrum but in terms of electromagnetic radiation in millimeter/submillimeter-wave frequency range (for which EHT is designed to probe the black hole M87*). It is this bright glow—the accretion disc (Fig. 3.6b)—that was imaged by EHT at 230 GHz (1.3 mm wavelength). The center of the image is observed to be a kind of ghostly, backlit shadow as expected (which marks the event horizon from where going toward the center of the black hole nothing is observable and ultimately we emerge into singularity at the heart of the black hole).

The image thus obtained matches well with what theoretical physicists worked out on the basis of general relativity (once again proving Einstein to be correct; the prediction of a black hole was from the general theory of relativity). The image obtained with EHT also surprisingly supports the imagination of the Hollywood film directors in the famous film *Interstellar* (2014), directed by Christopher Nolan with scientific guidance from Kip Thorne.

In 2021, EHT also helped to find the closest known black hole to Earth, a weirdly tiny object dubbed "The Unicorn," which lies a mere 1,500 lightyears from us and is three times more massive than our Sun. Few such super-lightweight black holes are known because they're incredibly hard to find, though EHT has now made such discovery possible.

Since we have mentioned above the name of the film *Interstellar*, based on physics/astrophysics, it will be worthwhile to mention here that there are also many documentary films based on physics/astrophysics—a brief discussion on that might inculcate thoughtfulness and imagination in the mind of young readers interested in physics and astrophysics. Esteemed physicists/astrophysicists and science popularizers like Carl Sagan, Stephen Hawking, and many others have taken this initiative to educate and provide food for thought to the young minds. We will mention here a few of them. *Cosmos: A personal Voyage* is a 13-episode series created by Carl Sagan, based on his bestselling book of the same name. It includes a number of scientific subjects, including the origin of life and a perspective of our place in the universe. Since 1980, it has been broadcasted in 60 countries and seen by more than 500 million people. In the documentary film *Into the Universe*, Stephen Hawking brings his vision of universe to the screen for the first time to delve into questions like how the universe began, whether life exists on other planets, and whether time travel is possible? Hawking appears on the show in linking scenes using his own synthesized voice while the voice over narration is provided in character as Hawking by renowned actor Benedict Cumberbatch. It was released on Discovery channel. In the documentary film *Secrets of Quantum Physics*, British physicist Jim Al-Khalili shows how quantum physics is in everyday life as Robins navigate using quantum entanglement, how our sense of smell is influenced by quantum vibrations, and that quantum physics might play a role in biological evolution. Particle physicist Brain Cox discusses various aspects of universe in the documentary film entitled *Wonders of the Universe* featuring a wonder related to each topic. The topics include nature of time, life cycle of stars, and the effect of gravity in the creation of the universe.

3.2.3.7 Black Hole versus White Hole

The *white hole* might appear to belong to science fiction at present. However, the origin of white holes may be traced back to Einstein's general theory of relativity. Some solutions of the theory's complex equations produce black holes, and, it turns out that other solutions produce white holes. The two are, mathematically at least, equally likely to exist, though Einstein's equations do not describe how black holes/white holes can form. Black holes and white holes can be thought to be intertwined. For its part, NASA astrophysicists have referred to the white hole as the time-inverse of black holes.

Unfortunately, whether or not white holes truly exist remains an open question, but that hasn't stopped researchers from thinking about what they'd be like if they did exist. Some theories suggest they might even be necessary to solve some longstanding problems, ranging from the nature of dark matter to the nature of the universe itself, but no plausible theory exists as yet. We all know about black holes, those gravitational monsters that swallow anything and everything (including light) that gets too close to it. But what about the opposite? Might the vast unending cosmos also feature white holes that emit matter and energy into the void, but can never be entered? Can we consider white holes to be the mirror image of black holes? Things often come in binaries and so it is that white holes also feel like a necessary balance to the finality of black holes: Where does all that swallowed stuff from the black hole go? Some astrophysicists think that "white holes" make a lot of sense. For them, they seem inevitable, even though we know nothing about them as yet.

One final interpretation of white holes to close the discussion for the moment: Can we think of the Big Bang as the white hole itself, the black hole being "cosmic sink" and the white hole possibly as a cosmic source!

3.2.4 Gravitational Waves

3.2.4.1 Introduction

Gravitational waves were predicted by Einstein from his general theory of relativity, and their practical existence was established with the LIGO experiment in 2015 on the basis of the collision of two black holes. Gravitational waves (ripples of space–time curvature) are generated when massive bodies, say two black holes, coalesce or collide. This collision is possible in regions of space where gravity is extremely intense (i.e., in the space where Newton's description of gravitation fails and only Einstein's chronogeometric gravitation rules firmly) and where a huge amount of matter or space–time curvature moves or vibrates or swirls at or near the speed of light. This can happen due to either the collision or coalescence of two black holes or due to the implosion of the core of a star that triggers a supernova or the merger of two neutron stars takes place that are orbiting each other.

Since the space–time curvature is the manifestation of gravity, the ripples produced in the space–time curvature due to the collision of two black holes or their coalescence are actually the waves of gravity or gravitational waves. Just as the sound waves carry their encoded symphony from the orchestra that produces it to the audience, so the space–time curvature ripples carry their encoded history from the colliding/coalescing black holes to the distant parts of the universe. One of the locations may be our Earth. Since the gravitational waves, unlike electromagnetic waves, are not absorbed or distorted by any of the intergalactic gases, matters of the stars, and so forth, we can monitor the ripples of the space–time curvature with an instrument installed at the Earth. However, that must be done with an extremely sensitive instrument. With this view in mind, the LIGO instrument was designed to act as a gravitational-wave detector.

Two questions naturally arise. First, what is the frequency of the oscillation of gravitational waves? Like the electromagnetic spectrum, the gravitational wave spectrum is extremely broad, and the frequency range of a gravitational wave signal provides information about its source; the lower the frequency, the larger the mass involved and vice versa. It also tells scientists which type of detector to use to look for which source, as the detector size should be comparable to the wavelength of the signal. The LIGO detector was designed to receive gravitational waves ranging from a few tens of hertz to a few kilohertz (i.e., typically in the audio frequency range of 20 Hz–20 kHz).

Second, what is the velocity with which the gravitational waves propagate from its source of origin? The answer is definitely "with the velocity of light." We know that when an electron is disturbed for some reason or, other (say on switching an electric light or fan) it creates a change in the electric field that eventually creates a changing magnetic field. Such time-varying electric and magnetic field pairs spread out as electromagnetic waves at the velocity of light. In the same way, the shaking of huge masses that change the gravitational field results in gravitational waves that also propagate at the speed of light. This is because the information transmission (with the shaking of electrons or with the shaking of mass) is limited by a maximum speed, the speed of light, as per Einstein's relativity theory.

Deciphering the ripples' symphony would yield a wealth of information, most important of which is the direct signature of the black hole. This was not possible with electromagnetic wave (radio, light, and X-ray) astronomy because these waves are produced far outside the hole's event horizon and are emitted by hot and high-speed electrons unlike the space–time curvature, whose extreme distortion is the unique signature of a black hole. The electromagnetic wave is also strongly absorbed and distorted by matter intervening between its source of origin and the Earth. The gravitational wave ripples' symphony can tell us how heavy each of the black holes were, how fast they were spinning, the shape of their orbit, the merging of the two holes' horizons, and the wild vibrations of the newly emerged hole, including the warpage of the fabric of space–time. These are just vaguely known on the basis of Einstein's general theory of relativity; deciphering the symphony of the ripples from gravitational waves would help us to gain all this information more accurately, and also for the first time we could even test the general relativity's predictions.

3.2.4.2 Gravitational Wave Detection with LIGO

On September 14, 2015, LIGO made the first-ever direct observation of gravitational waves: Ripples in the fabric of space–time predicted by Einstein over 100 years ago. The public announcement took place on February 11, 2016, in Washington, D.C. Each of the twin LIGO observatories—one in Hanford, Washington, and the other in Livingston, Louisiana—picked up the feeble signal of gravitational waves generated 1.3 billion years ago when two black holes spiraled together and collided. The information about that collision was carried to the Earth by gravitational waves that traveled for 1.3 billion years, riding on the shoulder of space–time curvature, that is, it traveled a distance of 1.3 billion light years with the velocity of light. Two additional detections of gravitational waves, once again from merging black hole pairs, were made on December 26, 2015 and January 4, 2017; and on August 14, 2017, a fourth event was detected by LIGO and Virgo's (a similar facility installed near Pisa, Italy, jointly sponsored by France and Italy) gravitational-wave detector. Rainer Weiss, Barry Clark Barish, and Kip Thorne received the Nobel Prize in 2017 for their "decisive contributions to the LIGO detector and the observation of gravitational waves." Upon hearing the news of the Nobel Prize award along with two others, Barish commented: "The detection of gravitational waves is truly a triumph of modern large-scale experimental physics." From the standpoint of modern astronomy, the detection of gravitational waves ushered in a new era of gravitational-wave astronomy. LIGO and Virgo provided astronomers with an entirely new set of tools with which to probe the cosmos with new vigor and with new expectations.

The basic structure of LIGO is remarkably similar to the Michelson interferometer used in 1897 in the Michelson-Morley Experiment which disproved the existence of luminiferous ether—a substance at that time thought to permeate the universe. An interferometer uses two sources of light to create an interference pattern that can be measured/analyzed—hence the name Interfere-o-meter. Interferometers come in different shapes and sizes, and they can be used to measure everything from the smallest variations on the surface of a microscopic organism to the structure of enormous expanses of gas and dust in the distant universe.

LIGO is a very complex instrument that incorporates state-of-the-art laser and sensor technology, for it has to detect extremely feeble gravitational waves. A simple sketch of LIGO is shown in Fig. 3.7a, together with the photograph of the actual site of the L-shaped LIGO observatory in Hanford, Washington/Livingston, Louisiana, in Fig. 3.7b, which consists of a laser source, a beam splitter, mirrors at the perpendicular arms that are of equal length, and a photodetector that records the interference pattern. Each of the perpendicularly placed arms of LIGO is 2.5 miles, that is, 4 km in length (to reduce the possible "kick" or "shake" due to the uncertainty principle, as discussed in Chapter 2, and to increase the sensitivity of measurement enormously, since gravitational waves demand a sensitivity on the order of 10^{-21}), separated by 1,865 miles (the distance from Hanford, Washington, to Livingston, Louisiana). In addition, it had to sense a miniscule distortion—just one-ten-thousandth the diameter of a proton, that is, 10^{-19} m—arising out of the passage of a gravitational wave; a true marvel of modern technology.

In the design of the LIGO instrument, much has been done to shield the interferometer from external noise that might be caused by earthquake or a nearby passing vehicle, so that the detectors have a better chance to pick up just the exceedingly subtle signal of gravitational waves. Subtler effects like quantum fluctuations inside the interferometer itself, and specifically, the quantum noise generated among the photons in LIGO's laser that might tilt the mirrors by minuscule angle, need to be taken care of in the design, enabling detection of the telltale flicker of light only due to gravitational waves with due sensitivity.

Gravitational waves cause space itself to stretch in one direction and simultaneously compress in a perpendicular direction (as per the tidal gravity concept of Einstein, which is believed to be the basis of his general theory of relativity, discussed in Chapter 1). In the LIGO instrument, this would cause "differential arm" motion, that is, one arm of the interferometer will get longer while the other gets shorter (of course, by a minuscule amount), then vice versa, back and forth, as long as the gravitational wave is passing/sensed by the

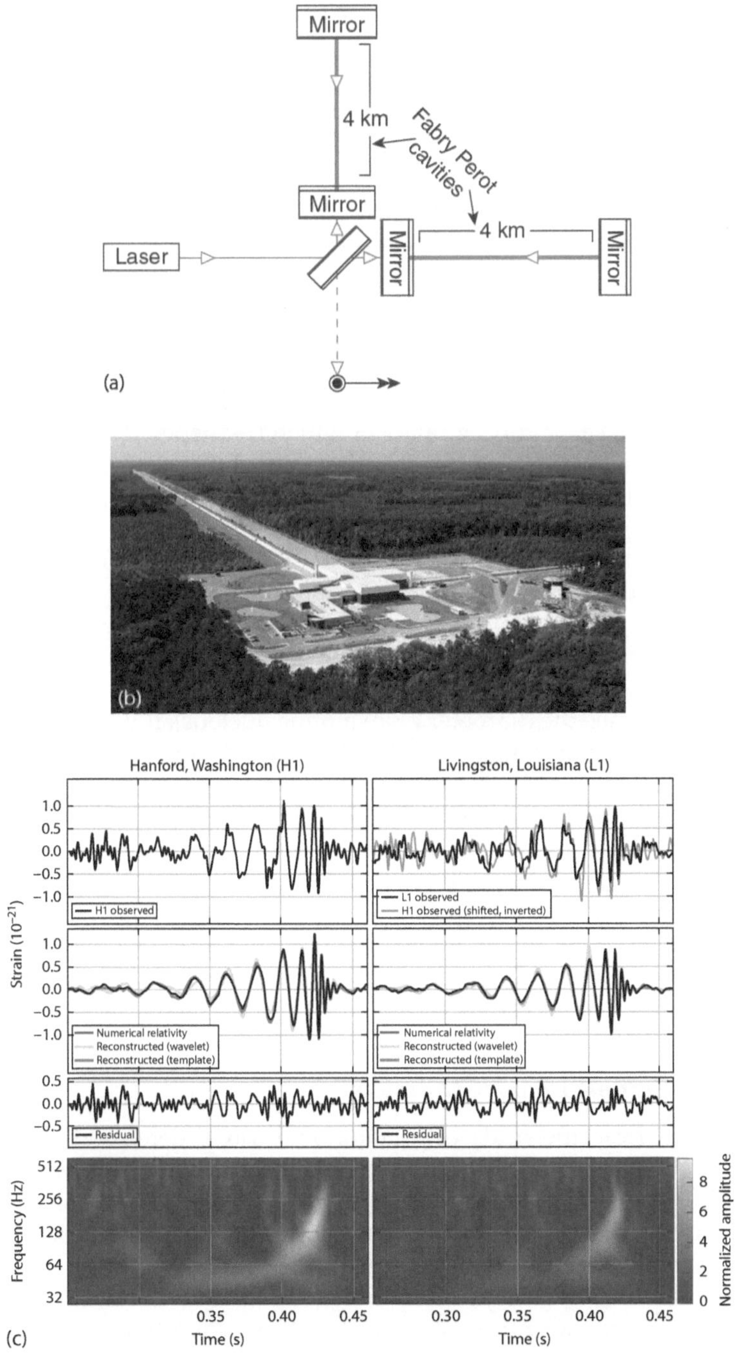

FIGURE 3.7
(a) LIGO instrument schematic; (b) the L-shaped LIGO observatory in Livingston, Louisiana (similar arrangement is also installed at Hanford, Washington); and (c) the gravitational wave signals received at the two LIGO observation sites [10].

instrument. As the length of the arms change, so too does the distance traveled by each laser beam in the perpendicular arms of LIGO. The beam in the shorter arm will return to beam splitter before the beam in a longer arm, and then the situation switches as the arms oscillate between being longer and shorter. Since the light waves in the two arms meet back at the beam splitter "out of phase" (i.e., not in exact time matching), a flicker of light is detected at the detector. The two arms of the interferometer are made exactly equal so that when no gravitational wave passes it, the two signals are reflected from the mirrors and received at the detector at exactly the same time, that is, in phase, and so no flicker of light takes place.

The gravitational wave signals received at the two LIGO observation sites are shown in Fig. 3.7c (two observation sites are used for "diversity reception," which is commonly used in communication engineering so that if the received signal fades out at one location, it will not fade out at the other location). The gravitational waves so received by the LIGO instrument was depicted as audio signal (over which frequency range the instrument was designed) and the sound resembled the chirping of bird (the said "chirp" of collision may be heard in the following video: Video | The Sound of Two Black Holes Colliding | LIGO Lab | Caltech).

Finally, gravitational waves could have been sensed with the change of tidal waves produced by the force of these gravitational waves (similar to that caused by space–time curvature resulting from the gravitation of the Moon on our Earth, the nature of the two of course being different), caused by the coalescing of black holes resulting in ripples in the space–time curvature. But this change in tidal wave is extremely small and is no larger than 10^{-14} m (only 10 times the diameter of an atomic nucleus). It is thus utterly hopeless to think of measuring such a tiny tide on the Earth's turbulent ocean. It may be noted that the gravitational waves reaching Earth are very feeble because the coalescing of black holes is taking place very far away, one or two or even more billion light years away from the Earth, and the strength of gravitational waves is known to die out inversely with the distance traveled similar to light waves. Thus, the effort to develop a gravitational wave detector with other novel principles was started by Joseph Weber in the late 1950s. Following a lot of changes in principle and technology, the outcome was LIGO instrument thanks to the efforts of Caltech and the MIT team. They worked very hard for the last two to three decades to realize this marvelous instrument.

3.2.4.3 A Recent Outcome from LIGO Results

One recent outcome from the study of gravitational waves detected by LIGO (in 2021) is excitingly important. The study proved the so-called *area theorem of the black hole,* which was proposed by Stephen Hawking in 1971 using Einstein's theory of general relativity as a springboard. The theorem states that "it is not possible for the surface area of a black hole to become smaller overtime, that is $(dA/dt) \geq 0$." The theorem parallels the second law of thermodynamics, which says that the entropy (disorder) of a closed system cannot decrease over time. Since the entropy of a black hole is proportional to its surface area, its surface area must continue to increase.

The black hole's monstrous gravitational attraction swallows anything and everything in its surroundings, whether it be stars, solar systems, or galaxies. As a black hole gobbles up more and more matter, its mass and surface area grow. But as the black hole grows, it also spins faster, which is known to decreases its surface area. Hawking's area theorem maintains that the increase in surface area that comes from the added mass would always be larger than the decrease in surface area caused by the added spin.

In the data analysis of gravitational waves, detected by LIGO in 2015, the researchers separated the signal into two parts depending on whether it was from before or after the merger of the black holes. This allowed them to figure out the mass and spin of the original black holes, as well as the mass and spin of the merged black hole. With this information, they calculated the surface areas of the black holes before and after the merger. The surface area of the resulting black holes was found to be larger than the combined area of the original black holes. This conformed to Hawking's area law. Thus, scientists believe that this might lead to uncovering more underlying laws of the universe in the coming days.

However, in 1974 Hawking also proposed his striking concept, Hawking radiation, with quantum mechanical analysis of black hole. According to this concept, because of strange quantum effects near the hole's event horizon, with the release of these radiations a black hole would shrink over a longer period of time and eventually get evaporated. Thus, his work of 1971 based on Einstein's general relativity and that of 1974 based on quantum mechanics appear to be paradoxical. Although the area theorem is understood to be applicable over a short to medium timeframe, Hawking Radiation is applicable for very large timeframe. No substantial evidence in favor of Hawking radiation has yet been recorded which needs extensive study on older black holes.

3.2.4.4 A Comment on the Fruits of Collaborative Research

In the design of LIGO to detect the gravitational waves, EHT to image the black hole, the Large Hadron Collider (LHC) to study the properties of subatomic particles, and for such technological marvels around the world, an international collaborative effort is a must to ensure the success of the project. This is because such machine design and the interpretation of data received from them require a multidimensional knowledge bank and enormous funding too. In this respect, we may quote John Bardeen, the only person who received the Nobel Prize in Physics twice: "Science is collaborative effort. The combined results of several people working together are often much more effective than could be that of an individual scientist working alone." This statement may be debatable in some respects because we have seen that Newtonian mechanics and Einstein's relativity, the bedrocks of classical physics, were developed single handedly by Newton and Einstein, respectively. However, we may point out that those scientific developments came at a different period of scientific developments. But in the age of modern technological developments, a group effort on a worldwide collaborative basis is a must. Otherwise LIGO, EHT, and LHC developments would not have been possible. This is the reason why present-day research in science, and especially in technology, is mostly based on international collaborations.

3.3 Unification Efforts Beyond the Fundamental Developments of Relativity and Quantum Mechanics

3.3.1 Einstein's Unified Field Theory

Having succeeded with his new concept of space–time and his chronogeometric vision of gravitation in his general theory of relativity, Einstein engaged in the more challenging game of unifying gravitation with light (more specifically electromagnetism). Because both gravitation and electromagnetism follow force equations that are strikingly similar (Section 1.5, Chapter 1), Einstein may have been prompted to hunt for their unification in his unified field theory. However, most physicists of his time were skeptical about such an effort. They thought that it was premature or even impossible to do that, for he was ignoring the fruits of the then emerging topic of atomic and nuclear physics based on quantum mechanics.

Einstein's simple yet elegant vision led him to propose the theory of relativity. For the first three decades of the 20th century, he was a towering figure in physics because his physical pictures and conceptual ability were unerringly correct. The mathematics, no matter how abstract and complex, however, came later and always as a tool for him to translate those physical pictures into a precise analytical language of science. The pictures he used to form even a complex physical problem were so simple and elegant that even the general public could grasp it. But ironically, in the last three decades of his life, Einstein abandoned his conceptual approach, resorting to obscure mathematics without any clear visual picture. Moreover, he unfortunately disregarded the inclusion of quantum mechanics in his unified field theory. He gradually became obsessed with purely mathematical concepts, such as "twisted" geometries, which are bizarre mathematical structures devoid of physical content. As Einstein once wrote, "I believe that in order to make real progress one must again ferret out some general principle from nature."

Throughout the 1930–1950s, the dominant activity in physics was not relativity or the unified field theory but developments in quantum theory. However, throughout that era, Einstein was practically alone in his pursuit of the unification of light with gravity. In the 1940s and 1950s, many physicists claimed that Einstein was over the hill. They described him as isolated, out of touch, and ignorant of the new developments in quantum physics that were being announced every new morning it seems. Even Robert Oppenheimer reportedly often told one of his colleagues that Einstein's quest for a unified field theory was futile. Einstein himself confessed, "I am generally regarded as a sort of petrified object, rendered blind and deaf by the years." In 1954, he remarked, "I seem like an ostrich who forever buries its head in the relativistic sand in order not to face the evil quanta." Although the world recognizes that Albert Einstein ranks alongside Isaac Newton in his quest to unlock the secrets of the universe, he spent the last 30 years of his life on a solitary, frustrating, and ultimately futile quest for the unified field theory.

When Einstein was working on the unified field theory, only two of the fundamental forces (gravity and electromagnetism) were known, whereas the third and fourth forces, the strong and weak nuclear forces, were under development. However, he consciously chose to ignore the nuclear forces, which is perhaps understandable because it was the most mysterious of the four forces at that time. Einstein was actually uncomfortable with the theory that describes the nuclear force, the quantum mechanics. (Scientists are in general agreement that Einstein made the biggest blunder of his life by rejecting quantum mechanics.) This may have been why Einstein failed in his untiring efforts to establish the unified field theory. He died while pursuing the notion that the forces of nature ultimately must be united by some physical principle of symmetry. The superstring theorists believe that Einstein's dream of unification may have been consonant with their so-called TOE.

3.3.2 Quantum Field Theory

QFT is a theoretical framework and bedrock of modern particle physics that combines classical electromagnetic field theory, special relativity, and quantum mechanics. In the four-dimensional space–time, QFT is considered to be the most fundamental theory of nature known today. It is the basis for explaining the behavior of subatomic particles and their interactions with a variety of force fields. Two examples of QFT are the QED (quantum electrodynamics), which describes the interaction of electrically charged particles and the electromagnetic force, and the quantum chromodynamics (QCD), which represents the interactions of the quarks and the strong nuclear force. According to QFT, all elementary particles exhibit wave-particle duality, acting as both a particle and a wave. We all know that in modern terminology, "light" is constituted of a quantum packet of energy ($h\nu$), photon, and the light/photon behaves both as a particle and as a wave. However, according to QFT, not only photon but all elementary particles exhibit the wave-particle duality. QFT understands that all elementary particles are localized vibrations in their respective quantum fields, which propagate in a wave-like manner. Each particle's wave-like and particle-like qualities are two complementary observable aspects of a single phenomenon that can be described in terms of excitations in a quantum field. For example, the Higgs boson particle discovered in LHC of CERN (Conseil Européen pour la Recherche Nucléaire or European Council for Nuclear Research), Geneva, in 2012, is the quantum excited form of the Higgs field.

Dirac and Feynman both contributed significantly to the development of QFT and QED. Shortly after completing the general formalism of quantum mechanics, Dirac applied the quantum theory to the electromagnetic field (leading to QFT) and made it consistent with the special theory of relativity (leading to QED). Dirac visualized that the cloud of probability that accompanies electrons between one interaction and another resembles the field of electrons or quantum field. He wrote equations for the field of electrons and other elementary particles. He concluded that particles are quanta (the packets of energy) of the so-called quantum field, just as photons are quanta of the electromagnetic field (light). All fields display a granular structure at the end of an interaction; "photon" in the case of light (the electromagnetic field) and "electron" (or other particles) in the case of the quantum field. The quantum field is unique in the sense that their quanta are particles (electron, proton, etc.) that appear when they interact with something else; left alone, they unfurl into a "cloud of probability." Thus, the notion of fields and particles, introduced by Faraday and Maxwell as entirely separate entities, disappear; and in QFT manifests either as a particle or a field depending on the state of an interaction—an illuminating philosophy and realistic practicality of quantum mechanics.

An electron is a quanta of the quantum field that does not in reality follow a trajectory in space (as we understand with a classical view) but appears in a given place at a given time as a particle when colliding with something else (i.e., taking part in an interaction). From the Schrödinger equation we know that within an atom, electron is in no particular place but is diffused in a cloud of probability in all places of the said cloud. Though it is not possible to know with certainty (in the classical sense) when and where it will appear, however, only in terms of its probability cloud one can determine its appearance in the next interaction. The elementary particle, more precisely the wave function of the elementary particle, is the excited state (or quanta) of the underlying quantum field, and each interaction can be visually represented by the Feynman diagram. This is the crux of QFT.

The standard model of physics that was finalized in the 1970s with only one missing field/particle—the Higgs field/Higgs boson (finally discovered in 2012 in LHC)—is based on QFT, QCD, and so on, and is discussed in Chapter 4.

3.3.3 Quantum Gravity—Theory of Loop Quantum Gravity

The most difficult part of the effort to unify two dissimilar entities in physics has been quantum mechanics with general relativity. Einstein's general relativity states that space–time is curved where everything is continuous, while quantum mechanics maintains that the world is flat in which discrete quanta of energy leap and interact. Modern cosmology as well as astrophysics have developed from the general theory of relativity that predicted gravitational waves, black hole, and expanding universe some 100 years back (all of which are now established experimentally). Quantum mechanics provided the foundation for atomic physics, nuclear physics, and physics of elementary particles, including condensed matter physics and so on. Both theories work remarkably well, one for comprehending the mysteries of the universe in the cosmic scale and the other for forming the basis of modern technology and works on the atomic scale. However, the paradox is that both cannot be true (in the form in which they existed before the development of the theory of quantum gravity) for any particular phenomenon in the sense that they appear to contradict each other. To be clearer, the gravitational field is described without taking quantum mechanics into account, while quantum mechanics was formulated without taking into account the fact that space–time is curved. (Due to the effect of mass/energy as given by Einstein's relativity theory, space–time bends and curves under the weight of matter and plunges into black holes when matter is too concentrated.)

But every experiment and test in the cosmic world say "relativity you are true," and every experiment and test in the micro-world say "quantum mechanics you are true." At the same time, we can neglect one or the other in particular situations. For example, the Moon is too large to be sensitive to the minute granularity of quantum mechanics (so we can forget quanta when describing its movement around the Earth), while an atom is too light to curve the space–time around it to any significant degree, so we can neglect the curvature of space–time when describing it. But there are situations—at the minute dimension of the so-called *Planck length* (= 1.62×10^{-35} m or 10^{-20} times the size of proton)—where both curvature of space–time and quantum granularity matter. Here was the *quantum gravity* that has given birth to the theory of loop quantum gravity or, in short, the loop theory.

The loop theory combines general relativity with quantum mechanics in a rather conservative way because it does not employ any other hypothesis apart from those two theories themselves, suitably written to render them compatible. But the consequences are radical. Einstein understood that space and time are manifestations of a physical field, the gravitational field. Bohr, Heisenberg, and Dirac understood that physical fields have a quantum character: Granular, probabilistic, manifesting through interactions. The quantum gravity theorists thus visualize that space and time must also possess the strange properties of quantum mechanics: Granularity, probabilistic and relational characteristics. Thus, the concept of quantum space and relational time appeared in quantum gravity.

It was Matevi Bronstein, a young theoretical physicist of the former Soviet Union, who showed that the gravitational field was not well defined at a point when quantum mechanics was taken into consideration. Bronstein worked out the mathematics of the problem and stated that quantum mechanics and gravity can be made compatible in a region of space smaller than the *Planck length* (more fittingly it should have been called the *Bronstein length*) given by $l_p = (hG/2\pi c^3)^{1/2}$. In this equation, h is Planck's Constant, G is the universal gravitational constant, and c is the velocity of light. Thus, this relation relates gravity (G) to quantum mechanics (h) in the realm of relativity (c). The value of l_p is 1.62×10^{-35} m. Similar to Planck length is Planck time (see Table 2.1), which is defined as the time required for light (i.e., photon) to travel a distance of 1 Planck length in a vacuum, which is a time interval of approximately 5.39×10^{-44} s.

The marriage of quantum mechanics with gravity has been the most difficult task for physicists. The list of scientists trying to solve the puzzle of quantum gravity would read like an honor roll. However, the scientist who has contributed significantly to develop the current form of quantum gravity was John Wheeler. He was more intuitive than mathematical (he was the first to use the term *black hole* for dying star in the cosmos and he popularized the term). Searching for the visionary aspect of Bronstein's quantum properties of the gravitational field in an extremely small scale (Planck length) of space, Wheeler looked for novel ideas and with his fervid imagination eventually derived the concept of *quantum space*. (To hear the metaphor spoken by John Wheeler himself as he explains the quantum space, go to the following website: http://www.webof stories.com/play/9542?o=MS.)

Following Wheeler's account, let us imagine that we are looking at the surface of an ocean from a great height. What do we see? We glimpse a vast and flat cerulean table. As we descend, we begin to observe great waves of ocean swollen by wind blowing over it. On further descending we find that the waves break up and that the surface of the ocean is a turbulent frothing. Yes, this is the pictorial vision of quantum space as visualized by Wheeler. Though in our gross vision (in a scale immensely larger than the Planck length) the space around us appears to be continuous and of course nongranular, in reality (down to the Planck length), in terms of the theory of quantum gravity, the same space is quantized and shatters like ocean generating foam. Wheeler terms these quantum elements of space "spin foam."

In the theory of loop quantum gravity, the quantum of space (i.e., of the gravitational field) is represented by a "spin network," as shown in Fig. 3.8. Every node stands for a volume element (to which the space is understood to be divisible to), and every line connecting the node is a half-integer (1/2) or the multiple of half-integer (half-integer/multiple of half integer in particle physics are called "spin" associated with the quantum-mechanical spinning of particles—electron and proton have half spin while the particle of light, photon, has spin 1)—hence the name "spin network." Again a loop will be completed if one moves from one volume element of space to another via the link lines and comes back to the initial volume element in a spin network—hence the name "loop quantum gravity." The mathematics of the theory determines the curvature for every closed circuit on a spin network; this makes it possible to evaluate the curvature of space–time, and hence the force of the gravitational field. The nodes of the graph represent the discrete packets of volume, and, as in the case of photons, they can only have certain sizes, which can be computed by

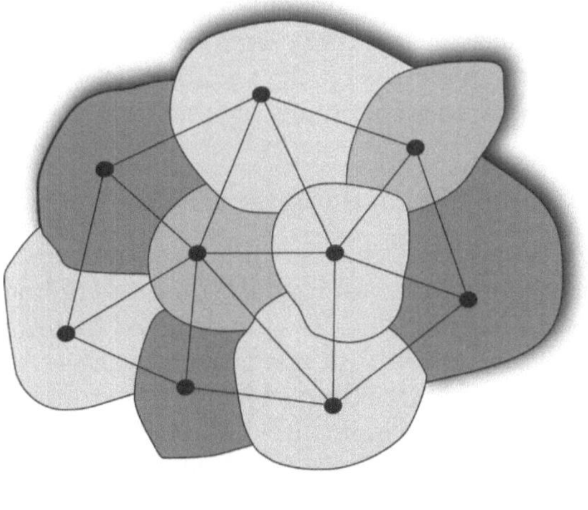

FIGURE 3.8
Representative diagram of spin network of the theory of loop quantum gravity.

using Dirac's general quantum equation (the eigenvalue equation for the volume operator). The nodes are the elementary quanta of which the physical space is made of. Every node of the graph is a "quantum particle of the space." A link is an individual quantum of a Faraday line. Space appears continuous to us only because we cannot perceive the extremely small scale (the Planck length at which the quantum gravity dominates) of these individual quanta of space. Just as when we look closely at the cloth of a T-shirt, we see that it is woven from small threads. On our scale, which is immensely larger than the Planck length, the space is smooth. If we move down to the Planck scale, it shatters and foams—just as the light coming to us from the Sun appears to provide continuous rays of light waves, but in reality they are streams of electromagnetic particles, the photon.

Just as the Faraday lines led to the concept of field in electromagnetism, the closed loops of the theory of loop quantum gravity may be interpreted as the lines or loops (i.e., lines closed on themselves) of the gravitational field. However, the infinitely fine, continuous spiderweb of Faraday lines is now a real spiderweb of finite numbers of distinct threads represented by quantum loops. Further, the so-called Faraday-type lines (closed loops) of the quantum gravitational field are threads of which the space is woven, unlike the Faraday/Maxwell and Einstein concept that field is immersed in space. The loop theory deals with how the quantum loops "weave" the three-dimensional physical space with a probability cloud of superimposed geometries leading to the pictorial view of the quantum space–time.

The central prediction around which the theory of loop quantum gravity revolves is that the space is formed of the quanta of gravitational field (the loops), just as quanta of light (photon) are the quanta of electromagnetic field. This granular quantum structure of space is a precise mathematical form obtained by applying the general equations of quantum mechanics written by Dirac to Einstein's gravitational field and developed through the Wheeler–DeWitt equation (developed by John Wheeler and Bryce DeWitt), modified by Abhay Ashtekar, and solved by Lee Smolin and Ted Jacobson. The essential difference between photon (the quanta of the electromagnetic field) and the loops of the loop quantum gravity is that the photon exists in the space while the quanta of gravity (loops) constitute the space itself. The quanta of gravity is not something that is laid in space; they are the minuscule elements of space and thus themselves constitute the space, thereby making the space constituted of space quanta. Further, it may be noted that in the space quanta (i.e., in the loops), one cannot talk about time instants (as we conventionally use for time variation of a physical quantity/variable), but rather in a quantum sense only the relational aspect prevails among the variables. In other words for variables P, Q, R in quantum loops, the behavior of variables is expressed as P(Q,R), Q(P,R), R(Q,P), ... and not as P(t), Q(t), R(t), ... though in a fashionable way some populists say that time does not exist in those loops!

3.3.4 Evolution of the Connotation of the Term *Time*

The connotation of the term *time* has changed significantly from the time of Aristotle to the time of Stephen Hawking and beyond with the progress of knowledge in the domain of physics and astrophysics. Aristotle and Newton believed that time is an absolute entity. Their understanding was that the events in this universe succeed each other in an orderly way—pasts, presents, and futures. Since the past is fixed, the future is open and the present is the instant of time we are in at the moment. However, with Einstein's relativity theory, the notion of time changed significantly—whose proof has been established firmly and understandable even to the average mind when applied to the cosmic domain (as discussed earlier). According to the general relativity theory, time passes faster in the mountains than it does at sea level (i.e., close to the Earth's surface due to the larger strength of gravity). Even a clock in the floor runs a little more slowly (i.e., time is dilated) than a clock on a table; the difference is extremely small (perhaps microseconds or even nanoseconds), but it is measurable by the present day's accurate clocks. Thus, after Einstein's theory of relativity, time no longer remained an absolute entity but was understood to be a relative one, that is, *time is elastic.*

This might appear surprising to an average mind, but it is how the phenomenon in nature works in reality; it could not be visualized in the past because of the inability to measure time with the level of precision possible today. (In addition, this revolutionary idea about time was not being thought about by any one before Einstein.) Because of time dilation close to the strong gravity of a body, the time is said to stop near the black hole (from the viewpoint of an observer outside the black hole) due to the stupendous gravitational strength of a black hole. The concept of "now" is also understood to be relative. For example, it differs by 4.2 years when we refer to the so-called instant of time at the Earth and at our closest star (apart from Sun), the Proxima Centauri. In addition, that difference is discernible for the so-called same moment at the Moon and Earth. Further, with regard to the present notion of the origin of the universe in terms of the Big Bang theory, "time" is understood to have started at the moment of the Big Bang. In looking into such varied notions of time from the time of Aristotle to that of Einstein, the cosmology expert Stephen Hawking may have been inspired to write his best-selling book, *Brief History of Time*, incorporating many fundamental issues of physics.

But the most striking challenge to the meaning of the term *time* has been raised by the latest theory of quantum gravity or the loop theory, which seeks to combine quantum mechanics with gravity and proclaims *Time does not exist*! What a shock to our understanding of the notion of time, though our average mind could somehow digest the so-called elasticity of time according to Einstein's relativity theory. But this appears to be a blow head-on! This seemingly absurd issue about time has been discussed in brief earlier in this chapter, while discussing the theory of quantum gravity. Although it appears to be incomprehensible to the average mind but true to the facts, it is understood to be intimately related to the intricacies of the latest theory of physics. This issue of time as a relational entity, and not as an instant of time (which is conventionally used as a function of any physical quantity, for example, for voltage [V] varying with time [t] is represented by $V[t]$, etc.), is, however, applicable only to the extremely small length of space called Planck length ($= 1.63 \times 10^{-35}$ m). But such issues are not of any importance to our day-to-day lives and for the phenomena around us. Thus, let us be cool and feel happy that our age-old proverbs such as "Time and tide waits for no one" and "Time is short but art is long" still hold important lessons of life for us. From time immemorial, these proverbs have told us to act in the present so that we can make our future bright.

3.3.5 A Digression: Big Bang versus Big Bounce as the Theory of the Origin of the Universe

We have already discussed the Big Bang theory and have also mentioned that it has strong astronomical support based on CMB radiation. Big Bang postulates that 13.8 billion years ago, our universe emerged from a singularity—a point of infinite density and gravity—and that before this event, space and time did not exist. In fact, it is understood that space and time emerged after the moment of the Big Bang. The Big Bang theory and other theories like the Grand Unified Theory (GUT) and superstring theories give a pictorial view of how the universe evolved once it began (details of which are given in Section 3.5). Creation of the universe is highly controversial through and through, although from ancient times, mystics and scientists alike have been proposing their views for unraveling the mysteries of the origin of the universe. Mystics believe that God chose how the universe began (the reasons for that are not understandable scientifically), while scientists have an explanation of how our universe evolved after Big Bang, but that theory cannot explain what the state of the universe was before Big Bang—that is, how the universe really began. This is where quantum gravity enters into the picture, and the big bounce theory has emerged as a competitor to understand the origin of the universe.

Quantum cosmology or, more specifically, quantum gravity and the theory of loop quantum gravity predict that the primordial state of the universe was not just the *big bang* but more specifically the *big bounce,* which may be understood as follows. When the universe was an infinitely dense fireball of extremely small size on the order of Planck length ($= 1.6 \times 10^{-35}$ m or million trillion trillionth of a centimeter), at that state of it the equations of general relativity were no longer valid and at the same time it was not possible to ignore quantum mechanics. In other words, under such circumstances, we enter into the realm of quantum gravity. If we consider only relativity, in such a nonquantum mechanical situation, with infinite density of the fireball, we are in a state of singularity where all physical laws including relativity collapse. Thus, Einstein's theory of relativity predicts that in that state of the universe, it is expected to be squashed ad infinitum and at a certain point would disappear altogether (resembling the classical situation of an electron falling into the nucleus, as discussed in Section 2.7, Chapter 2). This was the Big Bang predicted by Einstein's equations, if we ignore quantum mechanics.

But if we take quantum mechanics into account, the universe cannot be indefinitely squashed. A quantum repulsion is understood to make it rebound—just as a quantum repulsion pushes away the electron when it gets too close to the nucleus, as per Heisenberg's uncertainty principle, and prevents it from falling into the nucleus, forcing it into an orbit (the so-called stationary orbit of Bohr's atomic model) of the size determined by the de Broglie wavelength, as discussed in Chapter 2. Similarly, a contracting universe does not collapse down to a point; it bounces back and begins to expand as if it were emerging from a cosmic explosion (Fig. 3.9). The past of our universe may therefore well be the result of just a rebound—a gigantic rebound known as big bounce instead of Big Bang. This is what seems to emerge from the equations of loop quantum gravity when they are applied to the expansion of the universe.

FIGURE 3.9
An artist's view of the big bounce model of the universe [11].

The image of the bounce must not be taken literally. Going back to the example of the electron, recall that if we want to place an electron as close as possible to the nucleus of an atom, the electron is no longer a particle. We can think of it instead as a cloud of probability. An exact position does not make any sense for the electron. The same concept is possibly true for the universe. In the crucial passage through the big bounce, the space and time cannot be thought of as a single entity but wildly fluctuates in granular form as a spread-out cloud of probabilities. However, this concept is still at an infant and exploratory stage and is understandable only in terms of some mathematical equations to predict the possibility of such an event—in which our observable universe could be the result of the collapse of a previous contracting universe passing across a quantum phase, where space and time are dissolved into a swarming cloud of probabilities.

3.4 New Thoughts beyond Relativity and Quantum Mechanics: The String Theory or the Theory of Everything

3.4.1 Introduction

The 20th century was gifted with two jewels: *Relativity* and *Quantum Mechanics*. Each of these two theories, in its particular domain, has scored spectacularly and has become triumphant in its own right. General relativity, for example, has shown brilliant success in the cosmic scale of stars, galaxies, and so on, explaining gravitation with a chronogeometric vision. The black hole, which physicists believe is the ultimate state of a massive cold star (with death-time mass more than three times the solar mass) is a well-known prediction of general relativity. Further, the prediction of general relativity is that the universe originally started in a Big Bang that sent the galaxies hurtling from one another at enormous speeds, leading to an expanding universe (experimentally verified by the so-called red-shift of galaxies observed experimentally by Hubble and his co-researchers) and so on. However, the theory of general relativity cannot explain the behavior of the atoms and molecules of the micro-world. On the other hand, quantum mechanics has no rival to explain the secrets of the atom and subatomic particles. Quantum mechanics has unraveled the secrets of nuclear physics and explained the working of everything starting from transistors to lasers, conductivity to superconductivity, and so on. Modern chemistry and biology have also been extremely benefited by quantum mechanics for understanding many of the mysteries of their micro-phenomena. Although quantum mechanics has been undeniably successful in explaining the micro-world phenomena, the theory fails when we try to describe the gravitational force.

So physicists were faced with two distinct theories, each employing a different set of mathematics and with a totally different physical visualization of fundamental phenomena in cosmic and atomic scales. But each made astonishingly accurate predictions, with experimental support within its own realm. This is where the *string theory* and its competitor/alternative theory of loop quantum gravity entered into the picture. The first marriage of quantum mechanics with relativity was a disaster, creating a crazy theory—QFT—that for decades produced only a series of meaningless results. Every time physicists tried to calculate, for example, what happens when electrons collide, QFT would predict infinite values for collision! The theory was plagued with the riddle of infinite because quantum mechanics (including gravitation) assumed the particle nature of the fundamental building blocks of matter that follow "inverse square law." That is, the farther one distances oneself from a particle (electron) or a celestial body (the Sun), the weaker the field becomes. In other words, as one approaches the particle/celestial body, the force rises dramatically, and at its surface the force field becomes practically infinite (1/0, an expression that is ill defined or even undefined) making all physical quantities like energy and so on infinite! There were efforts, lasting for almost half a century, to take care of this riddle of infinity using the renormalization technique and so on. However, ultimately QFT has been very successful in combining quantum mechanics with relativity (of course, without gravity), which is the foundation of the standard model of physics for elementary particles. But to combine quantum mechanics with the gravitation of relativity, physicists thought to take an absolutely different approach, leading to the theory of loop quantum gravity on the one hand and string theory or, more specifically, superstring theory on the other.

3.4.2 Concept of String Theory

The mechanism by which string theory manages to marry both special and general theory of relativity and quantum mechanics and also resolves the problem of the riddle of infinity is fascinating. String theory assumes that the ultimate building blocks of nature consist of tiny vibrating strings that are one-dimensional extended objects, rather than the zero-dimensional point particles that form the basis for the standard model of particle physics propounded during the 1970s. By replacing point-like particles with strings, an apparently consistent quantum theory of gravity emerges. String theory is a quantum theory in that the mass spectrum of strings is discrete, so string theory may be considered an example of a quantum theory of gravity too. It may be possible to unify all the known natural forces (gravitational, electromagnetic, weak nuclear and strong nuclear forces) with the string theory. Particles like the electron, proton, and neutron in all matter that constitutes our body, the laptop, and even the farthest stars and galaxies are ultimately supposed to be the manifestation of the strings of the string theory. Such strings are not observable to us because they are extremely small (they are about 100 billion times smaller than a proton). According to string theory, our world only appears to be made of point particles because our measuring devices are too crude to see those tiny strings.

Knowledge of the physics of a violin string gives us a comprehensive theory of musical tones and allows us to predict new harmonies and chords. Today the diverse rules of music that have developed through thousands of years of trial and error can be easily derived from a single picture—that is, a single string that can resonate with different frequencies, each one creating a separate tone of the musical scale. The tones created by the vibrating strings, such as C sharp or B flat, are not in themselves any more fundamental than any other tone. What is fundamental, however, is the fact that a single concept, vibrating strings, can explain the laws of harmony. Similarly in the string theory, the fundamental forces and various particles found in nature are nothing more than the different modes of vibrating strings. The gravitational interaction, for example, is caused by the lowest vibratory mode of a circular string (a loop). Higher excitation of the string creates different particles of matter. From the point of view of string theory, all particles are just different vibratory resonances of vibrating string, and no force or particle is more fundamental than any other. The string theory can thus provide a coherent and all-inclusive picture of the nature similar to the way a violin string can be used to "unite" all the musical tones and rules of harmony. Thus, John Schwarz and Michael Green, the creators of the string or superstring theory, called it a TOE. Even *Science Magazine*, always careful not to exaggerate the claims of scientists, compared the birth of the string theory to the discovery of the Holy Grail.

Physicists are particularly excited because string theory demands that everything from the motion of galaxies down to the dynamics within the nucleus of the atom can be explained with a single theory/equation of string, with its different vibrational modes manifesting as different fundamental particles. It appears to solve the most challenging scientific problem of the century: How to unite the four forces of nature into one comprehensive theory. The theory even makes startling predictions concerning the origin of the universe, the beginning of time, and the existence of a multidimensional universe. Excitement is also because it forces us to revise our understanding of the nature of matter. Since the time of the ancient Greek philosophers, scientists have assumed that the building blocks of the universe were tiny point particles. Democritus coined the word *atomos* to describe these ultimate, indestructible units of matter which today we know as atoms. All the theories of physics and experiments support the presence of atoms and subatomic particles, but this new theory says that the ultimate building blocks of nature are more fundamentally not particles but tiny vibrating strings!

The mathematical analysis of string theory and its corresponding physical interpretation are quite involved. However, string theory uses some sort of interactions and Feynman diagrams that resemble respectively the Tinker toys played by children and the famous origami art of Japan. Only five types of interactions (joints) are required to describe string theory (see Fig. 3.11a) resembling Tinker toys. The Feynman diagrams for colliding open strings (resembling the Möbius strip) and closed strings (resembling the Klein bottle of topology) are shown in Fig. 3.11b—which reminds us of Japanese origami art.

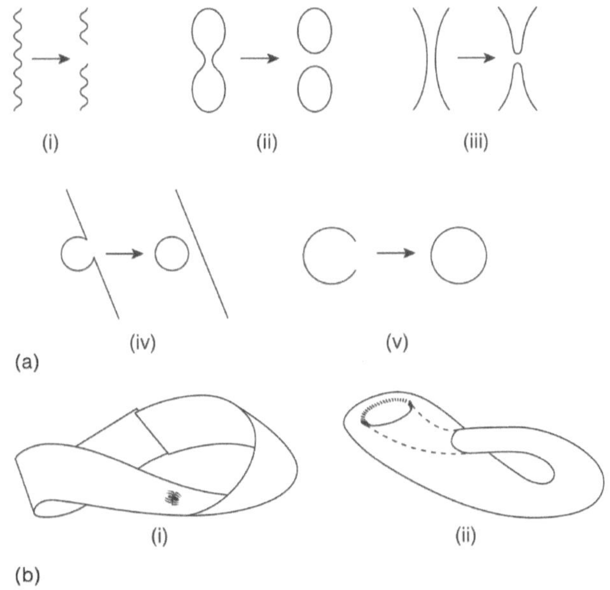

FIGURE 3.10
Representative diagrams of string theory. (a) Five types of string interactions: (i) A string spits and creates two smaller strings, (ii) a closed string pinches and forms two smaller strings, (iii) two strings collide and re-form into two new strings, (iv) a single open string re-forms and creates a closed and an open string, and (v) the ends of an open string touch and form a closed string; (b) single-loop Feynman diagrams for (i) colliding open strings and (ii) colliding closed strings.

3.4.3 History of the Development of String Theory

The fascinating history of string theory is as interesting as the startlingly different physical picture on which the theory is based. The origin of this theory can be traced back to the 1960s, when it was applied possibly for a wrong problem of strong interaction theory (and not as a theory of the universe). It burst forth as a sheer accident (!) and not as a logical sequence of ideas like QFT, which was based on years of patient developments. Yoichiro Nambu first wrote down the basic equations of string theory. He proposed the idea of the strings to make some sense out of the chaos of the hundreds of hadrons (a class of heavy mass elementary particles like proton, the light mass ones being known as leptons such as the electron) that were discovered in atom smashers in the 1950s. Nambu's seminal idea was to assume that the hadron consisted of a vibrating string, with each mode of vibration corresponding to a separate particle. According to this theory, the infinite variety of particles found in nature is simply different modes of resonances of the same string, with no particle any more fundamental than any other. However, the original Nambu theory was self-consistent in only 26 dimensions; that sounded more like science fiction than true science and was thus not taken seriously when it was proposed. Also, this theory was applicable only to bosons (which follow Bose–Einstein statistics, having an integer-valued spin).

The Ramond–Nevue–Schwarz (RNS) model of string theory proposed by Pierre Ramond, André Neveu, and John Schwarz made it self-consistent only in ten dimensions. Their theory is known as the *superstring theory*, for it is applicable for both bosons and fermions (that follow Fermi-Dirac statistics, having a half integer spin), satisfying the supersymmetry (SUSY). Nambu's string was just bosonic string, while RNS string is fermionic accompanied by the bosonic one. Quantum mechanics divided all the particles of the world into just two types: Bosons and Fermions; 50% of fundamental particles are bosons. Although Nambu may be considered one of the founding fathers of the string theory, his main contribution is in particle physics. He received the Nobel Prize in Physics in 2008 "for the discovery of the mechanism of spontaneous broken symmetry in subatomic physics."

String theory is the original and generic name of the entire theoretical framework where the point-like particle in particle physics is replaced by a hypothetical subatomic object made up of a one-dimensional entity with dynamic properties similar to those of a flexible elastic string. A version of string theory called supersymmetric string theory or, in short, superstring theory, was developed to account for both fermions and bosons, as mentioned above, incorporating SUSY. In SUSY, each particle from one group would have an associated particle in the other, which is known as its super partner, the spin of which differs by a half-integer. Modern string theory is almost entirely superstrings. Niels Bohr famously said that any great theory should be "crazy enough," but possibly string/superstring theory stretched beyond the limits of our scientific imagination to believe that the universe could be 26 or even 10 dimensions! Anyway, it is known that a higher dimension provides more information about an object. For example, compare the two-and three-dimensional views of the same object, the latter being capable of giving a better perspective view of the object.

The 1970s through 1984 were lean years for the string model, with most physicists preferring to work on the fast-paced developments in electro-weak and GUT-type theories. GUT is a model in particle physics in which electromagnetic, weak, and strong nuclear forces are merged into a single force at high energies. However, in 1984 Green and Schwarz observed that the string model possesses enough symmetry to eliminate all infinites and anomalies that were plaguing the QFT used to describe gravity interacting with other particles. This realization touched off an explosion in string theory. Nobel Laureate Steven Weinberg, who with Abdus Salam and Sheldon Lee Glashow is known for his contributions to the unification of the weak force and electromagnetic interaction between elementary particles (leading to electro-weak force), hearing the excitement about string theory, immediately switched to working on this theory. "I dropped everything I was doing," he recalls, "Including several books I was working on, and started learning everything I could about string theory." It was not easy, however, to learn an entirely new mathematics. "The mathematics is very difficult," he conceded.

3.4.4 How to Verify String Theory Experimentally

The world of physics, which is closing in on the unified description of the four fundamental fields/forces of nature, has since the 1990s focused on designing, developing, and experimenting with particle accelerators/atom smasher/colliders in tera electronvolt (TeV) range to probe deep into the atom's nucleus. The objective is to test certain aspects of the standard model of physics and also, if possible, to reach to the fringes of the GUT theory. The goal is to try to unify electromagnetic, weak, and strong interactions, or forces into one single force, including the string theory. Physicists were hoping that locked deep within the nucleus of the atom was the crucial data necessary to verify some aspects of these theories. Building a colossal atom smasher called the Superconducting Super Collider (SSC) was planned to study the phenomena associated with high-energy physics. This would have been the largest scientific machine ever built, costing billions of dollars; however, the project was abandoned by the U.S. Congress in 1993.

The LHC of CERN, Geneva (a European effort), started functioning in September 2008 and made a successful run in 2012 that produced some important experimental signatures in support of some of the theories for unifying the four fundamental forces of nature. The signature of the particles received from the run of LHC, in 2012, was the *Higgs* (more specifically, *Higgs-Boson*) *particle*—the quantum of the Higgs field, predicted by Peter Higgs and others in 1964. It was found to have the properties precisely predicted by the standard model of physics. Peter Higgs is a British theoretical physicist, emeritus professor at the University of Edinburgh, who received the Nobel Prize in Physics in 2013 for the "theoretical discovery of a mechanism that contributes to our understanding of the origin of mass of subatomic particles, and which recently was confirmed through the discovery of the predicted fundamental particle, by the ATLAS and CMS experiments at CERN's Large Hadron Collider." The Higgs particle was termed the God Particle, an interesting term. It may have been so named because it experimentally established the last missing link of the standard model of physics (which made physicists of the standard model school excited now that their decades of hard work had finally been rewarded through LHC experimentation). Actually, Leon M Lederman (an American experimental physicist who received the Nobel Prize in Physics in 1988, along with Melvin Schwartz and Jack Steinberger, for research on neutrinos) wanted to call it "that goddamn particle" (in his book entitled *The God Particle*, published in 1993) to popularize the Higgs boson. However, the editor of the said book did not agree to that term, and instead opted for the God particle. This is how the term originated. But Peter Higgs himself didn't endorse the name.

As it turned out, the supporters of GUT and superstring theory were disappointed with the outcome of the 2012 experiment with LHC, as their expectation of getting the signature of the so-called *SUSY particle* was missing from the results of the LHC experiment. This is because the energies needed to probe the GUT and superstring theories are so fabulously large that the ultimate verification is expected to come from the field of cosmology. The energy scale in which the unification of four force/fields of nature is understood to take place can be found only at the beginning of time when the universe evolved from the Big Bang. In this sense, solving the puzzle of the unified field theory with the possible experimental verification of the predictions (unbroken symmetry) of string theory may well solve the riddle of the origin of universe too. Let us hope that future atom smashers of much higher energy level will bring excitement for GUT and superstring theory supporters.

3.4.5 Few More Words about String Theory

Let us now summarize some of the aspects of string theory that makes it a novel theory.

According to string theory, the universe originally existed as a superdense ball in 10 dimensions, not the four dimensions (three space dimensions and one time dimension) visualized by Einstein's relativity theory and the convention of the day in physics and astrophysics. However, because the superdense ball was unstable in ten dimensions, it "cracked" into two pieces, with a small, four-dimensional universe peeling off from the said superdense ball. This process may be imagined in terms of the fission of a soap bubble (originally, a ten-dimensional superdense ball), caused by a strong enough vibration, into two more soap bubbles (one of the smaller soap bubbles being our universe). In terms of string theory then, we have a "sister universe" that coexists with our universe (which, however, we cannot reach as it shrank to an incredibly small size; billion times smaller than the nucleus of an atom). It also means that the original fissioning of our universe was so violent that it created the explosion that we know as Big Bang. Thus, the string theory can give an original view about the origin of Big Bang and also touches on the state of the universe before Big Bang. Einstein's theory, however, is unable to answer these questions. Einstein's theory only states that the universe was born from a gigantic explosion called the Big Bang some 10–20 billion years ago (the actual age as per the present study is 13.8 billion years). All the matters in the universe, including the stars, galaxies, and planets, were originally concentrated in one superdense ball, which exploded violently, creating our current universe. But why did the superdense ball explode? What was the state of the universe before Big Bang (the explosion of the superdense ball)? None of these questions was addressed in Einstein's theory, but string theory stands as a new candidate with its own viewpoints.

SUSY is central to superstring theory. Although SUSY was first found in a ten-dimensional theory, it also could be applied to four-dimensional theories, and it became quite fashionable by the late 1970s. GUTs, it turned out, suffered from certain anomalies that SUSY could cure. The significant advantage of superstring theory (because of its sufficient symmetry) over the QFT is that renormalization is not required. All the loop diagrams of superstrings at each level are finite by themselves, requiring no artificial sleights of hand like renormalization to remove the infinities. Symmetry explains why all the potentially harmful divergence and anomalies, sufficient to kill other theories, cancel each other perfectly in the superstring theory. The superstring model has such a huge set of symmetries that the theory can include all the symmetries of the electro-weak and GUT-type theories, as well as Einstein's theory of general relativity. All the known symmetries of the universe, and many that have not yet been discovered, are found within the superstring theory. In retrospect, it is clear that the symmetries are the reason why the superstring theory works so well.

3.4.6 A Comment by the Author

Physicists are always cautious in their approach in accepting new theories that might revolutionize knowledge that has been logically established and experimentally verified. However, supporters of a new theory and skeptics (who are still not convinced about its worth) are always divided in the initial stage of its development when a new theory is propounded; naturally, no exception was made for string theory. Staunch supporters of string theory have commented that "string theory is a miracle, through and through"; that "it is probably going to lead to a new understanding of what space and time really are, the most dramatic (understanding) since general relativity"; and that "we may be witnessing a revolution in physics as great as the birth of the quantum mechanics." However, the opposite views are equally strong and worth noting: "years of intense effort by dozens of the best and the brightest have yielded not one verifiable prediction, nor soon anyone is expected"; "all advertisement and very little substance"; and "the ground of physics is littered with the corpses of unified theories."

Although the string theory opens up a new vista of mathematics and probably a new vision of the fundamental nature of the physical world that has even startled the mathematicians and excited physicists from around the world, it may even take many decades before any suitable machine can be made to test the theory conclusively. Thus, skeptics remain unconvinced by the so-called "tall talks" made by the string theorists, despite its beauty, elegance, and uniqueness. But the string theory's defenders point out that, although a decisive experiment that could prove the theory might take decades, there are no experiments that contradict the theory too. Thus, I (the author) believe that as long as we cannot disprove a theory, we have no right to discard it into the dustbin.

In this perspective, human tolerance should be exhibited, and we must have the courage to honor the viewpoint of others with patience and due honor. In the August gathering of the World Parliament of Religions in 1893, the Indian monk/social reformer Swami Vivekananda said, "I am proud to belong to a religion which has taught the world both tolerance and universal acceptance. We believe not only in universal toleration, but we accept all religions as true." To respect others' viewpoints with human generosity is extremely important and is essential for the progress of humankind. In truth, religion and science are never at odds; both are equally required for the progress of humankind, the one enriching our ethical values and the other enriching our rational values and viewpoints. There is nothing wrong with either religion or science by themselves. However, misunderstandings and misinterpretations are to be avoided at all costs to allow science and religion to shine with their own effulgence.

3.5 A Digression: The Scientific Model of the Evolution of the Universe and Four Fundamental Forces of Nature

3.5.1 The Scientific Model of the Evolution of the Universe

The scientific models of the evolution of the universe are highly speculative. Based on the Big Bang theory, theory of quantum gravity, and inflation and superstring theory, a plausible idea of the origin and subsequent evolution of the universe has been developed. It is understood that when the universe was perhaps only 10^{-43} seconds old and 10^{-33} meters across, it was an extremely (infinitely!) dense fireball from which the Big Bang started. In the heart of the Big Bang was the *Plank epoch* when matter and energy are understood to consist of unbroken superstring/unified fields or forces and all the four fundamental forces (gravity, electromagnetic, strong, and weak nuclear force) of present universe were but one force. Quantum gravity, as described by the superstring theory, was the dominant force in the universe. Since, at the Planck scale, the current physical theories are not applicable to calculate what happened, as of now, very little is known about the physics of the Planck epoch. After the Plank epoch, started the *GUT era* when at the incredible temperature of 10^{32} Kelvin (a thousand trillion, trillion times hotter than the temperature found in the Sun), one of the four fundamental forces, the force of gravity, split apart from the other three fundamental forces, i.e., the GUT forces (weak and strong nuclear forces and electromagnetic force) of the standard model of particle physics (Fig. 3.11a). At this point, according to the inflation theory (that has widespread support among the cosmologists), the universe inflated rapidly, under the sway of an exotic force that caused a brief, violent moment leading to a hyperexpansion

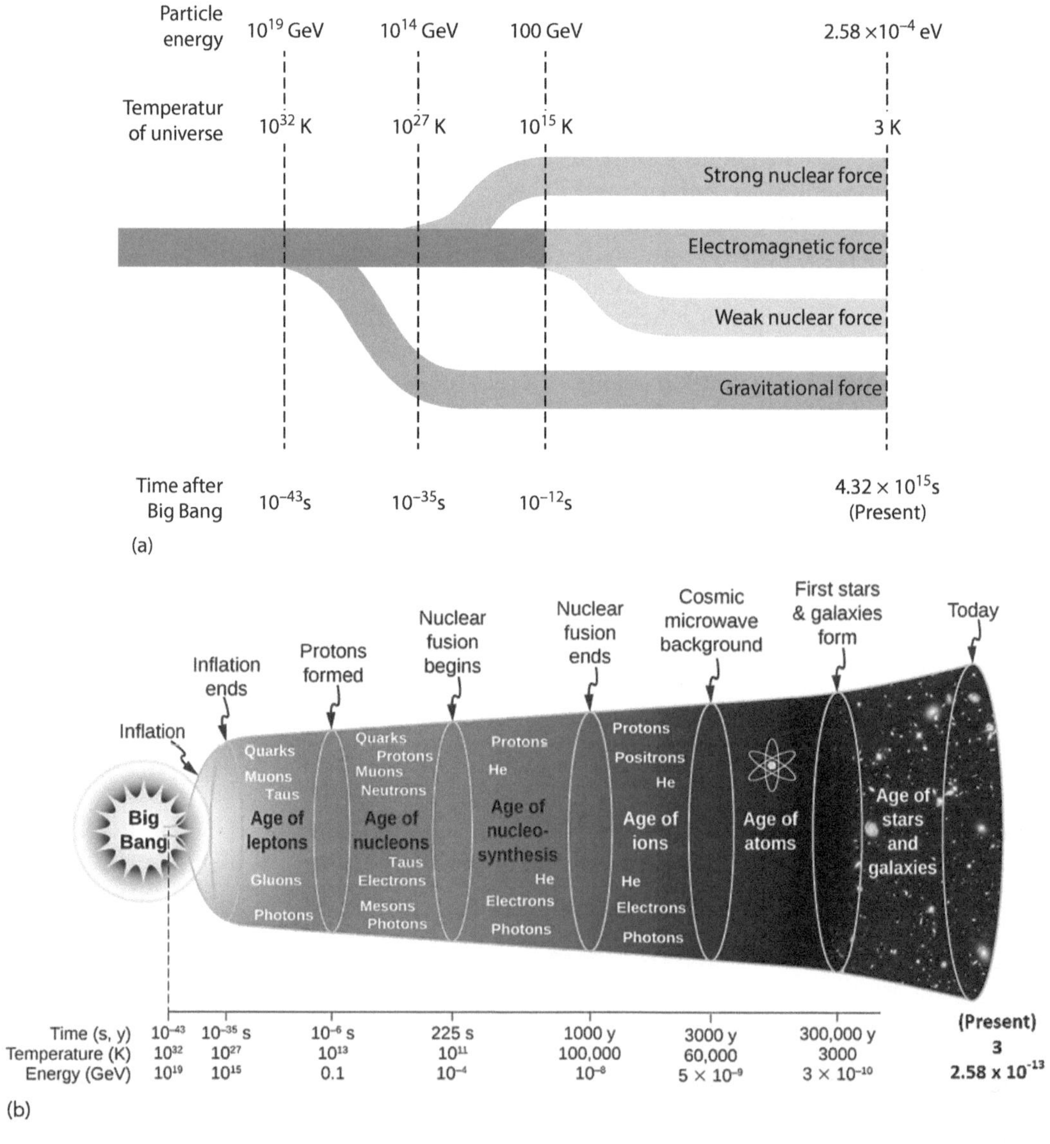

FIGURE 3.11
(a) The separation of the four fundamental forces in the early universe; (b) an approximate timeline for the evolution of the universe from the Big Bang to the present [12].

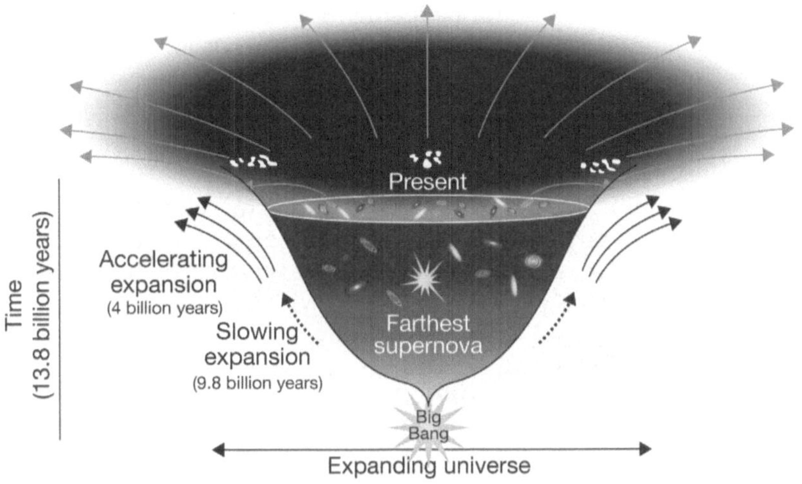

FIGURE 3.12
Accelerating expansion of the universe [13].

(imagine 1,000th of the thickness of a hair being expanded to the extent of trillions of miles) called *inflation era* (Fig. 3.11b). During inflation, the universe believed to expand at a rate faster than the speed of light; however, the exact physics of this sudden and stupendous expansion is also not very clear. Cosmic inflation ended very soon, and later, the universe started expanding normally. Now, the universe is 10^{-32} seconds old, the temperature has fallen to 1,000 trillion trillion (10^{27}) K. Following this, one by one the *age of leptons*, *age of nucleons*, and the *age of nucleo-synthesis* followed with the passage of time and falling of temperature of the original Big Bang fireball, as shown in Fig. 3.12b. Then subsequently, the atoms were formed followed by the formation of the stars and galaxies, and we ultimately reached the present state of the universe. The temperature of the original fireball has presently cooled down to 3 Kelvin (which is the measured temperature of the observed CMB radiation), i.e., close to absolute zero. Anyway, it may be mentioned that the first 3 minutes since Big Bang occurred is the period which gave us the most fundamental elements of our existence, i.e., hydrogen and helium, and set up the stage for the advanced processes of the formation of the present universe. Thus, the first 3 minutes after the Big Bang are undoubtedly the most crucial minutes in the history of the evolution of our universe.

Today, 13.8 billion years since the Big Bang explosion, the universe looks horribly unsymmetrical and broken—in terms of all the four forces being dramatically different from one another. But as mentioned above at the time of Big Bang, all the forces of nature were but one unbroken force entity. As the universe continued cooling, the gravitational force first separated out and then the GUT forces began to break, with the strong force peeling off from the electro-weak force. When the universe reached to a temperature of 10^{15} Kelvin—10^{-12} seconds after the beginning of time at the Big Bang—the electro-weak force broke into the electromagnetic force and the weak nuclear force. At this temperature, all the four forces had separated from one another, and the universe consisted of a soup of free leptons, quarks, and photons (details about all these subatomic particles will be discussed in Chapter 4).

From the days of Aristotle and, also, after the theory of Big Bang was proposed, the so-called steady-state theory of universe remained a strong contender of Big Bang theory. Aristotle, the most famous of the Greek philosophers and scientist of the ancient Greece, believed the universe had existed forever—something eternal is more perfect than something created! The motivation for believing in an eternal universe was the desire to avoid invoking divine intervention (God) to create the universe and set it going. If one believed that the universe had a beginning, the obvious question was who created it? What happened before the beginning of the universe? What was God doing before the creation of the universe? He made the world? Was He preparing Hell for those peoples who asked such questions?

The expansion of the universe based on Einstein's general relativity and Hubble's astronomical observations was one of the most important intellectual discoveries of the 20th century, or of any century. It settled the debate about whether the universe had a beginning. Many scientists, however, were still unhappy with the universe having a beginning with Big Bang because it seemed to imply that physics broke down at the instant of Big Bang. One would have to invoke an outside agency, which for convenience, one can call God, to determine how the universe began. They therefore advanced theories in which the universe was expanding at the present time but didn't have a beginning. One such theory was the *Steady-State theory*, about which we mentioned earlier, proposed by Fred Hoyle, Herman Bondi, and Thomas Gold, in 1948. In the Steady-State theory, as galaxies moved apart (according to expanding universe observation by Hubble), the idea was that new galaxies would form matter that was supposed to be continually being created throughout space. The universe would have existed forever and would have looked the same at all times. However, astronomical observations on the distribution of galaxies in the universe and most importantly the discovery of the CMB radiation, the leftover radiation from the early stages of the superhot fireball of the Big Bang that took place some 13.8 billion years ago, by Penzias and Wilson in 1965, went against the concept of Steady-State theory and ultimately that theory was abandoned.

3.5.2 How the Universe and Our Earth Might End Up and the Possible Means to Save It

Whether the universe was born of a Big Bang or a Big Bounce is the debate of concern for scientists, but we, the general public, are more concerned with the impending threats that might endanger the existence of our beautiful and blue Earth and our great human civilization. Let us first examine some of the threats that might endanger our existence on the Earth, beginning with astronomical and geological threats, and then turning to human-created threats.

The far-fetched astronomical threat that might lead to the wholesale destruction of the universe is expected to take place tens of billions of years from now; thus, there is no need to panic at all as a realistic and rational being. The proposed scenarios leading to this destruction of the universe includes a Big Crunch (re-collapse), a Big Rip (an infinite expansion rate tearing everything apart), a Big Chill (eternal expansion), a Big Snap (a fabric of space revealing a lethal granular nature when stretched too much), and Death Bubbles (space "freezing" in lethal bubbles that expand at the speed of light). What we know with certainty is that our nearest star, the Sun, which is 4.5 billion years old, will cause us problems much sooner. It keeps shining progressively more brightly because of the complex dynamics of the thermonuclear fusion reactions in its core as the hydrogen fuel gradually gets depleted, forming helium and heavy elements. Forecasts suggest that about a billion years from now, this solar brightening will start having a catastrophic effect on Earth's biosphere, added with the global warming caused by the human-made greenhouse effect—eventually boiling off our oceans, much like what has already happened in Venus. To save this precious and beautiful blue planet, the Earth, we have to make sincere efforts (with all of us checking the greenhouse effect) and taking possible technological approaches for maneuvering the orbit of the Earth around the Sun (by the scientific community).

The global warming caused by the greenhouse effect is a socioeconomic problem that can be solved with the adoption of new and green technologies if we have the political will and control over our lifestyle. The Inter-Governmental Panel for Climate Change (IPCC) of the United Nations is trying to generate awareness and to formulate climate policies to address the issue of global warming; hopefully, human conscience would win over petty (political) selfishness, making the problem addressable. To take care of the other problem (warming up of the Earth by the Sun), Earth can be kept at a constant temperature by gradually moving it out to a larger orbit around the warming Sun by clever use of asteroids. The basic idea is to nudge a large asteroid to fly very close to Earth every 6,000 years or so and give the Earth a gravitational tug in the right direction. Each such close encounter would be fine-tuned to send the asteroid passing near Jupiter and Saturn to get its energy and angular momentum reset to the required values for the next Earth encounter. Such "gravitational assists" have been successfully used before to send spacecraft such as NASA's Voyager probes into the outer solar system. If successful, this scheme could extend Earth's habitability from 1 billion to about 6 billion years. After that, our Sun will end its life as we know it, bloating into a red giant, and will engulf our Earth, including the planet Mars. Around the same time, a few billion years from now, our entire Milky Way galaxy will collide and merge with its nearest big neighbor, the Andromeda galaxy—forming a new galaxy. Fancifully, we may name it: "The Milkomeda"! All our imaginations, including even the wild ones, fail to guess what might happen in that situation of cosmic catastrophe.

Our fossil records reveal that five major extinction events occurred during the last 500 million years, with each event killing off more than 50% of all animal species on the Earth. The most recent of these extinction events took place 65 million years ago when an asteroid crashed into the Mexican coastline with impact energy of equivalent to many millions of hydrogen-bomb explosions. The most famous casualties were the nonavian dinosaurs. It blasted out a 180-km crater and engulfed our planet in a dark dust cloud that blocked sunlight for years, causing widespread ecosystem collapse. The extinction effect may also be caused by the massive floods of basalt lava that erupts from super-volcanoes in our own planet. They have the potential to create "volcanic winter" by enveloping Earth in a dark dust cloud, blocking sunlight for years much as a major asteroid impact would. They may also disrupt ecosystems globally by infusing the atmosphere with gases that produce toxicity, acid rain, and so on. Such a super-eruption in Siberia is widely blamed for the greatest recorded extinction of all, the "Great Dying" which wiped out 96% of all marine species about 250 million years ago.

Earth regularly gets hit by objects from space of various sizes, so the question is *not if* we will suffer another similarly deadly collision, *but when*. The answer is largely up to us; a good network of robotic telescopes should be able to give us decades of advance warning of dangerous inbound asteroids, which is ample time to develop, launch, and execute a mission to deflect them. If this is done sufficiently far in advance, only a gentle nudge is needed, which can be applied for example with a "gravity tractor" (a satellite whose gravitational pull nudges the asteroid toward it), a satellite based laser (which ablates material from the asteroid's surface and sends the asteroid recoiling in the opposite direction), or even by painting the asteroid so that the radiation pressure corresponding to solar heating will push it differently. If time is short, a riskier approach is required, such as a kinetic impactor (a satellite tackling the asteroid off course like football player) or nuclear explosion.

Well, small asteroids that approach toward Earth very frequently are of no concern because these small rocks usually burn up in the atmosphere on their way toward the ground. But the large asteroids approaching the Earth would hit the ground causing impact craters that might span miles in width and are the real danger for us. NASA and other space agencies around the world are cataloguing almost 90% of these large asteroids that could pose a danger to the Earth. This catalog of large asteroids that might hit or glide past the Earth in near future would help the scientists to take suitable action in time so that our lovable Earth can be saved from their dangerous impact. Just to have a feel, an asteroid having the size of a football field can easily wipe out a city like New York. The preferred method of dealing with an asteroid headed to Earth, as mentioned above, is to deflect it as one chunk using a nonexplosive kinetic impactor. But we need advance warning of the asteroid's approach to do that. If we know decades ahead of time that an asteroid is on an Earth-impacting trajectory, then we need only to launch a low-mass impactor. But what if an asteroid is heading straight to Earth and we don't have enough lead-up time? What if we have less than one year until impact? Then it appears that the nuclear devices can be the only available means in the humanity's arsenal in the struggle against asteroids. But attaching an explosive nuclear device to a rocket and launching it into space to hit and destroy the asteroid is not without risks. On November 23, 2021, NASA has launched the 560-kg Double Asteroid Redirection Test (DART) impactor that will crash into Dimorphos (a *moon* of a synchronous binary system with Didymos as the primary asteroid) over 480 million km away, at a speed of 6.6 km per second. By the end of 2022, the impactor is expected to hit the target asteroid and Didymos Reconnaissance and Asteroid Camera for Optical (DRACO) navigation system, which is carried in the DART impactor, will capture images until 20 seconds before impact while sensitive telescopes placed elsewhere in space will watch and see what happens after the impact. The "Double" in the name of DART refers to the double asteroid Didymos. Didymos is about 780 meters in diameter, but its little companion named Dimorphos/Didymoon is only about 160 meters in diameter. Most of the asteroids that pose a threat are of similar size as Dimorphos. Anyway, in 2026, NASA plans to launch the Near-Earth Object (NEO) Surveyor mission to find more asteroids in our neighborhood. But it's doubtful whether we'll ever have a complete picture of all the asteroids that could do us harm—as this vast and mysterious universe is full of surprises.

However, the danger of the human-created possibilities of the extinction of human civilization seems to me of more serious concern than the astronomical and geological dangers. The latter are very little under human control and happen in a considerable gap of time, but the former can happen at any time. However, the one that is under our control is the human conscience and value judgments, and if that can win over petty political and fringe religious agenda, then that will prove our worth as human beings. An accidental nuclear war launched because of a wrong political agenda of warmonger leaders or by religious fanatics might wipe out whole of human civilization. The consequences of the climate change that we are presently experiencing pale in comparison with a nuclear winter, when a global dust cloud would block sunlight for years, much as occurred when an asteroid or super-volcano caused mass extinction in the past. Concrete and straightforward steps can be taken to slash the risk of nuclear Armageddon as spelled out in numerous reports on nuclear disarmament. But warmongering leaders around the world just turn a blind eye to that, and interestingly these never become major election issues. The existing evidence suggests that there's no other lifeform as advanced as the human being in the entire universe; if so, then it is expected that human conscience would try to win over everything to save this beautiful blue Earth from the human-made possibilities of impending human suffering (global warming) or extinction of the whole civilization (with accidental nuclear war).

3.5.3 Fundamental Forces of Nature

With regard to string theory, four fundamental forces of nature were one and only one at the time of the creation of this universe, while with time the universe cooled down and the four forces with their individual characteristics were separated. In this perspective, it would be worthwhile to discuss the *four fundamental forces of nature* that we see around us today.

The most common of the four forces and discovered first was the *gravitational force*, initially formulated by Newton in the latter part of the 17th century and reformulated in a refined form by Einstein in the early part of 20th century. It is this force with which we are glued to the Earth; otherwise, we would be quickly spinning off to space. It is that force which binds together the solar system and keeps the Earth and the planets in their orbit, including the comets, asteroids, and other debris, all held together by the gravitational force of the Sun. This suggests that gravitational force is a long-range force; however, there is a limit to it because with increasing distance, its strength decreases by the square power of distance. The force of gravitation also prevents the stars from exploding with the energy of the thermonuclear fusion process going on within them. It is that huge force in a black hole that does not allow even light to escape from its unsurmountable grip.

The second force is the *electromagnetic force*. The term electromagnetism was first introduced by James Clerk Maxwell, who combined electricity and magnetism in the same framework. This is the force that holds together an atom, with the negatively charged electrons to orbit around the positively charged nucleus in the atom; everything in this universe is made of atoms of some material or the other. Because the electromagnetic force determines the structure of the orbits of the electrons in an atom, it also governs the laws of chemistry.

The electromagnetic force is enormously strong compared to the gravitational force. The electric force between two protons, for example, is 10^{36} (i.e., trillion, trillion, trillion) times the gravitational force between them, for a fixed distance. If we reflect a little, the enormous strength of the electromagnetic force is evident in our daily life. When you are holding this book in your hand, you are balancing the gravitational force on the book due to the huge mass of the Earth by the force provided by your hand (which is the electromagnetic force due to the charge constituents of your hand and the book, at the surface in contact). If electromagnetic force were not intrinsically so much stronger than gravity, the hand of the strongest man would have crumbled under the weight of a feather (caused by the force of gravity of the Earth). Indeed, to be more precise, in that circumstance we ourselves would have crumbled under our own weight!

Gravitational force plays a key role in the large-scale phenomena of the universe, and the so-called gravitational waves (predicted by Einstein in his general theory of relativity) was experimentally detected in 2015 on the basis of the collision of two black holes. Like gravitational force, the electromagnetic force also acts over large distances (including small distances of an atom) and does not need any intervening medium. The electromagnetic wave was hypothesized by Maxwell in the 1860s and was first used for wireless (radio) telegraphic communication by Marconi in 1901. From radio waves to gamma rays going through microwaves, terahertz radiation, light waves, infrared and ultraviolet rays, and X-rays are all electromagnetic waves (Fig. 3.4).

Human civilization has enormously benefited from electromagnetism since 1901, and in our everyday life we use radio, television, computers, and GPS, all of which work with wireless electromagnetic waves. With the advancement of technology, our devices became miniaturized, and now we have both smart phone and palm computer. With just the click of a button, we are connected with the world in a split-second via the Internet, which is again a gift of electromagnetism to the whole of humanity. With regard to medical treatment, a picture of brain activity can be generated with an MRI machine that uses radio waves, broken bones can be imaged with X-rays, and gamma rays are used to kill cancerous cells—all these are the boon of electromagnetic waves. The light wave that aids our vision and all the colors we see again fall in the visible spectrum of electromagnetic waves. For this reason, electromagnetic force is one of the most familiar force entities and is inseparable from our modern life whether in the selfie we click in our cell phone or the films we shoot.

But within the nucleus of an atom, the electromagnetic force is overpowered by the *strong and weak nuclear forces* that we have come to know with the advent of quantum mechanics.

The *strong nuclear force*, for example, is responsible for binding together the protons and neutrons in the nucleus as if in the form of glue. In any nucleus of an atom, all the protons are positively charged, and hence left to themselves; their enormous repulsive electric force would tear apart the nucleus. The strong force thus overcomes the repulsive force between the protons. Only a few elements can maintain the delicate balance between the repulsive electric force of protons (which tends to rip apart the nucleus) and the strong force (which tends to hold the nucleus). Thus, we have only about 100 known stable elements in nature (see the periodic table in Fig. 2.19, Chapter 2). Should a nucleus contain more than about 100 protons, even the strong force would experience difficulty in containing the repulsive electric force acting among the protons. The strong nuclear force is the strongest of all the 4 fundamental forces—about 100 times the electromagnetic force. Its range of action, however, is extremely small—of about nuclear dimensions (10^{-15} m)—and hence the electron does not experience this force.

If the strong force were the only force at work inside the nucleus, then most nuclei would have been stable—but we all know that certain nuclei, for example, uranium (with 92 protons) are so massive that they automatically break apart (with the so-called β-decay when the nucleus emits an electron and an uncharged particle called the neutrino), which we call *radioactivity*. The type of force responsible for the radioactive disintegration of very heavy nuclei is known as *weak nuclear force*. The weak nuclear force is not as weak as the gravitational force, but much weaker than the electromagnetic and the strong nuclear forces. The range of weak force is exceedingly small, of the order of 10^{-16} m.

When the strong force is unleashed, the effect can be catastrophic. For example, when the uranium nucleus in an atomic bomb is deliberately split, the enormous energy locked within the nucleus is released explosively in the form of a nuclear detonation. Strong force can yield significantly more energy than a chemical explosive, which is governed by the electromagnetic force. A nuclear bomb releases over a million times the energy contained in dynamite. The strong force explains why our Sun shines (including all other stars, the Sun being our closest star) so brightly and is able to give off so much light and heat for the last 4.5 billion years and will go on giving it for another 5 billion years or so. All the stars are basically huge nuclear furnaces in which the strong force within the nucleus is unleashed. If the Sun's energy, for example, were created by burning coal instead of nuclear fuel, only a minuscule fraction of the Sun's light and heat would be produced. Without the strong nuclear force, therefore, the stars would not shine, we would not have our life-saving Sun, and life on Earth would be impossible.

The *weak force* is so fleeting and ephemeral that we do not experience it directly in our lives. However, we feel its indirect effects—a Geiger counter placed near a piece of uranium measures the radioactivity of its nuclei caused by the weak force giving a clicking sound. The weak force also manifests in the form of heat; the intense heat found in the interior of the Earth is primarily caused by the decay of radioactive elements like uranium and thorium, which are plentifully available deep inside the Earth. These radioactive elements decay spontaneously, generating a stupendous amount of heat energy according to the conversion formula of mass into energy by Einstein's famous mass-energy relation, $E = mc^2$. This tremendous heat, in turn, results in volcanic fury when it reaches the Earth's surface. This enormous amount of heat inside the Earth also keeps the iron in melted condition which spins with the spinning motion of Earth; the free electron motion of this molten iron produces the Earth's magnetic field.

In this connection, it may be mentioned that different planets have different amounts of magnetic field around them because of the varied quantity of liquid iron in their cores. Jupiter, Saturn, Uranus, and Neptune have a strong magnetic field, while Earth has a decent magnetic field, but Mercury and Venus have almost no magnetic field at all. The Earth's magnetic field (i.e., geomagnetic field) on the Earth's surface ranges from 0.35 to 0.65 microteslas (i.e., 0.35–0.65 gauss). The Earth's magnetic field protects us from the blazing solar wind and the cosmic rays coming from deep space, which would otherwise have stripped away the ozone layer in the atmosphere that protects us from harmful ultraviolet radiation from the Sun. The Earth's magnetic field also helped the sailors for a long time in direction finding, and we use the compass even today, a feature that is also available in our smartphone. A recent study by Xu Jingjing et al. (published in June 2021 issue of *Nature* magazine) indicates that migratory birds may use Earth's magnetic field to navigate over long distances (day and night) and in different seasons. It is understood that the bird's retina has cryptochrome protein known as Cry4 that can sense Earth's magnetic field, helping them to navigate accurately both day and night. It has also been found that birds have increased Cry4 expression during the migratory season.

Weak nuclear force and electromagnetic forces were united by Sheldon Glashow, Abdus Salam, and Steven Weinberg into a single force category called *electro-weak force*, applicable for some elementary particles. However, gravity has long eluded physicists. Gravity is so unlike the other forces that its unification with other forces failed in the so-called standard model of physics. As noted earlier, Einstein tried to unite gravity with electromagnetic force in his unified field theory (ignoring strong and weak nuclear forces altogether), but he was unsuccessful. In order to unite gravity with other forces, two theoretical frameworks have been tried successfully: One is the theory of loop quantum gravity and the other is string theory, but till date, they have had no experimental support whatsoever.

Without the four forces of nature, life would be unimaginable: The atoms of our bodies would disintegrate, the fate of the Sun would be uncertain (without the balance of the implosive gravitational force and explosive nuclear force), and the atomic energy that fuels and lights the stars and galaxies would be snuffed out. The idea of forces, therefore, is an old and familiar one, dating back at least to the time of Isaac Newton. The physicists believe that if we can establish someday that all these four forces are in fact the manifestation of a single force, then The Holy Grail of physics will be attained. Table 3.1 summarizes the four forces along with their range, mediating particle, and spin.

TABLE 3.1

The Four Fundamental Forces of Nature and Their Carriers

Force	Range	Mediating Particle	Spin
Gravitation	Long	Graviton	2
Electromagnetism	Long	Photon	1
Weak Nuclear Force	Short	W and Z bosons	1
Strong Nuclear Force	Short	Gluons	1

3.6 Antimatter, Dark Energy, and Dark Matter

3.6.1 Introduction

Antimatter was discovered by Dirac upon investigating Einstein's mass energy relation in its totality, just as Einstein was propounding the time dilation of special relativity upon investigating Maxwell's equations. Every scientist tends to get engrossed in his or her theory when investigating the physical property he or she is looking for—as Maxwell was interested to unite electricity and magnetism in the same framework of electromagnetism and came up with his famous Maxwell's equations and hypothesized the electromagnetic wave and established that whatever the electromagnetic radiation was, radio wave, microwave, light wave, X-rays, gamma rays all travel with the same velocity, the velocity of light. Similarly, Einstein, with his mass energy relation, was interested in investigating the interchangeability of the two physical entities mass and energy in the framework of the special theory of relativity and in predicting the possible effect of mass (of Sun) on light energy (photon of star light). But when the perspective of investigation changes, the extra dimension of truth that is inherent in the philosophy and analytical equations of the concept emerges, leading to new discovery. Both the discovery of the special theory of relativity by Einstein from Maxwell's work and antimatter by Dirac from Einstein's work took place in this manner.

Dark energy and dark matter have become a hot topic of discussion in modern astronomy circles. Both of them can account for many missing links in the standard model of the universe, dependent primarily on the relativity theory. In fact, the possibility of the existence of dark energy was found necessary to account for the experimentally observed accelerating nature of the expansion of the universe. Also, dark matter is believed to be responsible for the extra gravitational pull exerted on the outer stars of a galaxy, so that those stars do not tear apart from the galaxy. To date, the results obtained regarding dark energy and dark matter have been mixed in nature, both supporting and opposing their existence. A recent experimental effort that started operation in late 2020, based on a huge machine named DESI, produced an initial map of dark matter in 2021.

3.6.2 Antimatter

In 1928, Paul Dirac, a 26-year-old electrical engineer with a BA in applied mathematics, encouraged by the exciting work of another physicist in his early 20s, Werner Heisenberg (who was creating a new theory of matter and radiation, quantum mechanics), discovered the theory of antimatter quite by accident. Dirac observed that Einstein's famous equation $E = mc^2$ needed to be interpreted as: $E = \pm mc^2$ (which Einstein might also have felt but did not concern himself with the minus sign because he was creating a theory of force exerted by mass on energy). But Dirac who was engaged in creating a new type of equation for electrons, known today as the Dirac Equation, could not ignore the possibility of matter with negative energy. The minus sign was puzzling, for it seemed to predict an entirely new form of matter.

Dirac found that matter with negative energy would look just like ordinary matter but would have the opposite charge. Such exotic matter is thus supposed to be composed of an anti-atom in which the anti-electron with positive charge would orbit around the anti-proton with negative charge. These anti-atoms in turn could combine to create anti-molecules and even anti-planets and anti-stars made of the so-called *antimatter*.

The possibility of opening the door to such an alternate universe or anti-universe on the basis of physics was first addressed by Dirac. Lewis Carroll's *Through the Looking Glass* (1865), where Alice walked through the mirror and entered into a mirror-reversed universe, reminds the world of antimatter in a fantasy world (the scientific theory of which is hardly 80 years old). In that alternate universe, everything seemed familiar, except there was a twist: In the Wonderland, logic and commonsense were reversed; people were left-handed, people's hearts were on the right side of their bodies, and clocks moved counter-clockwise. Carroll created Alice in the Wonderland series to amuse children with a twist of logic and sought to take them to the "other world" where rules were complete mirror images of our own world. Interestingly, superstring theory and GUT also talk of parallel universes like the antimatter universe, the mirror universe, the time-reversed universe, and so forth.

The existence of antimatter, first predicted by Dirac in 1928, was henceforth demonstrated conclusively by the discovery of the anti-electron by Carl Anderson. Anderson built a cloud chamber to determine the composition of cosmic rays, high-energy particles that rained down from space originating from a black hole or supernova explosion. To his surprise, in 1932, while analyzing the photograph of cosmic ray tracks, he observed that an electron was going in the "wrong way" in a magnetic field—which was termed a positron (an electron with positive charge)—the first particle of antimatter. For his work on antimatter, Dirac, at the age of 31, was awarded the Nobel Prize in 1933, and soon afterward, in 1936, Anderson received the Nobel Prize in Physics "for his discovery of the positron." Heisenberg was very impressed with Dirac's result, commenting with the highest praise: "I think that really the most decisive discovery in connection with the properties or the nature of elementary particles was the discovery of antimatter by Dirac."

In today's atom smashers around the world, pure anti-electron beams can be produced which, when colliding with beams of electrons, can generate GeV (i.e., giga or million electronvolt) to TeV (i.e., tera or trillion electronvolt) energies in such lepton colliders because the collision of matter and antimatter leads to their annihilation (neutralization), releasing enormous amounts of energy. The conversion of matter and antimatter into energy is much more efficient than the release of energy in a hydrogen bomb. This is because in the hydrogen bomb, the nuclear detonation process has only 1% conversion efficiency of matter into energy while in the antimatter bomb, if they could be built at all, the conversion efficiency would have been even up to 100%. However, the antimatter bomb is prohibitively expensive; although the annihilation of matter and antimatter might be useful as a possible energy fuel for space travel—but only if we could find large chunks of antimatter in the universe; which has not yet been possible.

Do we have an equal amount of matter and antimatter in our universe? The answer is an emphatic "NO." There is no evidence that any antimatter exists in our universe, except for those tiny bits created artificially by humans in particle accelerators and the tiny bits created by nature in collisions between high-energy matter particles in space. But the question is why? What accounts for the imbalance between matter and antimatter in our universe? Why should matter dominate over antimatter?

Over the decades, highly speculative mechanisms have been proposed hypothesizing that the matter–antimatter imbalance was present at the beginning of this universe, or perhaps the matter–antimatter in the universe could have been kept apart by some unknown force. The most plausible theory accounting for the nonabundance of antimatter in the present universe is based on *CP violation* supported by GUT and superstring theory. In particle physics, CP violation is a violation of CP symmetry: The combination of C-symmetry (C for charge conjugation, which transforms a particle into its antiparticle) and P-symmetry (P for parity, which creates the mirror image of a physical system). Discovery of CP violation in 1964 in the decays of neutral kaons resulted in the Nobel Prize in Physics in 1980 for James Watson Cronin and Val Logsdon Fitch, "for the discovery of violations of fundamental symmetry principles in the decay of neutral K-mesons." In 2013, LHC of CERN announced the discovery of CP violation in strange B meson decays and in 2019 in charmed decays.

The logic for CP violation for the nonavailability of antimatter in the observable universe goes like this. The Big Bang should have produced equal amounts of matter and antimatter. If CP-symmetry were preserved as such, there should have been total cancellation of both: Protons should have canceled with antiprotons, electrons with positrons, neutrons with antineutrons, and so on. This would have resulted in a sea of radiation in the universe with no matter at all, and our material universe would not have existed. Since this is not the case, after the Big Bang, physical laws must have acted differently for matter and antimatter, that is, violating CP-symmetry. According to GUT and the superstring theory, at the beginning of time, as a result of CP violation, there occurred a slight imbalance of matter over antimatter (roughly one part in a billion). That means that matter and antimatter in the universe annihilated each other at the Big Bang, creating radiation, but one billionth of the original matter was left over, which, ultimately, on the basis of the so-called inflation theory created our observable material universe as explained in Section 3.5.1.

3.6.3 Dark Energy and Dark Matter

It was mentioned earlier that the concept of the expanding universe which followed from Einstein's theory of general relativity was confirmed with the astronomical observation of the relative motion of galaxies by Hubble and his team in 1929. Before Hubble's experimental verification, Einstein himself was skeptical about the outcome of his own theory, as he was firm in the belief that the universe was a static one. To substantiate his view, he even introduced the so-called cosmological constant, which prevents the universe from expanding. However, with the experimental verification of the concept of the expanding universe by Hubble, as we have seen, Einstein changed his view and accepted the expansion of the universe.

We may ask: Has the universe been expanding uniformly since the Big Bang or has a rate of decrease/increase of the expansion occurred with time? Before 1998, physicists believed that the attractive force of gravity would slow down the expansion of the universe over time (and might one day in the distant future universe collapse, leading to the Big Crunch), for the universe is full of matter and the attractive force of gravity pulls all matter together in an implosive way. But in 1998, two separate teams of physicists measured the change in the universe's expansion rate, using distant supernovae as mileposts. If expansion was now slowing down under the influence of gravity, supernovae in the most distant galaxies should appear brighter and closer than their red-shifts would suggest. Instead, at high red-shifts, the most distant supernovae were found to be dimmer, thus suggesting that expansion of the universe is speeding up.

It was further learned that in the early stage, the universe, after the so-called inflation era, was actually expanding more slowly than it is today. Around 9.8 billion years after the Big Bang, the universe began expanding *faster.* That indicates some unknown force is fighting against the gravity's pull, causing galaxies to speed apart from one another. So the expansion of the universe has not been slowing down due to gravity, as everyone thought earlier; rather it has been accelerating (Fig. 3.12). Thus, the concept of an *accelerating universe* is a hot topic of astronomical research today. No one expected it to be so, no one knew how to explain it but some suspected that some mysterious force must be making this happen.

The accelerated expansion of the universe was understood to be driven by a kind of repulsive force that seems to be growing stronger as the universe expands; scientists, possibly for the lack of a better name, call this mysterious force *dark energy*. This so-called dark energy is thought to explain larger-than-expected values for the observed galactic red-shifts for distant galaxies. These red-shifts suggest that the universe is not only expanding but is doing so at an increasing rate. Virtually nothing more than this can be mentioned at present about nature and the properties of dark energy; it is only a repulsive force, and as for the rest, more is unknown than is known. However, it is known how much dark energy there is because we know how it affects the universe's expansion; other than that, it is a complete mystery. Eventually, theorists offered three explanations. (1) Maybe it was due to that long-discarded version of Einstein's theory of gravity, the one that contained what was called a cosmological constant (which Einstein added to his theory of general relativity to make his equations fit with the notion of a static universe); (2) maybe some strange kind of energy fluid filled the space; and (3) maybe there is something wrong with Einstein's theory of gravity, and a new theory could include some kind of field that would create this cosmic acceleration.

With regard to the first view, empty space possesses energy, and it would not be diluted as more space comes into existence rather with the expansion of the universe, causing more of this energy-of-space to appear. As a result, this form of energy would cause the universe to expand faster and faster with time. Unfortunately, no one understands why the cosmological constant should even be there, much less why it would have exactly the right value to cause the observed acceleration of the universe. The second viewpoint considers dark energy as being a new kind of dynamical energy fluid or field, something that fills all of space but something whose effect on the expansion of the universe is the opposite of that of matter and normal energy. Some theorists have named this "quintessence," after the fifth element of the Greek philosophers. But if quintessence is the answer, we still don't know what it is like, what it interacts with, or why exists at all. The third view, which demands a revision of Einstein's theory of gravity, would be faced with the question of what kind of theory it would be? How could the motion of the bodies in the solar system be correctly described, as Einstein's theory or even Newton's theory is known to do, and still give us the different prediction for the universe that we need? There are many candidate theories, but none are compelling. So the mystery continues.

By fitting the theoretical models of the composition of the universe to the combined set of cosmological observations, scientists have come up with the approximate composition of the universe as follows: Normal matter of the observable universe is 5%, dark energy is 69%, and dark matter is 26%, as shown in Fig. 3.13. Our observable universe is like the

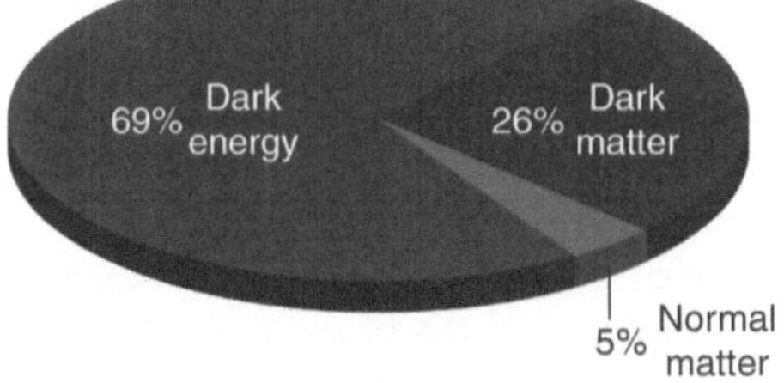

FIGURE 3.13
Composition of the universe in terms of normal matter of the observable universe, dark matter, and dark energy.

tip of the cosmic iceberg: It accounts for only about 5% of the total mass/energy, while the remaining 95% of the total universe is dark energy and dark matter!

We may ask: What is *dark matter*? Well, we are much more certain about what dark matter is not than we are about what it is. First, it is dark, meaning that it is not the normal matter with which the galaxies, stars, planets, and even ourselves in the observable universe is made of. Second, it is not made up of baryons of normal matter. We know this because we would be able to detect baryonic clouds by their absorption of radiation passing through them, but dark matter does not have any such signature. Third, dark matter is not antimatter because we do not see the unique gamma rays that are produced when antimatter annihilates with matter.

Because of the names, it's easy to confuse dark matter and dark energy. They may be related (which is not yet known), but their effects are quite different. In brief, dark matter like gravity attracts and pulls matter inward, whereas dark energy repels (as if an antigravity) and pushes matter outward. Also, while dark energy shows itself only on the largest cosmic scale, dark matter exerts its influence on individual galaxies as well as the universe at large. The word "dark" evokes black, meaning something that absorbs light by a significant amount with the least possible reflection, but that's not right. What we are seeking is something that does not interact at all with light but something invisible that generates an anomalous gravitational pull. Thus, it may be better to rename "dark matter" as "an entity for hidden gravity"; this, I believe, will give it a far superior description.

Astronomers discovered the possibility of the presence of dark matter while studying the outer regions of our galaxy, the Milky Way, as well as the galaxies beyond it. The Milky Way is shaped like a disc that is about 100,000 light years across. All the stars in this disc orbit the center of the galaxy very much like the planets orbit round the Sun in our solar system. The laws of gravity say that the stars at the edges of such a spinning, spiral galaxy should travel much more slowly than those near the galactic center, where a galaxy's visible matter is concentrated—just as Mercury, the closest planet to the Sun, races around the Sun much faster than Pluto, the farthest planet (to be more specific with the latest astronomical designation, the "dwarf planet") from the Sun. But astronomical observations have shown that all stars orbit at more or less the same speed regardless of where they are in the galactic disc. This puzzling result makes sense if one assumes that the boundary stars are feeling an extra gravitational pull toward the galactic center by an unseen mass—the mysterious dark matter. Otherwise, the gravitational pull of the stars would not be enough to hold the galaxy together, and the outer stars from the galaxy would fly apart into the space. Calculations showed that a galaxy can become stable if it is surrounded by a massive invisible "halo" (the so-called dark matter) that holds the galaxy together. The halo may be ten times as massive as the bright disc of our Milky Way galaxy, so it exerts a strong gravitational pull. The same effect is seen in many other galaxies. Clusters of galaxies show exactly the same thing; with exceptions observed by some astronomers.

Scientists have not yet observed dark matter directly. It doesn't interact with baryonic matter (consisting of protons, neutrons, and electrons bundled together into atoms), and it's completely invisible to light and other forms of electromagnetic radiation, making dark matter impossible to detect with our current instruments. But scientists are confident that it exists because of the gravitational effects it appears to have on galaxies and galaxy clusters. Dark matter could also explain certain optical illusions that astronomers see in the deep of the universe. For example, pictures of galaxies that include strange rings (the so-called Einstein ring) and arcs of light could be explained if the light from even more distant galaxies is being distorted and magnified by massive, invisible clouds of dark matter in the foreground—a phenomenon known as gravitational lensing.

Are dark matter and dark energy related? No one knows. The only thing they seem to have in common is that both were understood to be forged in the Big Bang, and both remain mysterious. The gravity of dark matter tries to pull the stars of the galaxy or galaxy cluster together, while dark energy tries to push the galaxies apart in the ever expanding (accelerating) universe. Dark matter dominated the early universe, but dark energy began to dominate about 4 billion years ago. As the universe gets larger, dark energy's domination increases.

In the 1930s, astronomer Jan Oort and physicist Fritz Zwicky deduced additional gravitational effects that seemed to indicate the presence of additional, unseen matter. Oort measured this matter within our Milky Way galaxy and called it "nebulous matter," whereas Zwicky measured it within clusters of distant galaxies and called it "dunkle Materie," or dark matter. These results were reported just a decade after Hubble's experimental observation of the characteristics of the motion of galaxies in the universe; but the astronomers of that time completely ignored them. However, in 1973, Jeremiah Ostriker and Jim Peebles at Princeton University resurrected Zwicky's theory by making rigorous calculations about the stability of a galaxy. But their result also received an icy response.

After decades of skepticism and derision, what finally turned the tide on dark matter were the careful, persistent results of astronomer Vera Rubin and her colleagues at the Carnegie Institution for Science in Washington, D.C. They analyzed hundreds of galaxies, and their results conclusively verified that the velocity of the outer stars in a galaxy did not vary much from that of the inner ones, contrary to the planets in our solar system. This meant that outer stars would fly away into the space, causing the galaxy to disintegrate into billions of individual stars, unless the gravitational pull of the invisible dark matter would have held them together in addition to the usual gravitational force. However, it was not an easy task for Rubin, a female scientist, to be accepted by her male peers. At every step of the way, her career came perilously close to being derailed by male hostility.

Scientists have a few ideas about what dark matter might be. One leading hypothesis is that dark matter consists of exotic particles that do not interact with normal matter or light but that still exert a gravitational pull. Astronomers believe that dark matter exists because visible matter doesn't have enough gravitational muster to hold galaxies together. What is it made of? There are two likely leading candidates: Superdense astronomical bodies called MACHOs and WIMPs (weakly interacting massive particles). However, the overwhelmingly predominant scientific view right now is that dark matter may consist of a fundamental particle or combination of particles. There are a lot of reasons for favoring this viewpoint. Physicists have detected 18 different types of elementary particles, along with a dizzying 200 or so composite particles, many of which do not fit the standard model of particle physics. It is highly likely that there are undiscovered particles out there, and it is entirely reasonable that some of them might not interact with light or other forms of radiation. We already know of one family of particles that is understood to behave as dark matter: The neutrinos. They do not engage with light or with the strong nuclear force, but they do contribute a gravitational effect.

Another possible particle responsible for dark matter may have been recently discovered, the X17 particle. The presence of such a particle was observed in the radioactive decay of the beryllium-8 isotope in 2016 and in the emission of light from the excited helium atom in an unusual angle in 2019, which was also found to be "protophobic." That is, it seems to be "scared" of the proton and was found to run away from it. Another very strong candidate accounting for dark matter is believed to be the axion, which is about a trillion times lighter than the electron; theorized in 1977 by R. Peccei and H.R.A. Quinn primarily as a candidate for solving the strong CP problem. The axion is a neutral particle that is extraordinarily weakly interacting. The CERN Axion Solar Telescope (CAST) is trying to investigate the possibility of axions produced at the core of our Sun. Axions are also believed to be produced within neutron stars, by nucleon–nucleon bremsstrahlung (a process that produces electromagnetic radiation when a sudden deceleration of charged particle, as of an atomic nucleus, takes place in an intense electric field). Scientists believe that there is a long way to go before we can have any solid understanding of the dark matter and dark energy. For the moment we can just quote Carl Sagan: "Somewhere out there, something astonishing is waiting to be known."

An international collaborative team from the United States, the United Kingdom, France, and other nations, with the Lawrence Berkeley National Laboratory as nodal agency, installed a massive machine named DESI in 2020, conceptualized in 2016 to probe the expansion history of the universe and the mysterious physics of dark energy. DESI sits at an elevation of 2,100 m on top of Kitt Peak in the Sonoran Desert, which is located 89 km from Tucson, Arizona. It will survey the cosmos for 5 years to gather information from 35 million galaxies and 2.4 million quasars to get a possible hint about the origin and nature of dark energy. The instrument has been mounted on a telescope and aims its robotic array of 5,000 fiber-optic "eyes," cycling through a new set of 5,000 galaxies every 20 minutes. DESI's components are designed to automatically point at preselected sets of galaxies, gather their light, and then split that light into narrow bands of color to precisely map their distance from Earth and gauge how much the universe expanded as this light traveled to us. It will allow us to go 11 billion years back in time and measure ancient light with unprecedented precision. DESI will peer deeply into the universe's infancy and early development to create the most detailed three-dimensional map of the universe. Its data will go beyond determining the rate of cosmic expansion, it will also shed more light on how the expansion affects the way galaxies and other astrophysical objects form and grow over time. The results are expected to provide a big leap in our understanding of dark energy. The DESI collaboration has participation from nearly 500 researchers at 75 institutions in 13 countries.

In May 2021, the DESI collaboration team analyzed the first set of images of 100 million galaxies, using artificial intelligence and obtained the map of dark matter in the sky of the southern hemisphere (an eighth of the total night sky visible from Earth). These initial results are interesting and might throw useful light on the future cosmological model of the universe. Hopefully, many more exciting news might be waiting for us, to be explored by the DESI team in the near future.

3.7 Some Comments Regarding the Present Scenario of Science and Technology Research After the Developments of Relativity and Quantum Mechanics

We must accept that science drives technology, and not the other way around. For example, the LHC at CERN, Geneva, is a technological marvel, but it was built primarily to discover the last missing particle of the standard model of physics, Higgs Boson (the so-called God particle), the creator of mass in subatomic particles and hence of matter with the elapse of time after Big Bang. It is known that the LHC project has many technological offshoots for the benefit of humankind, such as the World Wide Web (WWW) or the

Internet, PET (position emission tomography) scans (which is used to detect cancers) and so on. But because of CERN's active and single-minded pursuit for basic science, we have been able to discover the Higgs Boson and many other fundamental discoveries are also coming up every now and then. In short, science and technology are intrinsically related, and the progress of the latter is unequivocally dependent on the demand of the former. Technology, in other words, is neither sovereign nor autonomous, and if new discoveries of science cease to sustain technology, technology will collapse in time.

The first and middle part of the 20th century was the golden period of physics and astrophysics and, in general, of science. On the one hand, the relativity theory helped us to understand the cosmic world with deeper insight, and quantum mechanics enabled us to understand the mysterious behavior of the atomic and subatomic world. In the middle of the century, two new forces were discovered deep within the atom—the strong and weak nuclear forces. Finally, in the century's third half, we got the standard model of particle physics—an accounting of all the particles and forces (except gravitation) known to exist in our universe. But the 21st century has brought us a comparatively rough patch as far as new and fundamental discoveries of science are concerned.

Of course, there is no doubt or even debate about the fact that quite a few mysteries of science have been unlocked in the recent past based on technological marvels. The elusive Higgs Boson was pinned down in 2012 by the LHC at CERN, Geneva, and in 2015, LIGO detected the gravitational waves predicted by Einstein's theory of general relativity 100 years ago, using laser interferometry to measure the minute ripples in the fabric of space–time caused by passing gravitational waves from cataclysmic cosmic events of collision of a pair of black holes 1.3 billion light years away from Earth. Black hole M87*, which is 55 million light years away from Earth, was imaged with EHT in 2019. This was yet another technological marvel of the radio telescope. Further, the International Thermonuclear Experimental Reactor (ITER), the world's largest magnetic fusion device, is being built to mimic the way the Sun is powered. It will be the first fusion device to test integrated technologies, materials, and the physics regime necessary for the commercial production of fusion-based electricity. The neutrino factory is under development to study the properties of elusive subatomic particles having practically zero mass, but it is predicted to be a fundamental constituent of "dark matter" that is understood to constitute 26% of the matter of the universe (the observable universe being only 5%). Recent mapping of dark matter in 2021 by the DESI team is raising questions about rethinking the concept of gravity and the model of the universe based on the standard model of cosmology and so forth.

But all these are the technological triumphs that have established/modified to some extent the scientific theories developed decades earlier—a full century earlier in the case of gravitational waves and black holes. And new ideas like string theory (which hold that matter is made up of tiny vibrating strings) or the theory of loop quantum gravity (in which the spin-network of quantized loops represent the excited gravitational fields) are still lofty theoretical propositions with no visible possibility of experimental verification in the near future.

Physicists today write a lot of papers, build a lot of (theoretical) models, hold a lot of conferences, and cite each other, but of all of the theoretical work that has been done since the 1970s they have not produced a single successful prediction that might have a landmark or game-changing influence on future scientific thoughts and developments when compared with relativity and quantum mechanics. But true to its right spirit, physics is all about making successful predictions, and that has been unfortunately lacking in recent times. Physicists today have been seduced by complex mathematical equations that might be "beautiful" or "elegant," but lack the connection to the real development of physics based on original physical visualization of new concepts with deep insight. It is true that Galileo was the first scientist to detail the link between mathematics and the physical world, and indeed all of today's successful physical theories are expressed mathematically in some form or other. However, the great contributions of the 20th century, like relativity and quantum mechanics, are the most precious and timeless theories because they are at first the physical visualization of the macro- and micro-world, respectively, with deep insightfulness and later on they received a rigorous mathematical framework. But in the history of science, we can find many examples of great thinkers being led astray for being unnecessarily seduced by mathematics and possibly overlooking the physical insight—the latter being the real driving force for any fundamental discovery of science.

The last four to five decades may be called the decades of technological triumphs, which have experimentally established some of the most remarkable theories of science. But these technological advancements have greatly benefited the progress of astronomy, especially the radio astronomy. The latest technological marvel in radio astronomy is the EHT, which, as we mentioned earlier, obtained the first image of the event horizon of black hole M87*. Apart from that, the Atacama Large Millimeter/Submillimeter Array (ALMA), which is an astronomical interferometer of 66 radio telescopes in the Atacama Desert of northern Chile, has been used for many significant astronomical studies since 2013. Space-based astronomical observatories like the Hubble telescope, Planck satellite, and WMAP relate to the state-of-the art technology that has helped to obtain the latest information regarding innumerable galaxies in the observable universe, CMB radiation left behind by the Big Bang, and so on. We have also talked about LIGO, which established the existence of the gravitational waves predicted by Einstein's general relativity theory some 100 years back. The latest marvel of technology is DESI, which was used for some preliminary mapping of dark matter in 2021. Another space telescope, James Webb Space Telescope (JWST—jointly developed by NASA, the European Space Agency (ESA), and the Canadian Space Agency (CSA)—is expected to start functioning from 2022. The JWST, NASA's flagship astrophysics mission, is expected to provide improved infrared resolution and sensitivity over Hubble, and will enable a broad range of investigations across the fields of astronomy and cosmology, including observing some of the most distant events and objects in the universe, such as the formation of the first galaxies, and detailed atmospheric characterization of potentially habitable exoplanets.

But what about the further progress of science in a meaningful way beyond relativity and quantum mechanics? Is it that no further fundamental concepts of science need to be discovered, or there are none anymore? If we train our mind with that thought, then possibly we are making an utter mistake. Even today we can safely quote Sir Isaac Newton: "… whilst the great ocean of truth lay all undiscovered before me." The failure of science to make no new/fundamental discoveries during the last three to four decades is because our whole socioeconomic orientation and value judgments have changed, producing a serious negative impact in the progress of science. We are now overly dependent on technology and have actually been blinded by the marvels of technological developments. While technology is required for providing a good living standard, the absence of the new and fundamental developments of science will lead to a catastrophic halt in its progress. Thus, extensive and serious research in fundamental science is an absolute necessity. However, revival of the age-old spirit of scientific research based on *Curiosity*, *Courage*, and *Confidence* (popularly known as three C's) is needed. As Einstein once observed, "I am not particularly intelligent but I am curious." Marie Curie similarly commented: "Be less curious about peoples and more curious about ideas."

4

Zooming into the Subatomic World of Atomic Physics

Weirdness is the characteristics of the micro-world

4.1 Introduction

From the late 1890s with J. J. Thomson's discovery of the electron, all notions about atomic structure got revolutionized. It became increasingly clear that atoms are not featureless little billiard balls, but instead were made of complex structures consisting of *subatomic particles* that were individually far tinier than the intact atom. Whereas an atom as a whole has a diameter of the order of a hundred-millionth of a centimeter, the diameter of subatomic particles like the proton and neutron can be something like a ten-trillionth of a centimeter. The electron and quark are even smaller and typically are one-thousandth the size of the proton and neutron. To use a more easily grasped explanation, it would take 100,000 or more subatomic particles laid side by side to stretch across the diameter of a single atom. Since even the most complex atom contains only a little over 300 subatomic particles, it can clearly be observed that the intact atom has a vastly large empty space. Held by the electromagnetic forces, a few electrons orbit through the wide space around the tiny atomic nucleus, ten thousandth the size of an atom, which is at the center of the atom. If an atom could be broken up into its individual subatomic particles and compressed, the whole system could be made to shrink to a tiny fraction of its former self!

The concept of the atom as the smallest part of matter was first proposed by Democritus in the 5th century B.C. In 1897, J. J. Thomson, on the basis of his famous cathode ray experiment, first proposed that even the atom has constituent parts. With the discovery of the electron, he proposed the so-called plum-pudding model of the atom. In his model, the atom was considered to be a soup of positive charges, and negatively charged electrons were thought to be embedded within it to balance the total positive charge, as if the electrons were like plums and the atom as pudding. This concept was further modified by Ernest Rutherford who in 1911 introduced the idea of the nucleus in the atomic structure on the basis of his α-particle scattering experiment and proposed a solar system like model of the atom. Later, Niels Bohr proposed anatomic model that could resolve the problem of Rutherford's model of the fast spiraling fall of electron to the nucleus, jeopardizing the stability of the atom. In Bohr's atomic model, electrons rotate around the central nucleus in some specified stationary orbits and the electron is transferred from one stationary orbit to another with the emission/absorption of a photon. Finally, on the basis of the wave particle duality of the electron propounded by de Broglie and the uncertainty principle of Heisenberg, the stationary orbits of the atom proposed by Bohr's atomic model were justified. Ultimately, on the basis of the superposition property of the electron as per

DOI: 10.1201/9781003215721-4

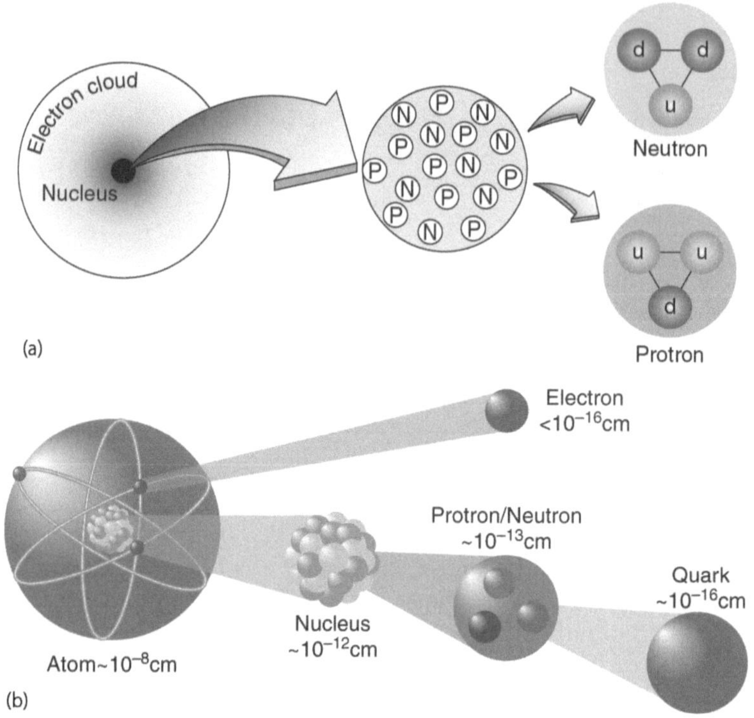

FIGURE 4.1
Modern atomic structure: (a) Atom with electron cloud and nucleus, the nucleus consisting of the proton and neutron, while the proton and neutron consist of up and down quarks; and (b) relative sizes of atom and its subatomic constituents.

quantum mechanics, the concept of the electron cloud was introduced by Heisenberg and Schrödinger in the modern atomic model.

Thus, until the 1930s, the basic atomic structure (as per quantum theory) consisted of a cloud of electrons surrounding a central massive nucleus (see the modern atomic structure, Fig. 4.1a). The electron cloud is roughly 10^{-8} cm in size (about a millionth of the diameter of a human hair), the nucleus at the core of the atom is 10,000 times smaller than that, that is, roughly 10^{-12} cm (Fig. 4.1b), and the atomic nucleus is several thousand times heavier than the tenuous electron cloud. Each atomic nucleus contains an equal number of protons and neutrons, and all the neutrons and protons are held together tightly in their tiny nucleus by a new type of force, neither electrical nor gravitational—a force called nuclear force (to be more specific, a strong nuclear force). Each neutron has the same mass as a proton but has no electrical charge, while the proton has a positive charge that is equal and opposite to that of the electron. The mass of the electron is $m_e = 9.1 \times 10^{-31}$ kg, and the mass of the proton is $m_p = 1.67 \times 10^{-27}$ kg (i.e., 1835 m_e), while the neutron has almost the same mass as a proton.

Until the late 1930s, only three subatomic particles were recognized: *Electrons, protons,* and *neutrons.* With the advent of the particle accelerator/atom smasher, following the invention of the cyclotron in 1939 by E. O. Lawrence, the domain of physics after the 1950s was flooded with the discovery of a host of subatomic particles in laboratories around the world. Nobel laureate Enrico Fermi, observing the plethora of new hadrons, each one with a strange-sounding Greek name, once lamented: "If I could remember the names of all these particles, I would have been a Botanist," Oppenheimer suggested in jest that the "Nobel Prize be given to the physicist who didn't discover a new (subatomic) particle that year." As Fig. 4.1a shows, the proton and neutron were subsequently found to consist of a more fundamental subatomic particle called the quark; the proton consists of two up quarks and one down quark, while the neutron consists of two down quarks and one up quark. The quarks are held together in the proton and neutron by the binding agent, the *gluon*. Details of all these are discussed in this chapter.

The *atom smasher/particle accelerator* is essential to probe into the mysterious world of subatomic particles within an atom. This is because to observe an object through radiation, the resolution (accuracy of determination of the object) improves if the wavelength of the irradiating signal becomes of the order of or smaller than the object itself. For example, we prefer to use the electron microscope and not the optical microscope to probe into atomic dimensions. In the electron microscope, the electrons are used to observe atoms, where an accelerated beam of electrons is used as a source of illumination, allowing us to observe objects a million times smaller than a human hair (~0.1 nm). This is just enough to "view" individual atoms. Thus, to probe into the subatomic level within an atom, we need to use extremely small-wavelength radiation of the signal, which in effect demands extremely high energy as per the quantum-mechanical relation $E = h\nu = (hc)/\lambda$, where λ is the wavelength of radiation, E is the energy of the radiation, and h is Planck's Constant and c is the velocity of light. Thus, the atom smashers operate in the domain of *high-energy physics*, where the energy level goes up to mega or million (10^6), giga or billion (10^9), and even to tera or trillion (10^{12}) electronvolts.

The universe is truly a *Lego-Land*. Everything in this universe, starting from stars to stereos, rocks to rabbits, is made of the largest Lego, the *atomic Lego*. Until 1930, it was known that atoms are made of merely three kinds of smaller Lego blocks: Electrons, protons, and neutrons. These were the *subatomic Legos* known to the scientists of that time. After the 1950s, the atom smashers around the world came up with newer subatomic particles, with exotic names such as pions and kaons, sigmas, omegas, muons, tauons, W-bosons, and Z-bosons (which are unstable and decay into more familiar stuff in a split-second). All except the last four are made of quarks. In fact, protons and neutrons are also found to be made of up and down quarks (but electrons have shown no sign of being made of anything smaller despite having been smashed at 99.999999999% of the speed of light at Conseil Européen pour la Recherche Nucléaire [CERN], Geneva—the European Organization for Nuclear Research). Now we have a zoo of subatomic Legos, with hundreds of species, including gluon, Higgs boson, and so on. So far, no evidence has been found that any of these bosons, quarks, leptons (the family name for electron, muon, etc.), or their antiparticles are made of any smaller or fundamental entity. Hence, all these subatomic Legos are known as elementary/subatomic particles in particle physics.

String theorists and loop quantum gravity theorists, with Sherlock Holmes' detective outlook, seek more fundamental Lego units like "vibrating minuscule strings" (as in string theory) or "spin-network of quantized loops of excited gravitational fields" (as in the theory of loop quantum gravity), which, they claim, can be observed with accelerators that have ten trillion times more energy than those of today. That's quite a claim. If we do not fully understand what that means, we need not worry at this moment, since even the most devoted practitioners of string theory and the theory of loop quantum gravity cannot claim to understand their theories fully yet. Only time will tell what can ultimately be termed as truly fundamental or elementary. We simply don't know yet. But there is good reason to suspect that everything we know so far—including the very fabric of space–time itself—might ultimately be made up of some more fundamental building blocks.

4.2 Classification of Subatomic Particles

4.2.1 Introduction

Subatomic particles are generally classed according to their mass, charge, and spin. By *mass* we mean the *rest mass* of a particle; that is, the mass in the frame of reference connected with the particle itself. The smallest mass is possessed by the electron ($m_e = 9.1 \times 10^{-31}$ kg); therefore, the mass of other particles is often expressed in terms of electron mass. Mass is also expressed in energy units (as per Einstein's equation $E = mc^2$) when the electron mass calculates out to be 0.511 MeV (mega electronvolts). The *electric charge* of particles is denoted by 0, +1, –1. In the first case, there is no charge (the particle is neutral, e.g., for the neutron). In the second case, the charge is equal to that of an electron, but unlike the electron, it is positive, as is the case with the proton. In the third case, the charge coincides with the electronic charge (1.6×10^{-19} C) in both magnitude and sign.

Our common experience is familiar, with spinning tops that spin about an axis or the spinning of Earth about its own axis that runs from the North Pole to the South Pole. But quantum mechanics tells us that the particles do not have any well-defined axis and are not little billiard balls spinning in space like planets. Here spin is a representative term. In fact, an electron's *spin* is related to the electron's inherent angular momentum, that is, the spin angular momentum. The spin multiplied by $\hbar$ (the reduced Planck's Constant) gives the value of the electron's intrinsic spin angular momentum. In 1920, Otto Stern and Walter Gerlach designed an experiment that unintentionally led to discovering an electron's spin.

Let us understand visually the concept of spin related to subatomic particles. By the spin of a particle, we mean what the particle looks like when it is viewed from different directions. A particle of spin $s = 0$ is like a dot: It looks the same from every direction, as in Fig. 4.2a. On the other hand, a particle of spin $s = 1$ can be compared with an arrow: It looks different when viewed from different directions. Only when it is turned around a complete revolution (360°) does the particle look the same, as for the club sign (say) in Fig. 4.2b. A particle of spin $s = 2$ is like a double-headed arrow: It looks the same if one turns it around half a revolution (180°), as for the picture of the King of diamonds (say) in Fig. 4.2c. Similarly, higher spin particles look the same if one turns them through smaller fractions of a complete revolution.

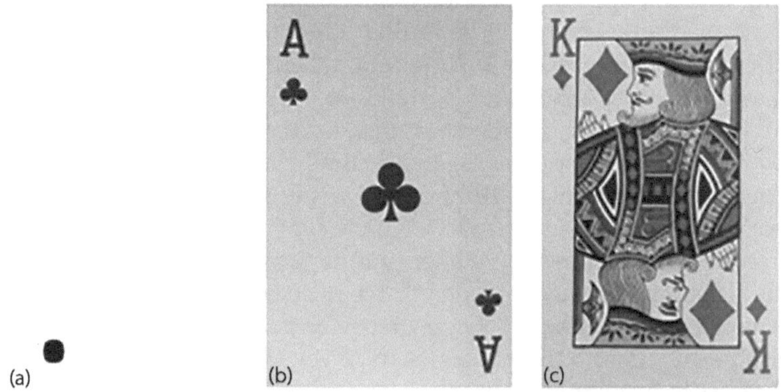

FIGURE 4.2
Representative diagrams showing spin 0, 1, and 2.

All these seem fairly straightforward, but the remarkable fact is that there are particles that do not look the same if one turns them through just one revolution: One has to turn them through two complete revolutions! Such particles are said to have spin $s = ½$. All the known particles in the universe can be divided into two groups: Particles of spin$s = ½$ (such as electrons, protons, and neutrons); which make up the matter in the universe (known as *fermions*), and those having spin $s = 0, 1, 2, \ldots$ This gives rise to forces between matter particles (known as *bosons*); almost 50% of the particles in this universe are bosons, while the rest, 50%, are fermions.

There are three families of subatomic particles. The *first family* is the smallest and consists of just one particle, the *photon* (the quantum of electromagnetic radiation). The rest mass and electric charge of photon is zero, while its spin $s = 1$. The *second family* consists of particles called *leptons*. The six leptons that are known are the electron (e^-), electron neutrino, muon, muon neutrino, tau or tauon, and tauon neutrino. Different leptons have different masses and charges, but all of them have half-integral spin, that is, $s = ½$. The *third family* consists of particles called *hadrons* (from the Greek word for large or massive). Hadrons are numerous: Several hundred of them are known. Typical examples of hadrons include the proton, neutron, and pion.

The hadron family is further subdivided into two *subfamilies*; the *mesons* and the *baryons*. Mesons have either no spin or integer spin, whereas the baryons have half-integral spins. Almost all subatomic particles have a companion *antiparticle*, with the exception of the photon, the neutral pion, and the eta meson. So, for the electron, we have its antiparticle, the positron (e^+); for the proton, the antiparticle is antiproton; and so on. Table 4.1 shows the properties of some of the subatomic particles.

Hadrons consist of more fundamental elementary particles called *quarks*. All matter around us is actually made of the elementary particles of two basic types: *Leptons* and *quarks*. Each of these elementary particles is found in nature in groups of six particles, which are related in pairs or "generations." The lightest and most stable particles make up the first generation, whereas the heavier and less stable particles belong to the second and third generations. All stable matter in the universe is made from particles that belong to the first generation; any heavier particles quickly decay to more stable ones.

The six leptons are paired in three generations: The electron and the electron neutrino, the muon and the muon neutrino, and the tauon and tauon neutrino. The electron, muon, and tauon all have an electric charge and a sizeable mass, whereas the neutrinos are electrically neutral and have very little mass. The six quarks are similarly arranged in three generations: The up quark and the down quark form the first generation, followed by the charm quark and strange quark, and then the top quark and bottom (or beauty) quark. Quarks also come in three different colors and only mix in such ways as to form colorless objects. Photons (as particles that make up light and other forms of electromagnetic radiation) are the mediators of interactions.

4.2.2 Photons

When you are reading a soft copy of a work on a screen or a hard copy of a book, streams of photons (i.e., light) carry the images of the words to your eyes and make it visible. In modern scientific terms, we now understand that light is emitted as a packet of fixed quantum energy ($h\nu$), called the photon, which also corresponds to energy transition in atomic states as per the quantum model of the atom. The term *photon* was coined by the chemist Gilbert Lewis in 1926 in a letter to the journal *Nature*. However, the concept that light (or any

TABLE 4.1

Properties of Some of the Elementary/Subatomic Particles

Particle Name	Particle Symbol	Anti-Particle Symbol	Mass	Spin Number	Electric Charge	Lifetime of Particle
Photon	γ	–	0	1	0	∞
Electron	e^-	e^+	1	1/2	–1	∞
Electron Neutrino	ν_e	$\overline{\nu_e}^-$	0	1/2	0	∞
Muon	μ^-	μ^+	207	1/2	–1	2.2×10^{-6}
Muon Neutrino	ν_μ	$\overline{\nu_\mu}$	0	1/2	0	∞
Tauon	τ^-	τ^+	3500	1/2	–1	$\leq 10^{-12}$
Tauon Neutrino	ν_τ	$\overline{\nu_\tau}$	0	1/2	0	$\leq 10^{-12}$
Pions	π^0	–	264	0	0	0.8×10^{-16}
	π^+	π^-	273	0	+1	2.6×10^{-8}
Kaons	K^+	K^-	966	0	+1	1.2×10^{-8}
	k^0	$\overline{k^0}$	974	0	0	K_s: 0.9×10^{-10} ------------ K_L: 5.4×10^{-8}
Eta Meson	η^0	–	1074	0	0	$\sim 10^{-17}$
Proton	P	$\overline{P}$	1836.1	1/2	+1	∞
Neutron	n	$\overline{n}$	1836.6	1/2	0	960
Lambda Hyperon	Λ^0	$\overline{\Lambda^0}$	2183	1/2	0	2.5×10^{-10}
Sigma Hyperons	Σ^+	$\overline{\Sigma^+}$	2328	1/2	+1	0.8×10^{-10}
	Σ^0	$\overline{\Sigma^0}$	2334	1/2	0	10^{-14}
	Σ^-	$\overline{\Sigma^-}$	2343	1/2	–1	1.5×10^{-10}
Xi Hyperon	Ξ^0	$\overline{\Xi^+}$	2573	1/2	0	3×10^{-10}
	Ξ^-	$\overline{\Xi^-}$	2586	1/2	–1	1.7×10^{-10}
Omega Hyperon	Ω^-	$\overline{\Omega^-}$	3273	3/2	–1	1.3×10^{-10}

other electromagnetic wave like microwaves, X-rays, and γ-rays; see Fig. 3.4) consists of photons has come a long way in the history of our understanding of light. In 1690, Christiaan Huygens published *Traité de la Lumière,* his treatise on light in which he described light as being made up of waves that moved through the so-called ether, which was thought to permeate space (prior to Michelson–Morley's experiment on ether and Einstein's special theory of relativity).

In 1704, however, Newton declared in *Opticks* that he disagreed with Huygens' concept of the wave property of light and introduced the corpuscular theory of light; that is, light consists of minuscule shining particles. But Huygens' model received strong support in 1801, when Young conducted the double-slit experiment showing the formation of the interference pattern of light, which is possible only if light has a wave nature. In 1850, Foucault compared the speed of light through air with the speed of light through water and found that, contrary to Newton's assertions, light did not move faster in the denser medium. Instead, just like a wave would, it slowed down. Eleven years later, Maxwell's work on electromagnetism established that light is a form of electromagnetic waves, and it seemed that Huygens' wave model had won the day.

But in 1900, Max Planck came up with a revolutionary idea that would spark a brand-new concept of light. Planck explained some puzzling behaviors of black body radiation by describing the energy of electromagnetic waves as being divided into individual packets of energy ($h\nu$) termed quanta. In 1905, Einstein built on Planck's concept of energy packets of light and finally settled the corpuscle-versus-wave debate through his photoelectric effect by declaring it a tie. That is, light has a dual nature, both wave and particle. In 1923, Arthur Compton provided additional support for Einstein's model of light. Compton aimed high-energy light (X-rays) at materials, and he successfully predicted the angles at which electrons released by the collisions would scatter. He did so by presuming that the light would act like tiny billiard balls. After just three years, the particle form of light—the quanta of light—got its name "photon" as coined by Gilbert Lewis. Today light/photon is understood to have both a wave and particle nature established firmly by quantum field theory (discussed in Chapter 3).

The term *photon* is derived from the Greek word for the prefix phot-meaning light, while the -on suffix is commonly used for subatomic particles. Sunlight warms us for a second with about sextillion (10^{21}) photons, which is why it feels like a continuous flow. Any electromagnetic signal is basically constituted of photons with different photon energy (see Fig. 3.4, where photon energy is given by $E = h\nu$; h is Planck's Constant, and ν is the frequency of the electromagnetic signal). Gamma rays have the highest photon energy (10^5 eV or more), while radio waves have very small photon energy (10^{-6} eV or less). The photon energy is expressed in the electronvolt (eV); it is the energy that an electron gains from an electric field of one volt. The more photons an object emits each second, the brighter it looks. That is, it is more intense, while the energy of the photons being emitted is dependent on the frequency of the light.

As noted earlier, the electromagnetic wave was first theoretically hypothesized by Maxwell and experimentally produced by Hertz during the latter part of the 19th century. Electromagnetic waves stretch over some 60 octaves—that is, with the wavelength doubling 60 times as one progresses from the shortest to the longest. The longest waves are therefore 2^{60}, or about a billion, billion times as long as the shortest. Of this vast range, the visible light spectrum covers only one octave with photon energy: Red (1.6 eV). . . . Violet (2.7 eV). (An interesting video lecture by Richard Feynman on photons can be found at https://cosmolearning.org/video-lectures/photons-corpuscles-light.)

For photon energy above 30 eV, the electromagnetic radiation becomes ionizing radiation that carries enough energy per quantum to ionize atoms or molecules, that is, to completely remove an electron from an atom or a molecule. Such radiation is harmful to our health, with long exposure leading even to cancer. The upper parts of UV-rays, X-rays, and γ-rays are ionizing radiation. But radio waves, microwaves, and terahertz (THz) waves all are nonionizing radiation and thus are practically harmless to us if their power level is kept within safe limits. Microwaves used for mobile communication do not cause any harm to human tissues, provided that they are kept within the specific absorption rate (SAR) limit. (We are familiar with this term in connection with our cell phone.) The present US standard for a safe SAR limit is 1.6 W/kg averaged over 1 gram of human tissue; the UK standard is 2.0 W/kg, averaged over 10 grams of human tissue. THz radiation that has penetrating capability (even in zero visibility conditions) of radio waves and spectroscopic (i.e., chemical-discriminating) capability of optical waves is now used for concealed weapon detection (CWD) and standoff detection of explosives and illegal drugs, all of which is important for airport and other security purposes. THz imaging has also become very important for early diagnosis of skin and breast cancer, as the absorptivity of the THz signal in malignant and unaffected tissues will be greatly different from that in healthy tissue. Also, the image resolution is pretty good with THz radiation compared to millimeter waves.

Charged particles interact through the exchange of photons—the carrier of the electromagnetic force. Whenever an electron repels another electron or an electron orbits a nucleus, a photon is responsible. Photons are massless and uncharged and have an unlimited range. The mathematical model used to describe the interaction of charged particles through the exchange of photons is known as quantum electrodynamics (QED), to which Paul Dirac made a significant contribution.

The photon has a *rest mass* of zero. That is, if it could be made to stand still, it would turn out to exhibit none of the properties associated with the possession of mass. It would have no inertia and would neither produce a gravitational field nor respond to one. It is therefore considered a *massless particle.* Such masslessness is purely theoretical, however, since a photon cannot be made to stand still. The moment it is formed, it moves away from its site of formation at 300,000 km per second in vacuum (which is the velocity of light *c*). While it is moving in this fashion, the photon does not possess the rest mass and exhibits some of the properties associated with the mass of a particles of matter that can respond to a gravitational field, for instance. Indeed, one prediction of the general theory of relativity that the light from stars would get bent while passing close to the Sun is an experimentally proved fact, establishing that the photon (the light quanta) showering out from stars behaves as a particle and possesses mass (while in motion) that is affected by the gravitation of the Sun.

The fact that photons always travel with the velocity of light *c* distinguishes it from conventional particles of matter, as all particles of matter can have any velocity but always less than *c*. At first sight, it seems to be in contradiction to the observed facts that the measured velocity of light in matter (water, glass, etc.) is less than *c*. But this is the velocity of the wave packet (i.e., group velocity) and not that of the individual photons. In any experiment where photons could be expected to be slowed down—for example in an encounter with electrons in the Compton Effect—it is found that energy and frequency, not velocity, decrease. The only slowing down that a photon can suffer is its complete annihilation, as happens in the photoelectric effect.

In the radio frequency and microwave frequency range (where the wavelength is from tens of kilometers down to submillimeters), electromagnetic radiation is associated with electric phenomena that occur in conductors. At such frequencies, the photon energy of the electromagnetic signal is extremely small. But starting from infrared and going through optical and X-ray regions and beyond, the generation of radiation involves atomic and molecular vibrations and transitions between energy states of various particles. In such cases, the electromagnetic signal has to be treated as a stream of photons (where photon energy also becomes significant). This has led to the development of the discipline of *photonics,* with a host of applications emerging every day for both commercial and industrial applications, including medical.

The 21st century can be called the Century of Photonics, just as the 20th century can be called the Century of Electronics. Photonics became a technically popular term during the late 1980s when fiber-optic data transmission was adopted by telecommunications network operators—though photonics as an independent domain of research and development actually began with the invention of the laser in the 1960s, followed by the development of laser diodes, optical fiber, and erbium-doped fiber amplifiers in the 1970s and 1980s. Photonics is the physical science of light (photon) generation, detection, and manipulation through emission, transmission, modulation, signal processing, switching, amplification, and sensing. Although photonics can be applied theoretically from infrared to gamma rays, most photonic applications today are in the range of the visible and near-infrared spectrum of electromagnetic waves. Photons are now at work all around us. They travel through optical fibers to deliver Internet and cell phone signals. They are used in hospitals, especially in beams that target and destroy cancerous tissues. In 2012, scientists at the Large Hadron Collider (LHC) discovered the Higgs boson by studying its decay into pairs of photons. Photonics, microwave-photonics, and nanophotonics are presently a promising field of research.

4.2.3 Leptons

Lepton is a name derived from the Greek word meaning thin, delicate, lightweight, or small (in ancient Greece, a kind of small coin known as a lepton also was used). Unlike hadrons, leptons are very lightweight, with the electron being the lightest ($m_e = 9.1 \times 10^{-31}$ kg). The lepton family started with the electron, followed by the muon and tau particle, each of which is identical to the electron except for mass. The muon is 207 times heavier than the electron, whereas the tau particle is 3,500 times heavier than the electron. Each of these three leptons has a neutrino as its partner, giving rise to another three leptons. There are thus six types of leptons: The electron (e^-), electron neutrino, muon, muon neutrino, tau particle, and tau particle neutrino. The lepton is truly an elementary particle (i.e., it has no internal structure), unlike hadrons, which are a composite of its fundamental unit called quarks. Leptons and quarks are the basic building blocks from which everything in this universe—ranging from the amoeba to the vast galaxies—is made. Originally, leptons were considered the "light" particles and hadrons the "heavy" ones, but the discovery of the tau lepton in 1975 broke that rule. The tau (the heaviest lepton) is almost twice as massive as a proton (the lightest hadron).

In 1889, J. J. Thomson was investigating a long-standing puzzle known as cathode rays. His experiments led him to propose that these mysterious rays are basically streams of particles much smaller than atoms, which he called corpuscles. In Thomson's landmark experiment, he subjected the cathode rays emanating from the cathode to the transverse magnetic and electric field. It was observed that the cathode ray always became deflected toward the positive and away from the negative, indicating that they constituted a negative charge. Thomson also determined the mass-to-charge ratio of the cathode ray particles, which was found to be much smaller than that of any known atom. Thus, the corpuscles of cathode rays eventually came to be known as electrons; very small, negatively charged particles that are the fundamental parts of every atom.

In 1937, the next member of the lepton, the *muon*, was discovered through cosmic ray photographs. It looked like the electron but was more than 200 times heavier than the electron. Physicists were surprised to find this twin/carbon (except heavier) copy of the electron. Making the issue more dramatic, physicists in 1962, using the atom smasher at the Brookhaven Lab, Long Island, showed that the muon, too, had its own distinct partner, the *muon neutrino*. In 1977–1978, experiments at Stanford University in the United States and work conducted in Hamburg, Germany, showed that there was yet another redundant electron, with mass 3,500 times that of electron. It was dubbed the *tau particle*, with its own separate partner, the *tau-neutrino*. Physicists' faith in the simplicity of nature was shaken by the existence of redundant pairs or "families" of leptons (which is also true with quarks). Strangely, nature may prefer redundancy in the construction of the universe.

Of all the particles in the universe, the neutrino is perhaps the most elusive: It has no charge, it probably has no mass, and it is exceedingly difficult to detect. The neutrino is so elusive a particle that it is also known as the ghost particle. This particle hardly ever interacts with matter; for example, about 100 billion neutrinos pass through your thumbnail every second without you even noticing. The Sun is the source of most of the neutrinos we receive on Earth. The particle was first postulated by Nobel Laureate Wolfgang Pauli, purely on theoretical grounds; he predicted that it explained the strange loss of energy found in radioactive decay. Pauli conjectured that the missing energy was carried off by a new particle that could not be seen in the experiments. In 1933, Enrico Fermi published the first comprehensive theory of this elusive particle, calling it the neutrino ("little neutral one" in Italian). Because the entire idea of the neutrino was so speculative, however, *Nature* rejected publication of his paper. Neutrino experiments were notoriously difficult to conduct because these particles are very penetrating and leave no traces of their presence. Indeed, they can easily penetrate through the Earth. The existence of the neutrino was proved in 1956 by Frederic Reines.

Neutrinos are an important subgroup within the leptons and are made up of the electron neutrino, muon neutrino, and tau neutrino. The Nobel Prize in Physics in 1995 was awarded to Martin L. Perl "for the discovery of the tau lepton" and to Frederick Reines "for the detection of the neutrino."

In 2021, the Breakthrough Physics Experiment of FASER (ForwArd Search ExpeRiment) at CERN's LHC has reported that they have detected the elusive neutrino candidates. This is definitely a major milestone in particle physics research as prior to this project, no sign of neutrinos has ever been seen at a particle collider. This significant breakthrough is definitely a step toward developing a deeper understanding of these elusive particles (that have a very small mass like an electron but have no electrical charge—a characteristic that has made them extremely challenging to detect) and the role they play in gaining further knowledge about the universe. Study of collider neutrinos is expected to shed new light on the still enigmatic nature of these fundamental particles, not least because collider neutrinos are produced at high energy. Their weak interactions with matter have been little studied.

All lepton interactions are mediated by an electro-weak force, unlike the hadron where the strong (nuclear) force is the dominant player. The electro-weak force stands for the electromagnetic and weak (nuclear) force combined in the same framework by Steven Weinberg, Abdus Salam, and Sheldon Glashow. In 1967–1968, they observed an amazing similarity between the photon and the W-particle (W for "weak"), the particle (wave) that is supposed to be exchanged when the force acts between electrons and neutrino. Thus, they combined the photon of electromagnetic interaction and the W-particle of weak interaction, leading to the new W-particle theory that came to be known as the *electro-weak theory*. The 1979 Nobel Prize in Physics was awarded jointly to Sheldon Lee Glashow, Abdus Salam, and Steven Weinberg "for their contributions to the theory of the unified weak and electromagnetic interaction between elementary particles, including, inter alia, the prediction of the weak neutral current."

The weak (nuclear) force that is responsible for radioactivity acts on all matter particles of spin ½ (as for all leptons), but not on particles of spin $s = 0, 1, 2$, such as the photon (with $s = 1$) and hadrons. The electro-weak theory used the most sophisticated form of gauge symmetry available at that time, the Yang–Mills theory. This theory, formulated in 1954, possessed more symmetries than is possible with Maxwell's equations that describe the electromagnetic force. The electro-weak theory treated the electron and neutrino symmetrically as one "family," considering the electron and neutrino as two sides of the same coin. Anyway, even electro-weak theory (or any other theory as of today) could not account for the redundancy in the lepton family.

4.2.4 Hadrons and Quarks

Hadron is a Greek term meaning large or massive. The proton, neutron, and so on are examples of hadrons, with the mass of the proton being 1,836 times that of the electron. Hadrons are characterized by a strong (nuclear) interaction, unlike leptons which participate only in electromagnetic and weak (nuclear) interactions. Hadrons are composites of more fundamental particles known as *quarks*. The most stable hadrons are the protons and the neutrons, the constituents of atomic nuclei.

Quarks come in three pairs or generations: (1) The lowest mass pair is called the up and down; (2) the next heavier mass pair is called the strange and charmed; and (3) the heaviest mass pair is the bottom and top. Each of the six "flavors" of quarks can have three different colors: Red, green, and blue. Color has nothing to do with the familiar concept of color in optics, which is determined in terms of a particular frequency. In fact, quarks are much smaller than the wavelength of visible light. Modern physicists seem to have more imaginative ways of naming new particles and phenomena; they no longer restrict themselves to Greek terms! Up and down quarks are generally stable and the most common in the universe, whereas strange, charmed, bottom, and top quarks can only be produced in high-energy collisions, such as those involving cosmic rays and those in particle accelerators. The heavier quarks rapidly change into up and down quarks through a process of particle decay: The transformation from a higher mass state to a lower mass state. For every quark flavor there is a corresponding type of antiparticle known as an *antiquark* that differs from the quark only in that some of its properties (such as the electric charge) have equal magnitude but of opposite sign. When we add up the pairs of quarks, their color varieties and the antiquark twins, the total number of quarks is 36.

Until the 1960s, protons and neutrons were thought to be elementary particles. In 1968, however, deep inelastic scattering experiments at the Stanford Linear Accelerator Center (SLAC), US showed that the proton contained much smaller, point-like objects, which Feynman named partons. The objects observed at SLAC would later be identified as up and down quarks as the other flavors of quarks were discovered. Nevertheless, "parton" remains in use as a collective term for the constituents of hadrons—quarks, antiquarks, and gluons. (Under certain circumstances, a Yang–Mills particle called gluon can act as if it were a sticky, glue-like substance that binds the quarks.) The strong (nuclear) force mediated by the gluon (spin $s = 1$) particle holds the quarks together in the proton and neutron and also holds the protons and neutrons together in the nucleus of an atom. Gluon interacts only with itself and with the quarks.

The quark model was independently proposed by Murray Gell-Mann and George Zweig in 1964 as part of an ordering scheme for hadrons. However, there was little evidence for their physical existence at the time of their proposal. Murray Gell-Mann won Nobel Prize in 1969 for "his contributions and discoveries concerning the classification of elementary particles and their interactions." The quark model was formed by a simple combination of three quarks: Up, down, and strange quark. The model could miraculously explain the hundreds of hadrons discovered every new morning in that period in laboratories around the world and, more importantly, predict the existence of new ones. This period is reminiscent of Mendeleev's formation of the periodic table in 1869 when chemists were confused as to how to account for hundreds and thousands of chemical compounds on the basis of less than 100 elements!

The name "quark" comes from a single line in Irish writer James Joyce's novel *Finnegan's Wake*: "Three quarks (quarts?) for Muster Mark." Since the pre-standard model theory of particle physics was complete with only three quarks, the name made some sense. The full standard model today, however, needs six quarks, which has not made the word "quark" any less fun to say. Now we know that the proton and neutron are made of three quarks: A proton contains two up quarks and one down quark, while a neutron contains two down quarks and one up quark, the binding glue being the gluon. Particles may be made of other quarks too (strange, charmed, bottom, and top), but these all have a much greater mass and decay very rapidly into protons and neutrons. Quarks are known to bind into triplets and doublets. The triplets are called baryons, a term derived from the Greek word meaning heavy. The doublets are called mesons, a term that comes from the Greek word for medium. Baryons found in the nucleus (the proton and neutron) are called nucleons. The Latin word for kernel is *nucleus*. Nucleons are found in the metaphorical kernel of the atom. Baryons that contain at least one strange quark but no charm, bottom, or top quarks are called hyperons—the Greek word for beyond, which morphed into the English prefix *hyper-*. Hyperons are particles that are "way out" in a certain sense. Quarks are the only elementary particles in the standard model of particle physics to experience all four fundamental interactions, also known as *fundamental forces* (electromagnetism, gravitation, strong interaction, and weak interaction), as well as the only known particles whose electric charges are not integer multiples of the elementary electronic charge.

The "strange" quark's existence was indirectly validated by SLAC's scattering experiments: Not only was it a necessary component of Gell-Mann and Zweig's three-quark model, but it provided an explanation for the kaon (K) and pion (π) hadrons discovered in cosmic rays in 1947. "Charm" quarks were produced almost simultaneously by two teams in November 1974—one at SLAC and one at Brookhaven National Laboratory (BNL), United States. The discovery finally convinced the physics community of the quark model's validity.

Physicists had been searching for the elusive top quark since 1977, soon after the bottom quark was discovered at the accelerator of the Fermi National Laboratory (FNL or Fermilab) near Chicago. However, since then and for the next 16 years, searches had failed to detect the presence of the even heavier "top quark." Physicists were becoming apprehensive: If the top quark did not exist, they asked, then elementary particle physics would collapse like a house of cards. Noble Laureate Steve Weinberg commented: "There was tremendous expectation that the top quark is there. A lot of us would have been embarrassed if it were not." In July 1994, the top quark was finally discovered at FNL's Tevatron accelerator facility. The evidence was found by smashing protons into antiprotons at energies of almost 2 TeV. The colossal energy released by this sudden collision produced a torrent of subatomic debris. Based on a trillion photographs of such debris, a team of particle physicists was able to select just 12 collisions that had the "fingerprints" of a top quark collision. Estimations revealed that top quarks had a mass of 174 billion eV, making it the heaviest elementary particle ever discovered. It was nearly as massive as a gold atom (which contains 197 neutrons and protons). By contrast, the bottom quark had a mass of 5 billion eV. It took 8 months for the group of FNL physicists and a rival group (using similar accelerator) to finally declare the proof of existence for the top quark on the basis of 38 photographs of top quark collisions.

Almost immediately, the *New York Times* trumpeted this discovery on its front page: The discovery of a new subatomic particle had never received such prominent coverage in a major newspaper in the history of the discovery of subatomic particles. All of a sudden, millions of peoples who didn't have the slightest understanding of (or even interest in) atomic physics were asking the question, "What is a top quark?" This was the last crowning achievement of a half century of painstaking effort to decode the mysteries of the subatomic world. *Top quark* was the last quark in the quark family necessary to complete the standard model of physics—the most successful theory of particle interactions.

Quarks stick to other quarks because they possess a characteristic known as color (or color charge). Quarks come in one of three colors: Red, green, and blue. (Actually, quarks are much too small to be visible and thus could never have a perceptual property like color as we commonly know it, and so the names were chosen only to have a convenient analogy.) The colors of quarks in the standard model combine like the colors of light in human vision. Red light plus green light plus blue light appear to the eye as "colorless" white light. A baryon is a triplet of one red, one green, and one blue quark, and putting them together, we get a color-neutral particle. A color plus its opposite color also gives white light. Red light plus cyan light appear to the human eye as white light, for example. A meson is a doublet of one colored quark and one anticolored antiquark, and putting them together, one gets another color-neutral particle.

Quarks can't stand being apart from one another. They have to join up, and they always do so in a way that hides their color from the outside world. One color is never favored over another when quarks get together. Matter is color-neutral down to the very small scale. Colored quark particles are bound together by the appropriately named gluons—the particles that mediate the strong force. Gluons are also colored but in a more complicated way than the quarks are. Gluons glue quarks together, but they also stick to themselves. One consequence is that they can't reach out and do much beyond the nucleus. The mathematical model used to describe the interaction of colored particles through the exchange of gluons is known as quantum chromodynamics (QCD). The whole sticky mess is called the strong force or the strong interaction since it results in forces in the nucleus that are stronger than the electromagnetic force. Without the strong force, every nucleus would blow itself to smithereens.

Before 1964, physicists were bewildered by the disagreement between the abundance of hadrons and a modest number of lepton types. Perhaps that is why a hypothesis advanced in 1964 appeared to be so attractive. According to this hypothesis, all hadrons consist of several elementary building blocks called *quarks*. With the passage of time, the quark hypothesis gained ground and eventually became formulated as a rule: *The number of quark types must be equal to the lepton types.* This rule ultimately became established with experimental support that reflects the *quark-lepton symmetry*, which remains enigmatic to this day: Possibly nature works with symmetry, and symmetry is at the heart of everything.

4.3 The Standard Model of Physics

Since the 1930s, with the advent of particle accelerators and the development of quantum mechanics (which deals with micro-particles), a large number of elementary particles have been discovered. The number of these elementary particles became so large that researchers needed to organize them, just as Mendeleev did for chemical elements through his periodic table. This process is summarized in a concise theoretical framework known as the *standard model* of physics. It incorporated all that was known about subatomic/elementary particles until 1970, and it predicted the existence of additional particles as well. There are 19 named particles in the standard model, organized as shown in Table 4.2. The last particles discovered were the W- and Z-bosons in 1983, the top quark in 1995, the tau neutrino in 2000, and the Higgs boson in 2012. (The W-bosons are named after the weak force. The physicist Steven Weinberg named the additional particle the Z-particle and later explained that it was the last additional particle needed by the model.)

TABLE 4.2
List of Subatomic/Elementary Particles in the Standard Model of Physics

Family		Particle	Discovered/ (Predicted)	Spin Number	Charge (e)	Color	Mass (MeV/c^2)
Fermions	Quarks	Up quark (u)	1968 / (1964)	1/2	+2/3	R, G, B	2.2
		Down quark (d)	1968 / (1964)	1/2	−1/3	R, G, B	4.7
		Charm quark (c)	1974 / (1970)	1/2	+2/3	R, G, B	1275
		Strange quark (s)	1968 / (1964)	1/2	−1/3	R, G, B	95
		Top quark (t)	1995 / (1973)	1/2	+2/3	R, G, B	173,000
		Bottom quark (b)	1977 / (1973)	1/2	−1/3	R, G, B	4,180
	Leptons	Electron (e)	1874 / (1897)	1/2	−1	None	0.51099895
		Muon (μ)	1936	1/2	−1	None	105.658375
		Tau (T)	1975	1/2	−1	None	1776.86
		Electron neutrino (V_e)	1956 / (1930)	1/2	0	None	$< 10^{-7}$
		Muon neutrino (V_μ)	1962 / (1940)	1/2	0	None	$< 10^{-7}$
		Tau neutrino (V_T)	2000 / (1970)	1/2	0	None	$< 10^{-7}$
	[a]	Proton (p)	1917 / (1815)	1/2	+1	None	938.272081
		Neutron (n)	1932 / (1920)	1/2	0	None	939.565413
Bosons	Vectors	Gluon (g)	1978 / (1962)	1	0	8 colors	0
		Photon (γ)	1899	1	0	None	0
		W-boson (W)	1983 / (1968)	1	±1	None	80,379
		Z-boson (Z)	1983 / (1968)	1	0	None	91,187.6
	[b]	Higgs boson (H)	2012 / (1964)	0	0	None	125,180

Notes:

[a] Proton (2 up quarks + 1 down quark) and Neutron (1 up quark + two down quarks).

[b] Higgs boson is the quantum excited particle form of the Higgs field.

Today, we have a very good idea of what matter is made of, how it all holds together, and how these particles interact with each other. Developed in the early 1970s, the standard model has successfully explained almost all experimental results and has precisely predicted a wide variety of phenomena. Over time and through many experiments, it has become established as a well-tested physics theory.

The subatomic/elementary particles are either the building blocks of matter, called fermions, or the mediators of interactions, called bosons (i.e., force carrier particles). There are 14 named fermions and 5 named bosons in the standard model. Twelve of the fermions are divided into two groups of six: Those that must bind together are the quarks, and those that can exist independently are the leptons. In other words, the difference between quarks and leptons is that quarks interact with the strong nuclear force that binds the nuclei of atoms together, whereas leptons do not. Protons and neutrons are also fermions, and they are composed of quarks. The proton is composed of two up quarks and one down quark, while the neutron is composed of one up quark and two down quarks. In the nucleus, protons and neutrons are held together by the strong force of the carrier gluon. Fermions obey Fermi–Dirac statistics. The statistical rules that bosons obey were first described by Satyendra Nath Bose and Albert Einstein, and are known as Bose–Einstein statistics.

The ordinary matter that we experience around us is actually just made up of four particles, the up and down quarks that make up the protons and neutrons in the nuclei of atoms, electrons that form a cloud around the nucleus, and a near massless particle—the electron neutrino that is created during the fusion process in stars like the Sun. The other particles are rare and don't typically exist in ordinary matter.

In 1926, Paul Dirac connected the Bose–Einstein statistics (1924) and Fermi–Dirac statistics (1926) of particle physics with the symmetry properties of their wave functions and named the particles bosons and fermions. Bosons (e.g., photons) have integral spin and can occupy the same quantum state, whereas fermions (e.g., electrons) have half-integer spin, and no two electrons can occupy the same quantum state as per Pauli's exclusion principle. Dirac coined the term *boson* to honor Bose's fundamental contribution in formulating the Bose–Einstein statistics.

Satyendra Nath Bose of Bose–Einstein statistics was a self-taught scholar and polymath. While teaching at Dhaka University, Bangladesh, in the early 1920s, Bose concluded that Planck's derivation of his radiation law using Boltzmann's classical statistics needed a different mathematical treatment, a quantum–mechanical analysis, when applied to photons. His paper, entitled *Planck's Law and the Quantum Hypothesis of Light*, predicted the number and probability of photons proposed by Einstein in 1905. The statistics for elementary particles are also applicable to all particles with integer spin. Bose's derivation was initially rejected as a radical idea, but Bose was not disheartened. At this point, he sent the paper directly to Einstein in Germany in 1924. Einstein recognized the insight and importance of Bose's work and immediately translated Bose's paper into German, publishing it in 1924 in *Zeitschrift für Physik*. Einstein described Bose's work as "a beautiful step forward." This is the background fact how the Bose–Einstein statistics was born. Remarkably, however, Bose never received the Nobel Prize, though he was nominated for this coveted prize a number of times.

Leptons and *quarks* are fermions, as well as the things made from them, including protons, neutrons, atoms, and molecules, as verified by our macroscopic observations of matter in everyday life. People cannot walk through a wall unless the wall gets out of the way. Gluons, photons, and the W, Z, and Higgs are all bosons. Our most direct experience is with light (and other forms of electromagnetic radiation), which is made of mediating particles and photons that are bosons. In our everyday experience, we never see beams of light crash into one another. Photons are like phantoms: They pass through one another with no effect whatsoever observable in the macroscopic perception.

The *vector bosons* are known to be the force carriers of the fundamental interactions. For example, for strong interactions, it is *gluon* (which carries the strong force that binds the nuclei of atoms); for electromagnetism, it is *photon* (which carries the electromagnetic force responsible for all electricity, magnetism, and chemistry); and for weak interactions, it is the *W- and Z-bosons* (which carry the weak force responsible for some kinds of radiation). Whereas Fermi presented the first theory of weak force in the 1930s, Steven Weinberg, Abdus Salam, and Sheldon Glasgow redefined it a decade later and predicted the existence of *W*- and Z-particles in 1968, which were finally detected at the LHC in 1983. All vector bosons have spin equal to 1. *Higgs boson* is the manifestation of the Higgs field caused by quantum excitation; this field permeates space and endows all elementary subatomic particles with mass through its interaction with them. The Higgs boson particle is a massive scalar boson with zero spin, no electric charge, and no color charge.

Four basic forces are at work in the universe: Strong, weak, electromagnetic, and gravitational. They work over different ranges and have different strengths. Gravity is the weakest of the four, but it has an infinite range. Electromagnetic force also has an infinite range, but it is many times stronger than gravity. The weak and strong forces are effective only over a very short range and dominate only at the level of subatomic particles. Despite its name, the weak force is much stronger than gravity, though it is indeed the weakest of the other three forces. The strong force, as the name suggests, is the strongest of all four fundamental interactions.

Three of the fundamental forces result from the exchange of bosons. Particles of matter transfer discrete amounts of energy by exchanging bosons with each other. Each fundamental force has its own corresponding boson. The strong force is carried by the gluon, the electromagnetic force is carried by the photon, and the W- and Z-bosons are responsible for the weak force. Although not yet found, the graviton should be the corresponding force-carrying particle of gravity. The standard model includes the electromagnetic, strong, and weak forces, together with all their carrier particles, and explains how these forces act on all of the matter particles. However, the most familiar force in our everyday lives, gravity, is not part of the standard model, as fitting gravity comfortably into this framework has proved to be a difficult challenge. The quantum theory used to describe the micro-world and the general theory of relativity used to describe the macro-world are difficult to fit into a single framework. No one has managed to make the two mathematically compatible in the context of the standard model. But luckily for particle physics, when it comes to the minuscule scale of particles, the effect of gravity is so weak that it can be neglected. Only when matter is in bulk, at the scale of the human body or of the planets, for example, does the effect of gravity dominate. So, the standard model still works well despite its reluctant exclusion of one of the fundamental forces, the gravity.

For particle physicists, the 1970s was the decade when two of the fundamental forces of nature, the weak and the electromagnetic forces, could be described within the same theory, which formed the basis for the standard model. This "unification" implies that electricity, magnetism, light, and some types of radioactivity are all manifestations of a single underlying force known as the electro-weak force. The basic equations of this unified theory correctly described the electro-weak force and its associated force-carrying particles, the photon and the W- and Z-bosons, except for a major glitch. All of these particles emerge without a mass. While this is true for the photon, it is not for the W- and Z-bosons, which have mass. This problem was solved through the Brout–Englert–Higgs mechanism, which gives a mass to the W- and Z-boson (and of course all matter in this universe) when they interact with an invisible field, the Higgs field.

On July 4, 2012, Argonne Tandem Linac Accelerator System (ATLAS) and Compact Muon Solenoid (CMS) detectors at CERN's LHC announced that they had both observed a new particle in the mass region around 126 GeV. This particle is consistent with the Higgs boson proposed by Robert Brout, François Englert, and Peter Higgs. As already noted, this was the last missing puzzle of the standard model. Like all fundamental fields, the Higgs field has an associated particle—the *Higgs boson*. The Higgs boson is the visible manifestation of the Higgs field and is rather like a wave at the surface of the sea. The Higgs boson, unlike the other bosons (gluons, photons, and W- and Z-bosons), doesn't result in anything resembling a force (like the strong, electromagnetic, and weak forces). The Higgs field is a scalar field, and the Higgs boson is a particle with spin zero. The Higgs field is an invisible field that pervades the universe. Just after the Big Bang, the Higgs field was understood to be zero, but as the universe cooled and the temperature fell below a critical value, the field grew spontaneously, so that any particle interacting with it acquired a mass. The more a particle interacts with this field, the heavier it is. The electro-weak force's force-carrying particle photons do not interact with it and are thus left with no mass. But W- and Z-bosons do interact with the Higgs field and thus have a mass that is nearly 100 times that of a proton. Englert and Higgs jointly received the Nobel Prize in Physics in 2013 "for the theoretical discovery of a mechanism that contributes to our understanding of the origin of mass of subatomic particles."

Even though the standard model is currently the best description we have of the subatomic world, it does not give us the complete picture. Most physicists believe that the standard model is just an intermediate step toward achieving a true theory of everything. The theory incorporates only three of the four fundamental forces, omitting gravity. Also, important questions remain to be answered, such as, "Why are there three generations of quarks and leptons with such a different mass scale?" The standard model is obtained by combining the available theories regarding the electromagnetic, weak, and strong forces into one theory.

Although the standard model received exceptional experimental support, it still suffers from a lot of redundancy. Einstein used to think that if he were God, he would not have constructed the universe with 19 adjustable parameters and a horde of redundant particles. Possibly he would have used one (or no) adjustable parameter, and just one object out of which all the particles found in nature could be constructed. With such thinking, he would possibly be pointing toward a unified field theory or a theory of everything. Although string theory and the theory of loop quantum gravity aim toward such a theory, they still appear to be simply lofty theories with no experimental support whatsoever. By analogy, we can look at the Mendeleev periodic table with its collection of over 100 elements, which may be termed the "elementary particles" of the last century. No one could deny that the Mendeleev chart was particularly successful in describing the chemical building blocks of matter. But the fact that it had hundreds of arbitrary constants was unappealing. With the advent of quantum mechanics, the modern periodic table of elements can be explained by just three particles: The neutron, the proton, and the electron. Similarly, physicists believe that the standard model, with its odd-looking and redundant quarks and leptons, should be constructed from even simpler structures in days to come.

4.4 Physics Beyond the Standard Model—The Possibility of a New Particle, a New Force

When I was at Lawrence Berkeley National Laboratory (LBNL) at Berkeley during 1999–2000, quite often I used to overhear the phrase, "Physics beyond the standard model." I was working there as a visiting Fulbright scholar from Calcutta University, India, to develop radio frequency structures for the Muon Collider. This was a collaborative research project conducted by Berkeley Lab with Fermilab, and other US universities. As I was trained in radio physics and electronics and was just a new entrant in the field of accelerator physics and particle physics, I was not too excited about that particular phrase at that time, though I had some exposure to particle physics while studying for my BSc Physics (honors) degree at St. Edmunds College, Shillong, Meghalaya, India. But senior particle physicists of Berkeley Lab were seen to be thoroughly engrossed in a deep discussion of that newly coined phrase.

Studies on the radioactive decay of the beryllium-8 isotope in 2016 and emission of light from excited helium atoms at an unusual angle in 2019 point toward the existence of a new particle called X17 (so-named because its calculated mass is 17 mega electronvolts). The behavior of X17 particles cannot be explained through the standard model of particle physics. The only plausible way to explain it is through the protophobic force since the particle dubbed X17 appeared to be "scared" of protons and was found to run away from protons. X17 could be a particle, which connects our visible world with the mysterious dark matter (the strange substance seemingly holding the universe together in addition

FIGURE 4.3
The author at Lawrence Berkeley National Laboratory as senior Fulbright scholar, 1999–2000.

to our known force of gravity). If this proves to be so, we may be moving toward another game changer in physics and a new understanding of the physical world as a whole.

The real quest and planning to investigate the physics beyond the standard model started in the late 1990s. The LHC's Beauty experiment (LHCb) aimed toward studying asymmetries and charge conjugation/parity (CP) violations and the ATLAS experiment aimed to study many issues related to physics beyond the standard model since 2010. The FNL, affiliated with the University of Chicago, started the Muon g-2 experiment in 2015 with the abandoned and refurbished experimental setup of BNL in Upton, New York, to study issues related to physics beyond the standard model. The Muon g-2 experiment relates to measurement of the anomalous dipole moment of muon, a subatomic particle that is a sibling of the electron but has 207 times the mass of electrons. The Muon g-2 experiment was initiated at CERN, which was further taken up by BNL, but even BNL abandoned it in 2001 and a decade later Fermilab took it up again. In 2021, both the LHC and Fermilab reported intriguing experimental results that pointed to important clues, leading particle physicists to be cautiously excited to look beyond the standard model and possibly a new force (fifth force?) beyond the four known forces of nature. The excitement over this unexpected news in particle physics might seem to be an overreaction, but in the history of science small discrepancies have been known to lead to massive breakthroughs. In the 1850s, for example, astronomers measuring Mercury's orbit noticed that it was off a little from Newton's predictions. That anomaly, along with other evidence, firmly established the 20th century's most precious discovery—the theory of general relativity.

Starting from March 2021, a series of experiments in the LHCb (b stands for beauty), experiment of LHC of CERN, the most powerful accelerator in the world as of today, has led to a possibly new discovery (http://lhcb-public.web.cern.ch) that confounds particle physicists' existing theories based on standard model of physics—believed to be an incredible and seemingly unchallengeable theory of particle physics since 1970s. The so-called LHC beauty experiment is one of the four large experiments at the LHC that involves collaboration of 1,000 scientists and engineers from 19 countries and 85 institutes. The LHC scientists have been searching for the signs of something beyond the standard model for quite some time, and despite a decade of work at the LHC, nothing conclusive did emerge so far. First data taking of LHCb experiment investigating the nature of the decay of "beauty quark" started in 2014 but the data analysis of more than a trillion gathered till 2021 has led to some repetitive result with so-called "3 sigma" accuracy, as of today, with which the scientists are bursting with enthusiasm. However, when the result reaches the 5 sigma threshold, particle physicists consider that to be a genuine discovery. Incidentally, by "3 sigma" accuracy, we mean that the difference between the LHCb result and the standard model is about three units of experimental error, i.e., only around a one in a thousand chance of the result being obtained is caused by a statistical fluke. By "5 sigma," it is meant that one in a million chance of the result being down to a cruel statistical fluke.

The tantalizing result that the LHCb scientists have observed after nearly a decade's effort is the following. As per the standard model of physics, beauty quark ought to decay into electrons and muons at equal rates. Beauty quarks (also known as bottom quarks) are a kind of subatomic particle in an atom. It is known that there are six flavors of quarks that are dubbed up, down, strange, charm, beauty/bottom, and top/truth. Up and down quarks, for example, make up the protons and neutrons in the atomic nucleus. Beauty quarks are unstable, living on average just for about 1.5 trillionths of a second before decaying into other particles. When a beauty quark decays, it transforms into a set of lighter particles, such as electrons and muons, through the influence of the so-called weak force. The 2021 data analysis from LHCb experiment found that beauty quarks were decaying into electrons and muons at different proportions (around 85 muon decays for every 100 electron decays). This was truly surprising because, according to the standard model, the muon is basically a carbon copy of the electron—identical in every way except for being around 200 times heavier. To address this intriguing observation, the theorists believe in the existence of a brand new kind of particles that are influencing the ways in which the beauty quark is found to decay. One possibility they think is the existence of a fundamental particle called a "Z prime"—in essence a carrier of a brand new force of nature (fifth force?). This force would possibly be extremely weak, that is why any sign of it could not be seen until now, and presumably this is the force that is interacting with electrons and muons differently. Anyway, scientists believe that more data is needed with painstaking experimentation to ensure if the effect is natural and every effort should be made to see if the most sought for result of 5 sigma mark can be reached. To get there, it is necessary to reduce the size of the error, and to do this more data is needed. One way to achieve this is simply to run the experiment for longer and record more decays. The LHCb experiment is thus being currently in the state of upgrading to be able to record collisions at a much higher rate in future, which will allow making much more precise measurements. Meanwhile, other experiments at the LHC, as well at the Belle 2 experiment at KEK Japan, are closing in on the same measurements. It's exciting to think that in the next few months or years, a new window could be opened on the most fundamental ingredients of our universe if the existence of a "fifth force" is really discovered with confidence.

In April 2021, Fermilab, reported its Muon g-2 experiment which revealed that the wobbliness (i.e., precession) of the subatomic particle muon is greater than expected, while whipping around a magnetized ring (https://news.fnal.gov/2021/04). Muons occur naturally when cosmic rays strike Earth's atmosphere; but the particle accelerator at Fermilab can produce them artificially in large numbers. C.D. Anderson and S. Neddermeyer discovered the muon in 1936 while studying cosmic ray showers. Because of their negative charge and quantum-mechanical spin, muons, like electrons, behave as if they have a minuscule internal magnet. In a strong magnetic field, the direction of the muon's tiny magnet wobbles much like the axis of a spinning top or gyroscope.

The results of the Fermilab Muon g-2 experiment significantly differ from those regarding the prediction of the standard model of particle physics. The g-factor of wobbling (i.e., the magnetic moment) of muons in a magnetic field, according to the standard model, is supposed to be 2. Hence, the name of the experiment is Muon g-2—but the Fermilab experimental result found it to be significantly different. The new world g-factor average when the measurements of both Fermilab and BNL experimental values are taken together differs from the theoretical predictions with a standard deviation or the sigma factor of 4.2 (which means a probability of 1 in 40,000 that this is a statistical fluke). This 4.2 sigma factor has excited hopes for a truly new scientific discovery, especially in particle physics, which is considered to be 5 (i.e., a p-value or probability of 3×10^{-7} or 1 in 3.5 million). Improvements are in progress in experimental setup, measurement technique, data analysis, theoretical modeling for both CERN's LHCb experiment and Fermilab's Muon g-2 experiment to reach the 5-sigma mark for possible discoveries reported in 2021.

Evidence for a new physics in Fermilab's Muon g-2 experiment, in conjunction with anomalies recently observed through the LHCb experiment at CERN, is tantalizing. Each new discovery in physics and astrophysics promises a number of cracks in the standard model. The standard model has been unable to provide a theoretical framework for the relationship between matter and antimatter. It has also failed to accommodate the dark matter and dark energy that constitute 95% of all the stuff in the universe. Nor can it explain the neutrino masses. It appears that the physics of the future may rewrite the standard model in order to include new fundamental particles and/or new fundamental forces of nature.

4.5 Particle Accelerators/Atom Smashers

4.5.1 Introduction

Particle accelerators are the particle/nuclear physicist's and cosmologist's primary scientific tool in reaching a fundamental understanding of the elementary particles within atoms and the physical laws governing matter, energy, space, and time. Thus, particle accelerators help us to probe both the elementary particles of nature (in the micro-world) and the origin and evolution of all matter in the universe. As noted, the LHC discovered the last missing particle of the standard model of particle physics, Higgs boson, in 2012. The quark–gluon plasma (QGP), that existed in the first millionth of a second (i.e., microsecond) after the Big Bang, was detected for the first time in the laboratory at CERN in the year 2000 and in May 2021, the scientists of the LHC of CERN successfully generated/recreated QGP. The more powerful the accelerator, the further we can see into the infinitely small and the infinitely large, respectively, in the micro- and macro-world. Although particle accelerators may be best known for their role in these applications, they are also used in medical and scientific research and in the matters of national security. They are used in medical diagnosis and treatment (e.g., in creating tumor-destroying beams and in killing bacteria to prevent food-borne illnesses). Terahertz (THz) free electron laser (FEL)-based portable accelerators are used for security inspection at airports, in public facilities, and also in principal buildings that help in security purposes.

Tens of millions of patients receive accelerator-based diagnoses and therapy each year in hospitals and clinics around the world. Particle accelerators have two primary roles in medical applications: As producers of radioisotopes for medical diagnosis and therapy and as sources of beams of electrons, protons, and heavier charged particles for medical treatment. The US Energy Department's National Labs played a crucial role in the early development of technologies for medical diagnosis and treatment using particle accelerators. Los Alamos National Laboratory helped develop linear accelerators (LINAC) for electrons, which have become the workhorses of external-beam therapy. In addition, Oak Ridge and BNL contributed much of the present expertise in isotopes for diagnosis and therapy. LBNL pioneered the use of protons, alpha particles (helium nuclei), and other light ions for therapy and radiobiology.

Particle accelerators are also called *atom smashers* when they are used for the discovery and characterization of elementary particles that are present deep within the nucleus of an atom. The energy with which the particles collide in an atom smasher breaks apart the atoms of the particle and leaves behind the signature of the elementary particles thus produced, which are detected by the detector of the atom smasher/collider. Such a signature was left behind by Higgs boson in the LHC's ATLAS detector (see Fig. 4.4b), together with its simulation diagram by the CMS experiment (as in Fig. 4.4a). Only a team of expert particle physicists can decipher the secret hidden in the picture regarding the signature of the elementary particles.

4.5.2 The History of Particle Accelerators

The history of particle accelerators begins with the work of Ernest Orlando Lawrence in whose name the LBNL came into being. In 1930, inspired by Norwegian engineer Rolf Wideroe, Lawrence created the first circular particle accelerator at the University of California, Berkeley, with graduate student M. Stanley Livingston. The machine accelerated hydrogen ions up to energies of 80,000 electronvolts within a chamber less than 5 inches across. In 1931, Lawrence and Livingston set to work on an 11-inch accelerator. The machine managed to accelerate protons to just over 1 million electronvolts, a fact that Livingston reported to Lawrence by telegram with the added exuberant comment, "Whoopee!" Later, Lawrence went on to build even larger accelerators and founded Lawrence Berkeley and Lawrence Livermore Laboratories with accelerator facility. He was awarded the 1939 Nobel Prize in Physics "for the invention and development of the cyclotron and for results obtained with it, especially with regard to artificial radioactive elements."

Particle accelerators have come a long way since the 1930s, with much larger size and creating beams of particles with greater energies than were previously thought possible. As noted earlier, the LHC at CERN, Geneva, is the most powerful circular accelerator in the world today that has an enormous diameter of 5 mile (17 mile in circumference). LHC is able to accelerate protons to more than 6 trillion electronvolts. In addition to being a primary tool for studying the nature and properties of the elementary particles and cosmic phenomena, the LHC has also benefitted the mankind with its technological offshoots. The concept of hypertext developed by Physicist Tim Berners-Lee of CERN in 1989 made it possible to use the so-called Intranet in 1991 as a tool to allow scientists of CERN to share information among them. The World Wide Web (WWW) or the so-called information highway was made freely available to the world in 1993 so that anyone from anywhere in the world could connect to others via the Internet. From 1971 until 1999, Fermilab's Meson Laboratory was a key part of the laboratory's high-energy physics experiments. To learn more about the forces that hold our universe together, scientists there studied the subatomic particles mesons and protons. There are now hundreds of particle accelerators around the world including the linear collider at SLAC, United States; relativistic heavy

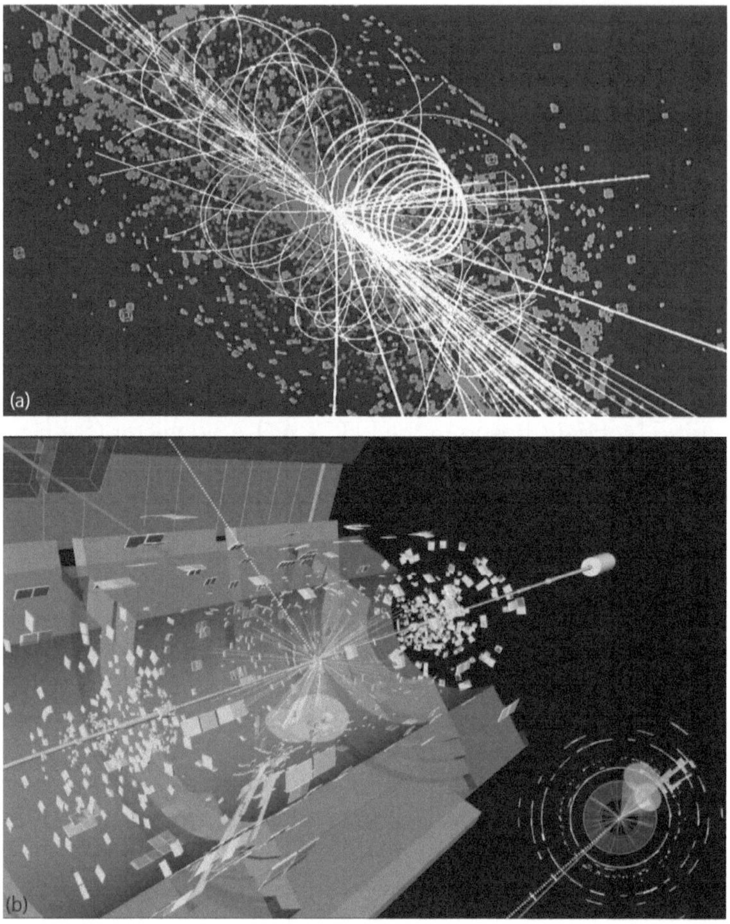

FIGURE 4.4
(a) Computer simulation of particle traces from an LHC collision in which a Higgs boson is produced in a CMS experiment [14]. (b) ATLAS observes elusive Higgs boson decay to a pair of bottom quarks [15].

ion collider (RHIC) at BNL, United States; the accelerator facility at KEK, Japan; Accelerator Science and Technology Centre at Cockroft Institute, United Kingdom; and so on.

An exhaustive list of different accelerators around the world may be found in the following website: https://en.wikipedia.org/wiki/List_of_accelerators_in_particle_physics.

Some scientists believe that the Grand Unified Theory (GUT) and also the string theory may be established experimentally in near future, provided we have accelerators capable of generating energy some billion times the energy being generated by today's most advanced accelerators. Whether that is possible remains a big question, for the energy it would require is close to the energy that was generated at the time of the Big Bang!

4.5.3 Principle of Operation of Particle Accelerators

The principle of operation of particle accelerators is dependent on concepts borrowed from both quantum mechanics and relativity. To probe into the micro-world at the atomic or subatomic level, we need to irradiate or shine the object with photons whose wavelength should be comparable to the atomic/subatomic dimension concerned. Such a small wavelength (λ) of incident radiation will have a very high frequency (since $\nu = c/\lambda$). The Planck–Einstein quantum relation $E = h\nu$, where ν is the frequency of incident radiation and h is Planck's Constant, indicates that a high value of ν implicitly means that a very high amount of energy E of the incident radiation is required to probe into the atomic or subatomic level. Thus, the need arises for an accelerator, where elementary particles (electrons, protons, etc.) are energized to very high levels—to the extent of giga (million) or tera (trillion) electronvolts. Thus, accelerators are said to fall into the regime of *high-energy physics*. In particle accelerators, particles are sped up very nearly to the speed of light. As the speed of a particle gets closer and closer to the speed of light, an accelerator gives more and more of its boost to the particle's kinetic energy, thus generating very-high-energy particles. However, we cannot speed up particles that are exactly equal or beyond the speed of light, according to Einstein's mass-energy relation $E = mc^2$, (see Chapter 2, Section 2.5) though we can get very close to the velocity of light. (In fact, the main injector at Fermi National Accelerator Laboratory accelerates protons to 0.99997 times the speed of light, while the LHC accelerates beams of particles, usually protons, around a 17-mile ring until they reach 99.9999991 percent the speed of light.)

A particle accelerator is a machine that accelerates elementary particles, such as electrons or protons, to very high energies. Electric fields are used to speed up and increase the energy of a beam of the charged particles, which are steered and focused by magnetic fields. The particle source provides the particles (protons or electrons) that are to be accelerated. The beam of particles travels inside a vacuum in the metal beam pipe known as the cavity. The vacuum is crucial to maintaining an air- and dust-free environment that will allow the beam of particles to travel unobstructed via the beam pipes/cavities. Electromagnets steer and focus the beam of particles while they travel through the vacuum pipe.

Electric fields spaced around the accelerator switch from positive to negative at a given frequency, creating radio waves that accelerate particles in bunches. Particles can be propelled along a linear or straight beam line (as in linear accelerators) and are directed at a fixed target, such as a thin piece of metal foil. Alternatively, two beams of particles can be propelled in opposite directions in a circular track and be made to collide with each other (as in circular accelerators/colliders). Particle detectors record and reveal the particles and radiation that are produced by the collision between a beam of particles and the target or particles colliding with each other. The accelerator at SLAC is the largest linear accelerator (LINAC) on the planet, while LHC is the world's largest circular accelerator.

Particles in circular accelerators gradually gain energy as they race through an accelerating structure at a certain position in the ring. For instance, the protons at LHC make 11,000 laps every second for 20 minutes before they reach their collision energy. During their journey, magnets guide the particles around the bends in the accelerator and keep them on course. But just like a car on a curvy mountain road, the particles' energy is limited by the curves in the accelerators. If the turns are too tight or if the magnets are too weak, the particles will eventually fly off course. Linear accelerators don't have this problem, but they face an equally challenging aspect: Particles in linear accelerators only have the length of the track where they pass through accelerating structures to reach their desired energy. So, if we want to look deeper into matter and further back toward the start of the universe—the fundamental questions searched with a particle accelerator—we have to go higher in energy, which means we need more powerful tools. One option is to build larger accelerators: Linear accelerators hundreds of miles long or giant circular accelerators with long, mellow turns. We can also invest in technology. We can develop accelerating structure techniques that will rapidly and effectively accelerate particles in linear accelerators over a short distance. We can also design and build incredibly strong magnets—stronger than anything that exists today—that can bend ultra-high-energy particles around the turns in circular accelerators. The ultimate decision depends on our requirements and the investment that can be made for the purpose.

4.5.4 Some Interesting Facts About the Particle Accelerators

Through modern particle accelerator technology, we have come across many impressive figures of high and low temperatures, size, and energy levels. In 2012, BNL's Relativistic Heavy Ion Collider achieved a Guinness World Record for producing the world's hottest human-made temperature: A blazing 7.2 trillion degrees Fahrenheit. The Long Island-based lab created a small amount of QGP—a state of matter thought to have dominated the universe's earliest moments of birth—that is so hot that it causes elementary particles called quarks to break apart from one another. However, it may be noted that quarks do not exist singly in nature but are combined with other quarks in stable hadrons; for example, proton consists of two up quarks and one down quark, while neutron consists of two down quarks and one up quark bound by gluon, as discussed earlier.

Scientists at CERN have also created QGP, at an even higher temperature. In order to conduct electricity practically without resistance, the LHC's superconductor-based electromagnets are cooled down to cryogenic temperatures. The LHC is the largest cryogenic system in the world, and it operates at a frosty –456.3 degrees Fahrenheit. It is one of the coldest places on Earth. It is even a few degrees colder than outer space, which tends to rest at about –454.9 degrees Fahrenheit. Interestingly, the SLAC's klystron (RF signal generator) gallery, a building that houses components that power the accelerator, sits atop the accelerator. It's one of the world's longest modern buildings. Overall, it's a little less than 2 miles long, a feature that prompts laboratory employees to hold an annual footrace around its perimeter. The energy level achievable with most modern accelerators today is on the order of tens of tera (i.e., trillion) electronvolts (TeV). However, it may be noted that nature produces cosmic rays that have energy some million times greater than the most powerful accelerator on Earth!

Scientists tend to construct large particle accelerators underground. This protects them from being bumped and destabilized, but it can also make them a little harder to find. For example, motorists driving down Interstate 280 in northern California may not notice it, but the main accelerator at SLAC's National Accelerator Laboratory runs underground just beneath their tires. Residents of the Swiss–French countryside live atop the highest-energy particle collider in the world, the LHC. And for decades, teams at Cornell University have played soccer and football on Robison Alumni Fields 40 feet above the Cornell Electron Storage Ring (CESR).

4.5.5 The Future Circular Collider

CERN has launched a very ambitious project: Building the Future Circular Collider (FCC). This project was conceptualized in 2019 and is expected to be completed by 2050. FCC would be approximately four times longer (100 km long) than the 27-km-long LHC. The FCC is expected to be nearly six times more powerful than the LHC (Fig. 4.5). While LHC accounts for proton–proton (p–p) collision at 14 Tev, the FCC would aim for p–p collisions at 100 Tev. The LHC is a hadron (p–p) collider; the FCC would be following three prospective tracks and types of particle collisions: Hadron (p–p) collisions, electron–positron (e^-–e^+) collisions, and proton–electron (p–e^-) collisions.

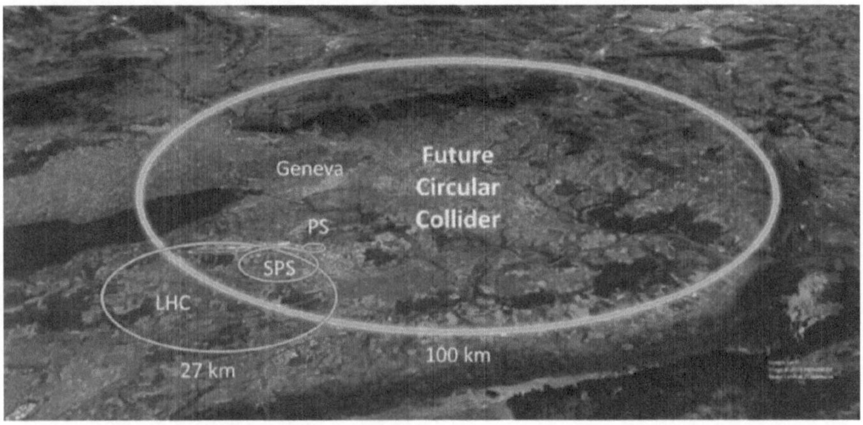

FIGURE 4.5
A view of the future circular collider (FCC) shown in the backdrop of LHC, SPS (super proton synchrotron), and PS (proton synchrotron) accelerators at CERN, Geneva [16].

The FCC aims to reach new and unprecedented levels of energy and luminosity and to replace the LHC once its lifespan ends. This will enable scientists to explore the standard model of particle physics in greater depth by studying known particles, whose parameters and characteristics still remain vague with higher precision than what proton–proton collisions offer. It is planned to be a collaborative effort of different universities, research facilities, and industries from all around the world, hosted by CERN. The cost of the whole project is very high, currently estimated at 20 billion Euros. Although the FCC's aim is to peer into the future of particle physics, many members of the scientific community feel that the money could be better spent elsewhere in the interest of furthering the progress of science and technology and bringing real benefit to humankind.

5

Zooming Out to the Cosmic World of Astrophysics

The cosmic world makes everyone awestruck by its immensity and variety

5.1 Introduction

The diameter of the observable universe is 92 billion lightyears, a total that quite surpasses our limits of comprehension. As Arthur Eddington observed: "Not only is the universe stranger than we imagine, it is stranger than we can imagine." Any common mind naturally receives such numerical figures with a good deal of skepticism and is flabbergasted by the enormity of the universe. Terms such as infinite and God have been coined since ancient times, and for countless people, they remain prevalent even today as a means of expressing our inability to comprehend the inconceivable. Actually, the only truly infinite thing in this universe is our ignorance. We generally use the term *infinite* to mean something that is beyond our comprehension. By doing so, we actually undermine our courage to earn knowledge about the unknown.

We study and decipher the mysteries of the cosmic world much as a detective approaches his work. Through cosmic detective work, useful clues have been obtained, and through use of scientific logic, humankind is unveiling the mysteries embedded in the seemingly infinite cosmos. Huge telescopes have been installed on Earth's surface from time to time, with the latest radio telescope being the Atacama Large Millimeter/Submillimeter Array (ALMA) telescope installed in the Atacama Desert of northern Chile. Also, space-based supertelescopes like the Hubble and Chandra X-ray Observatory, including the COsmic Background Explorer (COBE) telescope using electromagnetic signals in the infrared spectrum, followed by Wilkinson Microwave Anisotropy Probe (WMAP) and the Planck Satellite, help us to obtain reliable clues regarding distant galaxies and nebulae. In 2019, the world's most powerful radio telescope, the Event Horizon Telescope (EHT), imaged the black hole M87* in the constellation Virgo, which is some 55 million lightyears away from Earth. In 2022, NASA's James Webb telescope is expected to be operative, a space-based telescope, which is 100 times more powerful than Hubble telescope. It will be able to see galaxies and other celestial objects which are 13 billion years old, meaning that scientists will be able to see the astrophysical phenomena in the formative stage of the universe (as the age of universe according to the standard model of cosmology is 13.8 billion years).

DOI: 10.1201/9781003215721-5

5.2 The Sky

5.2.1 Introduction

Let us begin with a quote from my own poem entitled *The Sky*:

> The sky is always very dear to me—
> More I get lost in it, I am taken from deeper to the deepest of divine thoughts.
> Its joy is unbound in giving to others—the light, air and water.
> In the vast expanse of the sky—it's the playfulness of colors,
> To which I have got myself atoned once more today.

When we look at the sky in the light of day, we see its color as blue (though sunlight is white, which brightens the sky around our Earth in the daytime). At night our sky is dark, except for whatever brightness is provided when the Moon is visible, but there are innumerable stars in the absence of the Sun during night in a particular hemisphere (eastern or western) of Earth. The rising and setting Sun on Earth is also found to be reddish in color.

The sky in the Moon is always black (Fig. 5.1a, the picture was taken from the orbit of Apollo 11 in 1969), while the sky in the planet Mars is, in general, orange-red (see Fig. 5.1b).

When a rocket in outer space crosses the atmosphere around the Earth, the sky around it changes from blue to black. Thus, it appears that the atmosphere around the Earth with sunlight falling on it may be responsible for the colors we see in the sky; especially the blue color of Earth's daytime sky. The Moon has no atmosphere, and thus its sky is black, even in daytime, as there would be no scattered light. Again, owing to the red iron-rich dusts thrown up in the dust storms that are a frequent feature on Mars, the Martian sky is mostly orange-red in the day. However, in the absence of such a dust storm, the Martian sky is also expected to be blue in the day but darker than the Earth's daytime sky, due to Mars's thinner atmosphere.

FIGURE 5.1
(a) The black sky of the Moon with Earth on its sky [17]; and (b) the orange-red sky of the red planet Mars [18]. (Note: Colored pictures are available in the e-book.)

The immensities and verities of the sky above our heads have long mesmerized poets and scientists alike. Scientists, being naturally rational beings, have from time immemorial tried to understand the phenomena that take place in the sky. In ancient times, however, the sky was loosely thought to be synonymous with the space overhead. In his Queries appended to *The Opticks*, Newton observed that space is "God's Sensorium"—that is, the venue in which God exercises his divine will—in other words (with a more rational view), where the cosmic drama unfolds. Then we have the popular lullaby extolling the starry sky: "*Twinkle, Twinkle, Little Star,. . . Like a diamond in the sky.*" And in the Bible the sky is called "the firmament," the vast solid dome God created on the second day of his creation as a means of separating the Earth from the sky above and called it Heaven. In fact, it was understood to be scaffolding above the Earth, and in that canvas of the sky we see the stars and different constellations. However, following Einstein's formulation of the relativity theory, the sky is no longer mistakenly said to be synonymous with space (the new connotation becoming space–time) and scientists finally have a scientific understanding of various phenomena taking place day and night in our sky and in the skies of the other planets as well.

5.2.2 Why the Day Sky in the Earth Is Blue?

In premodern times, the sky was thought to be blue because of the sunlight that was reflected off the ocean and back into the sky, but the sky is blue even in the middle of the countryside, which of course is nowhere near the sea. The sky is also blue in places that are extremely dry, like the desert, for instance (see Fig. 5.2a).

The first steps in correctly explaining the color of the sky were taken by John Tyndall in 1859. He discovered that when light passes through a clear fluid holding small particles in suspension, the shorter blue wavelengths are scattered more strongly than the red. However, the more effective explanation is based on *Rayleigh scattering*—named after Lord Rayleigh, who studied it in more detail a few years later. Rayleigh showed that the amount of light scattered is inversely proportional to the fourth power of a wavelength for sufficiently small particles. It follows that blue light is scattered more than red light by a factor of $(700/400)^4 \approx 10$; where the wavelengths of red and blue colors are around 700 nm and 400 nm, respectively (1 nm = one billionth of a meter; see Table 5.1). As sunlight reaches our atmosphere, molecules in the air scatter the bluer light effectively but allow the red light to pass through practically unscattered.

FIGURE 5.2
(a) The blue sky over the Sahara Desert [19]; and (b) red sunlight illuminating the clouds at sunset [20]. (Note: Colored pictures are available in the e-book.)

TABLE 5.1

Colors of the Visible Spectrum with Corresponding Wavelengths and Frequencies

Color of Light	Wavelength Interval (Nanometer)	Frequency Interval (Terahertz)
Red	740–625	405–480
Orange	625–590	480–510
Yellow	590–565	510–530
Green	565–500	530–600
Blue	500–485	600–620
Indigo	485–450	620–680
Violet	450–380	680–790

The white light from the Sun is a mixture of all the colors of the rainbow. This was demonstrated by Newton, who used a prism to separate the different colors forming the spectrum. The colors of light are distinguished by their different wavelengths (the distance between two consecutive crests or trough) or frequency (the rate of vibration). The visible part of the electromagnetic spectrum (light is an electromagnetic wave that was established by Maxwell) constitutes red, orange, yellow, green, blue, indigo, and violet. In terms of increasing wavelength, the visible light spectrum reads VIBGYOR, while in terms of increasing frequency, it reads ROYGBIV or Roy G. Biv (see Fig. 5.3 and Table 5.1).

The *color of the sky* is caused by the scattering of sunlight passing through our atmosphere. "The sky" around our Earth is just the atmosphere that we see from underneath. The majority of our atmosphere extends about 16 km upward, and this is where most of the Rayleigh scattering happens, resulting in the blueness of our sky. When we look at the sky during the day, we mainly see the sunlight that has bounced off from the air molecules in the atmosphere and has scattered in all directions. Because blue light scatters more easily than red light, and because the light from the sky is mostly the scattered light, the sky appears to be blue during the daytime. But this extra bending, or scattering, applies just as much to the violet light too, so it is reasonable to ask why the sky isn't purple. The answer to this vital question lies in the fact that the receptors in the retina of the human eye respond most strongly to just three different colors: Red, green, and blue wavelengths. This gives us our color vision, making the sky appear blue rather than purple (see Fig. 5.2a). Interestingly, red is universally used as the "danger" signal light because red is least scattered by fog or smoke, making it visible in the same color even from a distance.

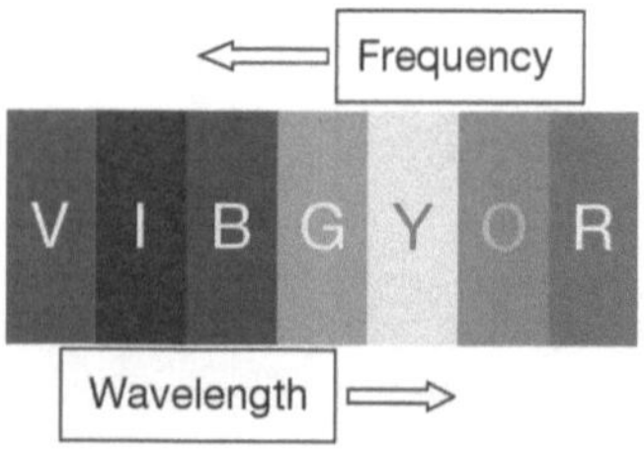

FIGURE 5.3
The seven colors of the visible part of the electromagnetic spectrum.

The "skyglow" that we observe in urban areas is caused by light pollution (due to excessive artificial lights all around the city), which reduces our ability to see stars, planets, and other celestial phenomena in the night. In the absence of light from human sources, skyglow may also be present due to a faint airglow in the upper atmosphere (a permanent, low-grade aurora) and starlight scattered in the atmosphere. Even beyond our atmosphere, faint skyglow is caused by sunlight that is reflected off interplanetary dust (zodiacal light) and background light from faint, unresolved stars and nebulosity. The appearance of streamers of reddish or greenish light in the sky, especially near the northern or southern magnetic pole, is caused by the interaction of charged particles from the Sun with atoms in the upper atmosphere. In northern and southern regions, it is, respectively, called the *aurora borealis* or Northern Lights and the *aurora australis* or Southern Lights.

5.2.3 Why the Night Sky Is Black/Dark—Olbers' Paradox?

Another question we need to address is why our night sky is black/dark (except in a moonlit night), though there are innumerable stars distributed all over the sky, which are much larger and brighter than our Sun. The answer to this question is not simple and has puzzled scientists since ancient times. Indeed, it was only after 1930s that the problem was resolved. Every physicist and astronomer before that, including Newton and even Einstein, believed that the universe was static until the experimental evidence by Edwin Hubble and his team in 1929 established the expanding nature of the universe.

This concept of a static universe suffers from the Olbers' paradox (or the dark night sky paradox), named after the German physician and astronomer Heinrich Olbers. In 1826, Olbers wrote a paper in which he argued that if there are an infinite number of stars, light coming from them should fill the night sky more brightly than the daytime sky. No matter where we look in the night sky, we should be blinded by the brilliant light. Olbers tried to justify the darkness of the night sky by suggesting that the light from distant stars and galaxies is absorbed by intergalactic dust and clouds, a theory that was subsequently found to be not the case.

The final logical explanation used to justify the darkness of the night sky is based on the concept of an expanding universe and the so-called Doppler Effect. According to the Doppler Effect, the light from a moving star changes frequency/color in the lower side of the electromagnetic spectrum (much as the whistle sound from a moving train changes pitch). This results in the observed red-shift causing light from receding stars and galaxies (in an expanding universe) to be shifted to the invisible (to the human eye) infrared and microwave frequency range—essentially causing them to disappear and casting no visible light in the night sky—making the night sky dark. In astronomy, by the term red-shift, we understand the displacement of spectral lines toward longer wavelengths/lower frequency (the red end of the electromagnetic spectrum, see Fig. 3.4 and Table 5.1) of radiation from distant galaxies and celestial objects.

Anyway, Olbers' paradox can be understood in simple mathematical terms as follows. The Olbers' model of the universe assumed a static and infinite universe with a homogeneous distribution of stars that would cause the adding up of the luminosity of all the stars according to the series $1 + 1 + 1 + \ldots . \infty$—a diverging series, making infinite luminosity of the sky even in the night with a star-studded sky. This infinite sum is the nub of the paradox. But when an expanding universe is considered with the red-shift of light received from the distant galaxies increasing with distance from the Earth, the series for overall luminosity calculation becomes 1 + 0.5 + 0.25 + 0.125 + ... This is a converging series that will always be less than 2 (as the increasing red-shift decreases luminosity, say for our discussion, by a factor of half successively).

In this context, we may mention a few words about the term "paradox." We have already come across a number of paradoxes like Bentley paradox (in connection with Newton's law of universal gravitation), Twin paradox (in connection with special theory of relativity), and so forth. In connection with the darkness of the night sky, we have discussed above the so-called Olbers' paradox. The word paradox is derived from the Greek prefix *para-*, meaning "contrary or opposed," and the word *doxos*, meaning "opinion." Thus "paradox" is a statement or tenet contrary to received opinion or belief, especially one that is difficult to believe. Physicist Richard Feynman once said: "The paradox is only a conflict between reality and your feeling of what reality ought to be."

5.3 The Sun and the Solar System

5.3.1 The Birth of Sun and Its Ultimate Fate

The Sun is our nearest star—it only takes 8 minutes for the light from the Sun to reach our Earth. But the light from other stars takes a few years, or even millions or billions of years, to reach us, depending on their distance from the Earth. The closest star to the Earth (next to the Sun), Proxima Centauri, is about 4.2 lightyears away from Earth and the farthest star (to date) from Earth. Icarus, whose official name is MACS J1149+2223, is about 5 billion lightyears away from Earth (it was discovered using the Hubble telescope in May 2016). In other words, the light from Proxima Centauri takes 4.2 years to reach Earth, while 5 billion years are needed for the light from Icarus to reach Earth.

A question naturally comes in the mind of general reader: How we have come to know that the Sun is also a star? With the advent of spectroscopy in 17th century, the German physicist Joseph von Fraunhofer in 1814, using the spectrometer invented by him, first identified dark (absorption) lines, known as Fraunhofer lines, in the spectrum of the Sun and other stars, caused by selective absorption of the Sun's/star's radiation at specific wavelengths by the various elements existing as gases in their atmosphere. It was thus observed that the Sun exhibits similar type of spectral lines (which represent the signature of their gaseous material contents) like all other stars in the universe. Further, in the late 1930s, Hans Bethe theorized that the energy generation in stars and also in our Sun is due to the thermonuclear fusion process that converts hydrogen into helium and other heavier elements. On the basis of these two key scientific features common between Sun and the rest of the stars in the universe, it has been confirmed that our Sun, the king of the solar system, is also a star.

When the Sun is high up in the sky, especially when it is overhead, it appears with its *true color*—yellowish white—as the sunlight then has to travel a relatively shorter distance and only a little blue and violet colors are scattered. But at sunrise and sunset, we see a redder Sun. This is because near the horizon the sunlight has to travel a larger thickness of the atmosphere. Hence, most of the blue and shorter wavelengths are scattered away from the air molecules of the atmosphere, and we are then looking mainly at the Sun and not its scattered light. Therefore, the light that reaches our eyes in such a situation is practically the unscattered red color of the solar spectrum that has a longer wavelength. Sometimes when we observe the clouds being illuminated in the sky during the setting of the Sun, we are rewarded with a beautiful array of red, orange, and pink colors (as shown in Fig. 5.2b).

A few questions naturally arise about the master of the solar system, the Sun. How was the Sun born, and what keeps the Sun with so much heat energy and light living so long? What is the age of the Sun and will the Sun ever die? What would happen to our Mother Earth, being a member of the solar system, should the Sun die?

The Sun is a second- or third-generation star that was born some 4.6 billion years ago from a huge, rotating, giant cloud of gas and dust that was present in space as debris of some supernova explosion. This swirling cloud of gas and dust was made mostly of the element hydrogen, with some percentage of helium and traces of heavier elements. As the cloud swirled, the law of gas physics and gravity set up a battle between two forces: Its gravity tended to crush it to a point, while its resulting pressure tried to blow it apart.

If gravity starts gaining the upper hand, compressing the cloud, it will get hotter; this is because as things get squeezed together, in this case due to gravity, they get hotter. This in turn boosts its pressure, halting gravity's advance. The cloud can remain stable for a long time as a *protostar,* while the squeezing gravity and blowing out pressure balance each other. But this uneasy truce is eventually upset. Because it's hot, the gas cloud glows, radiating away some of the heat energy that gave its pressure. This allows the gravity to compress the gas cloud further and so on. Eventually, the densest part of the cloud at the central portion gets so hot (maybe 15 million degrees centigrade) that thermonuclear fusion process sets in and it becomes much like a huge nuclear reactor where hydrogen is continuously converted to helium (with the intense gravity, however, preventing it from blowing apart) when the full-fledged star is born.

The basic equations for the burning of hydrogen and other elements in the Sun and other stars were worked out by Hans Bethe in 1939, for which he received the Nobel Prize in Physics in 1967 "for his contributions to the theory of nuclear reactions, especially his discoveries concerning the energy production in stars." Anyway, the outer parts of the nascent star are hot enough to shine intensely, and this starlight eventually blows away any remaining patches of the gas cloud, thus making it visible. The time it takes for a protostar to become a star depends on its mass. Our Sun took about 30 million years to become a star from its protostar state. Heavier stars contract faster and are born quickly. A star 15 times the mass of the Sun takes only about 1,60,000 years, while a star 5 times less massive than the Sun can take a billion years to be born as a full-fledged star from a protostar.

Unlike Earth, the Sun and all other stars are not solid but rather a gigantic ball of hot dense gas. The stability of the Sun (or any other star) is ensured by the balance between the energy created by the internal nuclear fire, which tends to be explosive (like hydrogen bomb) to blow it apart, and the gravitational force, which is implosive, that is, it tends to crush it down to a point. Thus, the stability of a star is based on the balance between the forces of gravity, which operates on a large scale, and nuclear fusion, which operates on a small scale. The generation of huge amount of energy with thermonuclear nuclear fusion process within the Sun keep it living with so much of heat energy and the light that it emits.

Life is a constant struggle for everyone, even for the Sun and all the stars alike. Indeed, mortality is the law of nature. A stable star, also called a main sequence star (e.g., our Sun), exists in the stable condition with a balancing act between contraction and expansion.

On the one hand, the force of gravity tries to contract the star (with an implosive force), while the huge amount of energy produced by the nuclear fusion process within the star tries to push everything in the star outwards (with an explosive force). This balance is maintained as long as there is sufficient hydrogen fuel is in the star. When all the hydrogen has been converted to helium, the star begins to die and gravity gains the upper hand (this takes billions of years for a small star like Sun, but only a few million years in case of massive stars). (In the next 5 billion years, the delicate balance between the explosive force caused because of nuclear energy generation in the Sun and the implosive gravitational force due to its mass will be destroyed and the gravitational force will take over.) Under such condition, the gravity forces the star to begin contracting and to fall in on itself. As a result of this immense contraction, the new energy produced at the center of the star gives the star a new vigor to put up a last battle to act against the inward attraction of gravity. In an average star like Sun, at this stage, the helium at the core of the star is converted to carbon (heavier elements, or even iron can be formed in massive stars) by the process of nuclear fusion.

During this dramatic phase of the star's life, nuclear fusion also begins to occur in whatever hydrogen is left in the shell surrounding the core of the star. The outer layer of the shell thus expand from the heat of this fusion energy, making the star to grow bigger. The core contracts further as its fusion fuel is gradually depleted. The contraction releases radiation energy causing the envelope to expand even more, making the star much bigger than it ever was—as much as 100 or more times bigger.

Because the size of the star has expanded, the surface cools down and goes from white-hot to red-hot. Now that the star (the Sun) is redder and physically larger than it was before, it will become a "red giant." In this situation, the surface of the Sun might expand out to the orbit of Mars, even engulfing our planet Earth in the process. One thing is certain: By that time, we most certainly won't be around. In fact, humanity only has about one billion years left on planet Earth. That's because the brightness of the Sun is increasing by about 10% every billion years owing to the burning of hydrogen. Increased brightness means increased heat on our planet. And as the planet heats up, the water on its surface will eventually begin to evaporate, making our planet uninhabitable.

5.3.2 The Solar System

This discussion of the Sun brings up the need to provide a qualitative and quantitative answer to one of our ancestors' most consequential question: *How and when was our solar system created?* This question has been posed throughout the ages and was finally resolved by 1950s. It is really a question about our own origin. To address this question, let us for a moment rewind and replay the birth process of our Sun. It has already been mentioned that our Sun was born from a huge (about 15 trillion miles across) swirling cloud of gas and dust. Suddenly, a disturbance, perhaps caused by the explosion of a nearby star, sent ripples through the cloud, stirring the gas and dust particles, jolting them closer and closer. Eventually, sufficient gravitational attraction occurred, when the particles began to clump together and interact. This interaction caused them to spin increasingly faster (to balance the attraction of gravity), and as the dense cloud spun, it flattened into a disc. The centrifugal force generated by the increasingly faster rotation of the cloud balanced the centripetal force of gravity, preventing the gas cloud from compressing down to a point. Instead, it was crushed into a disc shape.

Why the clumping together of gas and dust particles caused them to spin faster may be understood on the basis of the analogy of skaters spinning on ice. As the skater draws his or her arms closer, he or she spins ever faster. Why the fast spinning would make the gas and dust clumps flatten to a disc shape may be understood on the basis of the pizza baker's flattening of his pizza dough: When it is spun faster, it flattens to the disc shape of the pizza. In about 10 million years, the material at the center of the disc became the protostar, whose primary ingredients were hydrogen and helium gas. Eventually, it became a full-fledged star—which is our Sun. Within about 20 million years, the disc around the protostar will becomes a *protoplanetary disc,* whose primary ingredients are heavier elements such as carbon, oxygen, nitrogen, and silicon. When the hot central portion of the protostar disc forms into a star, the relatively colder outer part is blown away by the newborn star; it is this clump of colder stuff that formed our planets. Thus, planets are formed around stars as a natural by-product of the star formation process. It is to be noted that the spin (or in the physicist's language, *angular momentum*) of the original gas cloud has been taken up by Sun and planets after their formation. Our Earth thus spins about its axis in the anticlockwise direction (if we look down from the north pole of our Earth); and this spinning motion of the Earth about its axis (from west to east) takes approximately 24 hours. This prograde rotation of Earth makes Sun, Moon, and all other celestial bodies to appear to move with respect to Earth from east to west—the reason why we see the Sun rising every day in the east and setting in the west. The rotation of the Earth on its axis is thus responsible for day and night in Earth.

This explanation of the solar system's origin is supported not only by theoretical calculations, but also by the Hubble telescope observation of many other solar systems "caught in the act" of the birth process in various stages. Observations have revealed the presence of a protoplanetary disc around each of the protostars in many parts of our own galaxy and in other galaxies around the universe; as shown in Fig. 5.4.

This proves that the *formation of planets around stars is the rule rather than the exception.* The present astronomical observations confirm that a vast number of solar systems have been detected around stars in many galaxies, including our Milky Way galaxy. However, in case of the so-called binary star systems (where two stars are born together like twins), there is no planet around any of the stars. In fact, in such binary star system one star goes around the others all along their lives. Examples of binaries are Sirius and Cygnus X-1 (Cygnus X-1 being a well-known black hole).

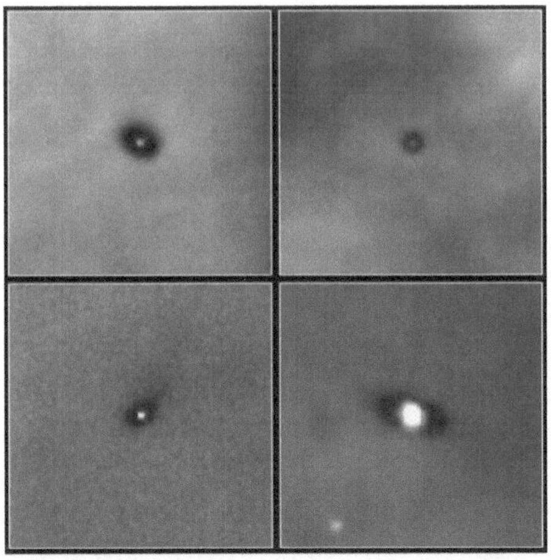

FIGURE 5.4
Four protostars are taking shape with a protoplanetary disc around each of them [21].

FIGURE 5.5
Our solar system [22].

The planets do not burn nuclear fuel and have no light of their own. They shine only through the reflected light of the Sun. This is why we are able to see the planets only in the darkness of the night sky. It is difficult to detect the planets in other solar systems around the universe, even when we use the most powerful telescopes; the reflected light from the planets of those solar systems is extremely faint compared to the light radiated by their respective Suns.

Our Earth and the rest of the planets––Mercury, Venus, Mars, Jupiter, Saturn, Uranus, and Neptune––all were formed from the grains of dust in the whirlwind of the protoplanetry disc around our Sun and in due time coalesced to form different planets. However, the dwarf planets like Pluto, Makemake, and Eris are believed to be formed from the Kuiper belt (a circumstellar ring of icy bodies beyond the planet Neptune). In addition to planets/dwarf planets, our solar system has an asteroid belt, Kuiper belt, comets, and so on, as shown in Fig. 5.5. In terms of the composition and bulk chemistry, we find the smaller rocky planets (Mercury, Venus, Earth, and Mars) in the inner part of the solar system, while the larger gas/ice giants (Jupiter, Saturn, Uranus, and Neptune) reside farther out. The density of the planets decreases with heliocentric distance (i.e., the distance from the Sun). There is an inner ring of small rocky bodies, that is, the *asteroid belt* (which has a chemical gradient) and an outer belt of icy bodies, that is, the *Kuiper belt*. In the center of the solar system is the blazing Sun. There are clearly more refractory materials close to the Sun and more volatiles farther away.

The *asteroids* are the debris left over from the formation of the solar system. While the plants were forming in the solar system, not everything managed to form something big enough to be called a planet. The astronomers call the bits of rock and metal that didn't make the cut, but still orbit the Sun, asteroids (the word "asteroid" means "star-like"), more technically they are known as planetesimals or planetoids. According to NASA's latest count, there are over a million asteroids zipping around the Sun. Many are less than 10 meters across. Some are pretty big, though. The largest, Vesta, is about twice the area of the state of California. Most asteroids are found in the main asteroid belt, orbiting the Sun between Mars and Jupiter. They don't always stay in a regular orbit, however. The effects of Jupiter's immense gravity, combined with the ever-present possibility of crashing into other orbiting objects, can sometimes hurl asteroids out of orbit and send them

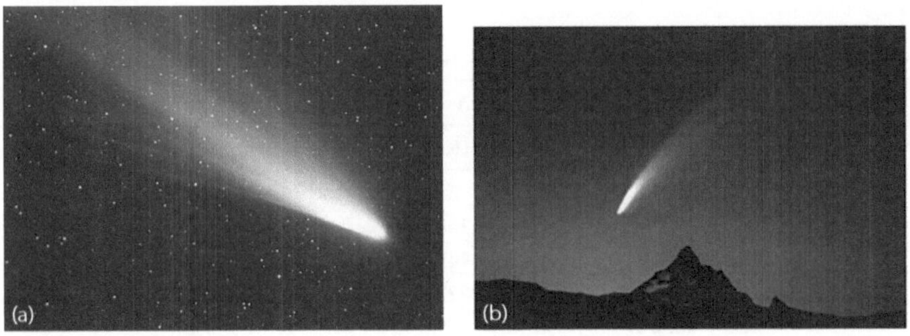

FIGURE 5.6
(a) Halley's Comet as seen in 1986 [23]. (b) Comet C/2020 F3 (NEOWISE) as seen in July 2020 [24].

careening wildly into space. Sometimes the asteroids crash into planets. Earth has been battered by asteroids plenty of times, perhaps most famously when an asteroid crashed into the Yucatán Peninsula and created a disaster that wiped out all nonavian dinosaurs—along with three quarters of the species on Earth. Apophis, an asteroid estimated to be about 340 meters across, caused some concern when it was discovered in 2004. However, after careful study of the object's trajectory, NASA announced in 2021 that we're safe from Apophis for at least the next 100 years. The Near Earth Object (NEO) study center of NASA's Jet Propulsion Laboratory (JPL) keeps an eye on any asteroids that look to be heading our way and might cause harm to our blue planet Earth.

In the solar system, we sometimes find the *comet*. Most famously, Halley's Comet (Fig. 5.6a), named after Edmond Halley, who predicted the orbit of this comet, is a short-period comet visible from Earth by the naked eye. It last appeared in 1986 and will appear next in mid-2061, with a periodicity of appearance in 75–76 years. Indeed, Halley's Comet is relatively close to the Sun. Some comets move around the Sun in such fantastically elongated orbits that they return only after intervals of many centuries or even millennia. The comet has a nucleus ranging from a few hundred meters to tens of kilometers across and is composed of loose collections of ice, dust, and small rocky particles and a tail that may stretch one astronomical unit (which is the average distance from Earth to Sun, i.e., approximately 93 million miles). As of July 2018, a total of 6,339 comets are known, a number that is steadily increasing. However, this represents only a tiny fraction of the total potential comet population possible in the whole universe. In July 2020, we saw the comet C/2020 F3 (NEOWISE), shown in Fig. 5.6b. Another comet, C/2021 A1, more commonly referred to as comet Leonard, discovered by G.J. Leonard at the Mount Lemmon Observatory in January 2021 when the comet was 5 AU from the Sun, became visible in the Earth's sky in December 2021 (on December 12, it came closest to the Earth). Unlike Halley's Comet (which is a short-period comet), the Leonard and NEOWISE both are long-period comets with orbital period of 80,000 years and 60,000 years, respectively.

The Kuiper belt and Oort cloud both are understood to be the origin of the comets. The Oort cloud was first described in 1950 by Dutch astronomer Jan Oort. The Oort cloud, consisting predominantly of icy planetesimals, surrounds the Sun at distances ranging from 2,000 to 200,000 AU (0.03–3.2 lightyears) and resides in the interstellar space beyond the heliosphere. Short-period comets (those with orbits of up to 200 years) are generally accepted to originate from Kuiper belt but the very long-period comets (whose orbits last for even millions of years) are believed to originate directly from the Oort cloud. However,

TABLE 5.2

Distances of the Planets/Dwarf planets of Our Solar System from the Sun

	Average Distance from Sun			
Planets/Dwarf planets	**Million (Miles)**	**Million (Kilometers)**	**Astronomical (Units)**	**Light (Hours)**
Mercury	35.98	57.9	0.386	0.0536
Venus	67.23	108.2	0.721	0.1001
Earth	92.97	149.6	0.997	0.1385
Mars	141.61	227.9	1.52	0.2111
Jupiter	483.74	778.5	5.19	0.7208
Saturn	891.04	1,434	9.56	1.3277
Uranus	1,783.95	2,871	19.14	2.6583
Neptune	2,793.06	4,495	29.96	4.1620
Pluto	3,670.00	5,858	39.05	5.4240
Eris	8,985.01	14,460	96.40	13.3888

Note: 1 km = 0.62137 mi, 1 astronomical unit (AU) = 150 million km. *Astronomical unit* stands for the average distance of the Sun from the Earth, while *light hours* stands for the time it takes light (or any electromagnetic radiation like the microwave and so on) to cross that distance, the velocity of light being 300,000 km per second.

Halley's Comet, even being a short-period comet, is hypothesized to have its origin in the Oort cloud.

In Table 5.2, the distances of the planets and dwarf planets of our solar system from the Sun are given.

Spectacular photographs are available for all eight planets of our solar system (see Fig. 5.7).

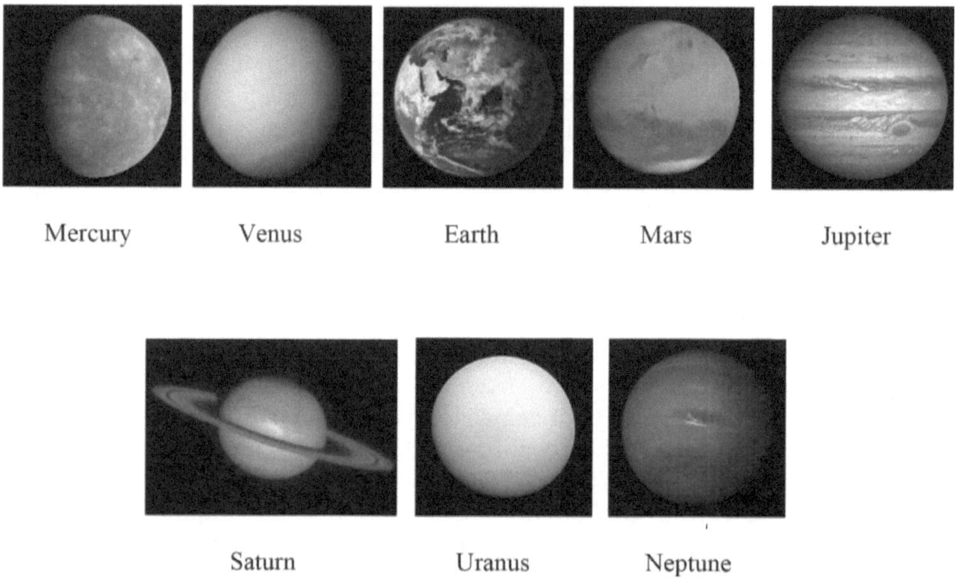

FIGURE 5.7
Photographs of eight planets of our solar system [25]. (Note: Colored pictures are available in the e-book.)

A rare view of Earth rising over the Moon captured by Japanese lunar orbiter Kaguya may be found at the website https://www.youtube.com/watch?v=H1KWtG66lEQ.

5.3.3 The Moon and the Satellites of Other Planets

The Moon is the only natural satellite of Earth. There are also a large number of artificial satellites revolving around the Earth launched in connection with astronomical investigations and for communication purposes. The Moon was born from Earth itself (all the natural satellites of planets are born from their respective mother planet). Shortly after the birth of Earth nearly 4.5 billion years ago in the protoplanetary solar disc, an object that may have been as big as a small planet collided with Earth and knocked a chunk of it loose. That chunk became our Moon. Scientists based this occurrence on their belief that the chemical composition of the Moon is quite similar to that of Earth's outer mantle. The Moon, however, has no atmosphere like our Earth, and the acceleration due to gravity in the Moon is one-sixth of that on the Earth (i.e., 6 kg in Earth will weigh only 1 kg on the Moon). Apart from Earth, other planets and dwarf planets of the solar system also have satellites: Mercury (0), Venus (0), Mars (2), Jupiter (79), Saturn (62), Uranus (27), Neptune (14), Pluto (5), and Eris (1). Recently, 20 more satellites of the planet Saturn were discovered, bringing the ring planets' total number of Moons to 82. Thus, Saturn is now the new Moon-king, surpassing Jupiter, which has 79 Moons.

A question that arises about our Moon is why can we see only one face of the Moon while the other face of it always remains invisible from everyone from the Earth? Also, how do the phases of the Moon take place? Just like the Earth, the Moon rotates about its own axis, and it takes the same amount of time for it to complete a revolution on its axis as it does to orbit around the Earth (about 27 days), exhibiting a trait known as synchronous rotation. It is for this reason that the same lunar hemisphere (the near side) always faces the Earth and we cannot see the other hemisphere (the far side) of the Moon. The Moon's synchronous rotation (or to use a more technical term, the "tidal locking") is not a coincidence. Rather, it is a consequence of Earth's gravity affecting the Moon in much the same way that the Moon's gravity causes tides on Earth.

About 4 billion years ago, the Moon became *tidally locked* with the Earth, and since then it has presented us the same face or the "nearside" (Fig. 5.8a), as seen from Earth, making the "farside" of it invisible from Earth. However, a number of lunar space missions by different countries of the world have made it possible to get nice pictures of the farside of our Moon (Fig. 5.8b). The Moon has striking asymmetries between its nearside and farside in topography, crustal thickness, and composition. Compared to the near side, the far side's terrain is rugged, with a multitude of impact craters and relatively few flat and dark lunar maria ("seas").

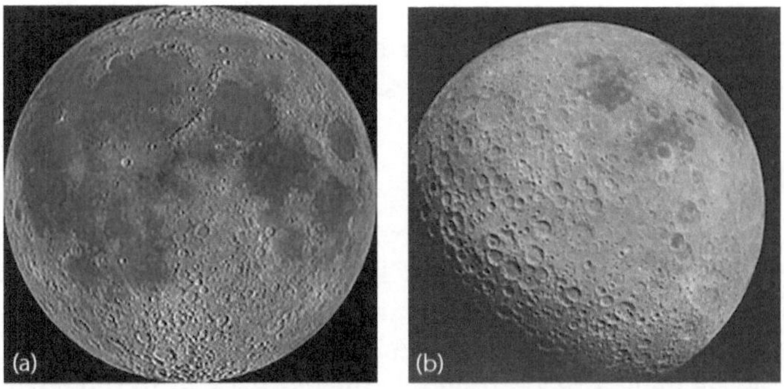

FIGURE 5.8
Photograph of the Moon's (a) nearside [26]; and (b) farside [27].

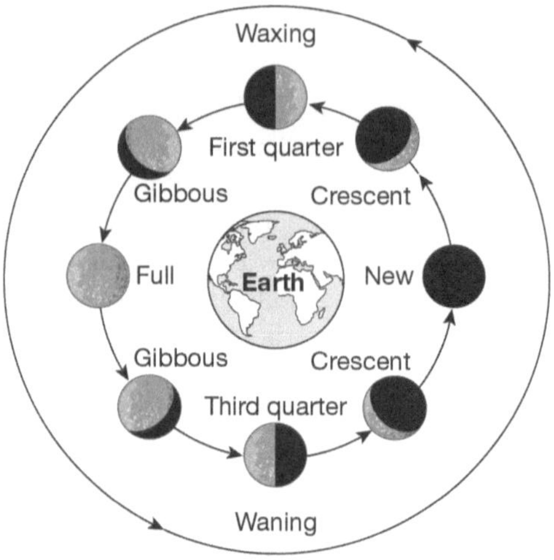

FIGURE 5.9
Phases of the Moon showing waxing and waning.

The phases of the Moon with the waxing and waning of its near side (visible from the Earth) have been well-known views since ancient times. The lunar phases change as the Moon orbits or circles the Earth (Fig. 5.9). The New Moon occurs when the Moon is between Earth and the Sun, when the near side of the Moon facing Earth does not get the sunlight (but its far side is still being lit by the Sun, which is not visible to us). Hence, it is fully dark and the Moon is not visible to us. As the Moon orbits the Earth, we can see more and more of the lighted up part of its near side. With the beginning of the waxing phase of the Moon, we gradually see the crescent, first quarter and the gibbous Moon, until finally the Moon is on the opposite side of the Earth from the Sun and we get a full Moon. As the Moon continues to orbit the Earth, we see less and less of the lighted up part of its near side when the waning phase starts, and we successively see the gibbous third quarter and crescent Moon—with the decreasingly visible part of the lit up nearside of it.

This is how the whole circle of the phases of the Moon visible from the Earth takes place. However, it must be noted that the phases of the Moon do not change owing to the shadow of the Earth falling on it, though some people have this wrong notion in mind. When the shadow of the Earth falls on the Moon due to the Earth coming between the Moon and the Sun (caused by the rotation of the Earth around the Sun), a lunar eclipse takes place. The Earth's shadow blocks sunlight from reflecting off the Moon, thus resulting in the lunar eclipse. The eclipse is either partial or total, depending on whether Earth's shadow partially or completely blocks the sunlight from falling onto the Moon.

5.3.4 Exoplanets

Exoplantes are the planets/worlds that are beyond our solar system and orbit around stars other than our Sun. As of today, more than 4,000 confirmed exoplanets in 3,063 systems, with 669 systems having more than one planet, have been discovered. The nearest exoplanet is Proxima Centauri b; it is located 4.2 lightyears from Earth and orbits Proxima Centauri, which, as we have already noted, is the closest star to the Earth next to the Sun. The first possible evidence of an exoplanet was reported in 1917 but was not recognized as such. An exoplanet orbiting a solar-type star in our home galaxy, the Milky Way, was first discovered in 1995 by Michel Mayor and Didier Queloz. At the Haute-Provence Observatory in southern France, using custom-made instruments, they saw planet 51 Pegasi b (an extrasolar planet orbiting around a main-sequence star and located approximately 50 lightyears away in the constellation of Pegasus), a gaseous ball comparable to our solar system's biggest gas giant, Jupiter. They won the Nobel Prize in Physics in 2019 "for the discovery of an exoplanet orbiting a solar-type star," along with James Peebles "for theoretical discoveries in physical cosmology."

The planet known as K2-18b lies in the constellation of Leo some 110 light-years away from Earth. It is the only world, other than our Earth, known to have water and to orbit its host star (a cold red-dwarf, which is a common type of star) in a habitable zone. The planet has a mass about eight times that of Earth and about one and a half times its pull of gravity, with an average temperature of 40°C. The planet's predicted temperature provides the right conditions for liquid water and complex organic molecules, a condition generally regarded as necessary for the existence of life as we know it on Earth. This discovery has opened up a sharp new focus in the hunt of extraterrestrials and the so-called alien civilization.

In October 2021, the Dutch Low-Frequency Array (LOFAR), the world's most powerful radio antenna, received radio signals from distant planets orbiting around stars outside our solar system. We have long known that the planets of our own solar system emit powerful radio waves as their magnetic fields interact with the solar wind (a stream of charged particles emitted from Sun)—the same process that drives the beautiful auroras we see at the poles of the Earth. The reception of such radio signals from exoplanets with possible biosignature in their atmosphere hints at the presence of "life" on them (may or may not be more diverse than that we see on the Earth). Such new discoveries of astrophysics naturally encourage us to dare to address the biggest question of mankind since ancient times, "Are we alone in this universe?" Or the question asked very frequently these days, "Are the Aliens out there?"

5.3.5 Atmosphere of Different Planets in Our Solar System

At the very beginning when Earth was formed, it had a thick toxic atmosphere resembling that of planet Venus until it cooled and became livable. Initially, Earth was very hot and lacked an atmosphere that could support the life form we find today. In the course of time, it cooled and acquired an atmosphere from the emission of gases from the rocks. The early atmosphere of Earth was such that we could not have survived. It contained no oxygen, but it did have a lot of other gases that would be toxic to humans, though other primitive forms of life could have survived under such ecological conditions. Over millions of years the atmospheric conditions changed, ultimately allowing the survival of some forms of life such as fish, reptiles, and mammals, as well as human beings.

The atmosphere in other planets is not at all conducive to life as we enjoy it on Earth. For a brief account of the constituents of the atmosphere in the eight major planets, see Fig. 5.10. The atmosphere of the dwarf planets Pluto, Makemake, and Eris is not clearly known. The planet closest to the Sun is Mercury, which experiences the full blast of the Sun's heat. The daytime temperature on Mercury reaches 450°C, and the nighttime temperature drops to 170°C. Venus is not suitable for either vacation or honeymoon: It's hostile atmosphere produces acid clouds and intermittent hurricanes with speeds up to 300 km per hour.

Mars, however, the outermost of the rocky/terrestrial planets, gives some hope as an alternative destination. There is strong evidence that Mars once had lakes and even floods

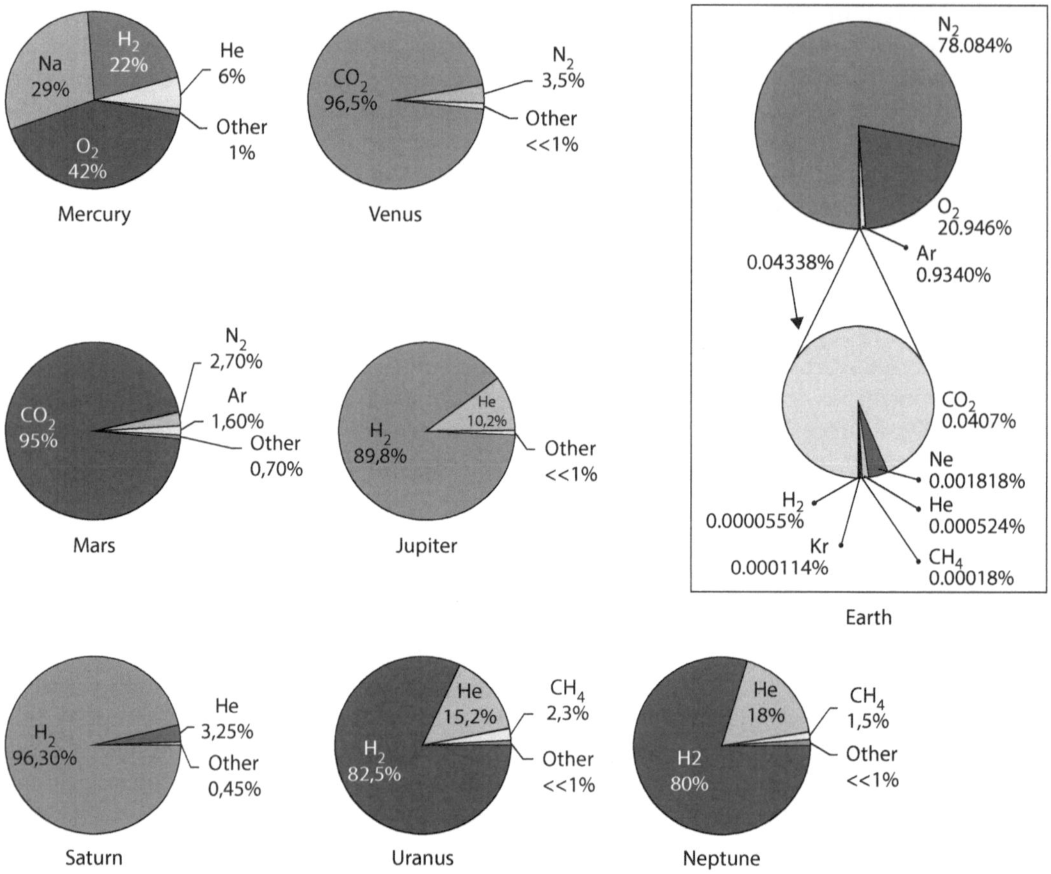

FIGURE 5.10
Atmospheric constituents of eight planets of our solar system [28].

of water. The surface water became vaporized and escaped via its thin atmosphere. The planet's present atmosphere has quite a bit of iron oxide (the chemical composition of rust), making it appear red from a distance—hence its named the red planet. The present space expeditions to Mars confirm that plenty of water is confined below the planet's surface or in the polar caps in the form of ice.

The fifth planet from the Sun, Jupiter, is the largest planet in our solar system: It is 2.5 times the combined mass of all the remaining planets and is thought to be a failed star (one whose mass never became great enough to trigger nuclear fusion process). From Jupiter onward, all the planets are called the gas giants because they have a solid core but no solid surfaces. Thus, such planets are not habitable. Following Jupiter is Saturn, the planet with magnificent rings, which were first spotted by Galileo in 1610.

The seventh member of the solar family, Uranus, is 2.871 billion km from the Sun, with wind speed on its surface of up to 725 km per hour. In 1986, the Voyager 2 spacecraft passed by Uranus and reported that its atmosphere is rich with methane. Sunlight scattered from clouds surrounding Uranus is reflected back through methane layers, giving it a blue-green color. The atmosphere of the eighth planet, Neptune, is composed of hydrogen, helium, and methane, giving it an aqua-blue appearance, which is befitting its namesake, the Roman god of the sea. No place in the solar system is as stormy as Neptune, making it a place unfit for any form of life.

Until 2006, tiny Pluto, named after the god of underworld, was accorded full status as the ninth and outermost planet in the solar family. It has now been demoted to the rank of dwarf planet. Scientists have discovered other rocky, icy, round bodies in the neighborhood of Pluto with names like Makemake, Eris and Sedan, some of which are also dwarf planets like Pluto. However, the secrets of the periphery of our solar system are yet to be explored fully.

Coming back to our Mother Earth, we observe that Earth's atmosphere is composed mainly of nitrogen (78.08%) and oxygen (20.95%); the remaining part of the atmosphere is divided into permanent and variable constituents. The permanent constituents are the five noble gases: Argon, neon, helium, krypton, and xenon and includes materials such as carbon dioxide, hydrogen, methane, nitrous oxide, and radon. The variable constituents of the atmosphere are water, carbon monoxide, and ozone.

Earth's atmosphere is stratified in five different major layers starting from its surface: The troposphere, stratosphere, mesosphere, thermosphere, and exosphere. These five principal layers are largely determined by temperature, and several secondary layers may be distinguished by other properties, as follows.

The *ionosphere* is a region of the atmosphere that is ionized by solar radiation. It is responsible for auroras. During daytime hours, it stretches from a height of 50 km from Earth's surface to 1,000 km and includes the mesosphere, thermosphere, and parts of the exosphere. However, ionization in the mesosphere largely ceases during the night, so auroras are normally seen only in the thermosphere and lower exosphere at night. The ionosphere is used for long-distance radio communication, which is affected differently during the day and at night due to differences in ionization in this layer by solar radiation. Well-known shortwave radio communication in the very high frequency (VHF) band (3–30 MHz), with maximum usable frequency (MUF) being 24 MHz, uses sky wave propagation via ionospheric reflection for long-distance radio communication.

The *ozone layer* is located in the lower portion of the stratosphere from about 15–35 km, and its thickness varies both seasonally and geographically. Ozone in the higher layers of the atmosphere is the product of ultraviolet (UV) radiation acting on oxygen (O_2) molecules. The higher-energy UV radiation from the Sun splits apart some molecular oxygen (O_2) into free oxygen (O) atoms. These atoms then combine with molecular oxygen to form ozone (O_3). At higher levels of the atmosphere, ozone performs an essential function. It shields the surface of the Earth from UV radiation coming from the Sun. This radiation is highly damaging to organisms, causing skin cancer in human beings. In 1980 it was discovered that the amount of ozone in the atmosphere was dropping sharply. The source of the decline was found to be linked to synthetic chemicals such as chlorofluorocarbons (CFCs), which are used as refrigerants and in fire extinguishers. In 1987, the United Nations Environment Programme (UNEP) successfully forged an agreement to curb CFCs' production significantly. The 1995 Nobel Prize in Chemistry was awarded jointly to Paul J. Crutzen, Mario J. Molina, and F. Sherwood Rowland for "their work in atmospheric chemistry, particularly concerning the formation and decomposition of ozone."

5.3.6 Global Warming and Moon's Wobbling

For the last two to three decades, the alarm has been sounded: *Global warming* is bringing ecological imbalance to our atmosphere. The melting of arctic glaciers, the rise of sea levels, frequent occurrences of typhoons and tornadoes, and many more natural catastrophes are already taking place and are widely understood to be due to so-called global warming. The Nobel Peace Prize 2007 was awarded jointly to Intergovernmental Panel on Climate Change (IPCC) and Albert Arnold (Al) Gore Jr. "for their efforts to build up and disseminate greater knowledge about man-made climate change, and to lay the foundations for the measures that are needed to counteract such change." IPCC predicted the scientific basis for the harmful effects of global warming, many of which are already occurring, there by endangering many different species. The IPCC report indicates that "it is extremely likely that human influence has been the dominant cause of the observed warming since the mid-20th century."

The largest human influence has been the emission of greenhouse gases such as carbon dioxide, methane, and nitrous oxide. Climate model projections indicate that during the 21st century, the global surface temperature is likely to increase a further 0.3–1.7°C or even 2.6–4.8°C, depending on the rate of greenhouse gas emissions (they should be kept below 1.5°C, otherwise it would be really dangerous for all of us). The anticipated effects of global warming include rising sea levels, changing precipitation, and expansion of deserts in the subtropics. Other likely changes include more frequent extreme weather events such as heat waves, droughts, wildfires, heavy rainfall with floods, and heavy snowfall. Massive extinctions of species due to shifting temperature regimes are expected. The effect on human civilization, including the threat to food security from decreased crop yields and the abandonment of populated areas due to rising sea levels, is expected to be significant. Because the climate system has a large "inertia" and because greenhouse gases will remain in the atmosphere for a long time, many of these effects will persist not for decades or centuries, but for tens of thousands of years.

The burning issue of climate change has been recognized by awarding the 2021 Nobel Prize in Physics to two climatologists who laid the foundation of our knowledge of the Earth's climate and how humanity influences it. The discoveries being recognized to demonstrate that our knowledge about the climate rests on a solid scientific foundation, based on a rigorous analysis of observations. One half of the prize has been given jointly to Syukuro Manabe and Klaus Hasselmann for "the physical modelling of Earth's climate, quantifying variability and reliably predicting global warming" and the other half to Giorgio Parisi for "the discovery of the interplay of disorder and fluctuations in physical systems from atomic to planetary scales."

Complex systems are characterized by randomness and disorder and are difficult to understand. One complex system of vital importance to humankind is Earth's climate. Syukuro Manabe demonstrated how increased levels of carbon dioxide in the atmosphere lead to increased temperatures at the surface of the Earth. In the 1960s, he led the development of physical models of the Earth's climate and was the first person to explore the interaction between radiation balance and the vertical transport of air masses. His work laid the foundation for the development of current climate models. About ten years later, Klaus Hasselmann created a model that links together weather and climate, thus answering the question of why climate models can be reliable despite weather being changeable and chaotic. He also developed methods for identifying specific signals, fingerprints, which both natural phenomena and human activities imprint in the climate. His methods have been used to prove that the increased temperature in the atmosphere is due to human emissions of carbon dioxide. The work of Giorgio Parisi in 1980s makes it possible to understand and describe many different and apparently entirely random materials and phenomena, not only in physics but also in other, very different areas, such as mathematics, biology, neuroscience, and machine learning.

It is worth mentioning herewith the comment made by Dr. Parisi just after announcement of the coveted prize: "it's very urgent that we take very strong decisions and move at a very strong pace" in tackling climate change. "It's clear for future generations that we have to act now," he added. However, a big question that haunts the mind of every individual of the general public, including the author, is that: "At the climate summit (that takes place almost every day in some part or the other around the world), can the world move from Talks to Action?" Also in this context, a comment by climate campaigner Bill McKibben may be mentioned, "Money is the oxygen on which the fires of global warming keep burning"—it is a serious thought we all must ponder upon, especially the national policy makers around the world (with a humanistic mind and not the materialistic one only).

Now let us discuss how the Moon's wobbling might add to the problem of global warming causing suffering to mankind in the decade of 2030s. A study, conducted in 2021, indicates that 2030s may be a disastrous decade for humankind, especially for those living near coastal areas. The satellite data on global sea-level rise by NASA show that it was only 3.4 mm in 1993, whereas data collected in March 2021 reveal that the figure has climbed to 96.7 mm—an increase by over 10 mm in the last 5 years. This rise in sea-level is expected to be accentuated by an inevitable natural phenomenon known as the Moon's wobbling.

The wobble is a regular undulation of the Moon in its elliptical orbit around the Earth, which occurs cyclically, a cycle that takes 18.6 years to complete. It was first documented in 1728. This phenomenon occurs when the Moon makes its elliptic orbit. Its velocity varies slightly, causing our perspective view of the lunar nearside to appear at a slightly different angle. The Moon's wobble thus impacts the gravitational pull of the Moon on Earth and influences the ebb and flow of the tides. One-half of the 18.6-year cycle of the Moon's wobbling suppresses the tides, which means that high tides get lower, while low tides become higher than normal. Once this cycle completes, the situation flips—in the subsequent cycle, the tides are amplified, with high tides getting higher and low tides, lower!

Though the Moon is in the tide-amplifying part of its cycle at present, however, the sea level rise due to global warming is not yet enough to cross the flooding thresholds with these lunar-assisted high tides to inundate larger areas in the coast lines around the world. But the National Oceanic and Atmospheric Administration (NOAA), United States, reported a total of more than 600 high-tide floods—also called *nuisance floods* or sunny day floods—around the US coastlines in 2019 and this will continue till 2022. Anyway, it will be a different story the next time the lunar wobbling-cycle will come around to amplify tides again, in the mid-2030s. World-wide sea level rise due to global warming is expected to take an alarming turn in another decade from now, if stern control is not brought into action by the nations around the world. The higher seas, amplified by the lunar cycle, will cause a leap in nuisance floods' numbers in almost all US mainland coastlines and many coastal areas of the United Kingdom, including coastal cities all around the world.

5.4 The Stars and the Stellar Evolution

5.4.1 Introduction

Some ancients speculated that stars were small holes in a black sphere (night sky) seen overhead through which the distant light of Heaven shone. Astronomer Giordano Bruno suggested that they were instead objects like our Sun, just much farther away and perhaps had their own planets and civilization. This concept did not go well with the Catholic Church, which had him burned at the stake in 1600 A.D.

Stars have been important to civilizations throughout the ages. They have been part of religious practices and have also been used for celestial navigation and orientation. The oldest accurately dated star chart was the work of ancient Egyptian astronomers in 1534 B.C. The first star catalogue in Greek astronomy was created by Aristyllus (3rd century B.C.). The star catalogue of Hipparchus (2nd century B.C.) included 1,020 stars and was used to assemble Ptolemy's star catalogue. Many of the constellations such as Capricornus, Aquarius, Pisces, Aries, Taurus, Gemini, Cancer, Leo, Virgo, Libra, Scorpius, Sagittarius; and star names in use today are derived from Greek astronomy. In spite of the apparent immutability of the Heavens (a belief of the ancients), Chinese astronomers were aware that new stars could appear. In 185 A.D., they were the first to observe and write about a newly watched star (supernova, now known as the SN 185). The brightest stellar event in recorded history was the SN 1006 supernova, which was observed in 1006 A.D. and described by the Egyptian astronomer Ali ibn Ridwan as well as several Chinese astronomers. The SN 1054 supernova, which gave birth to Crab Nebula, was also observed by Chinese and Islamic astronomers. The Indian astronomers of the 5–6th centuries A.D., including Āryabhaṭa, Varāhamihira, and Bhāskara, were also quite conversent with the astronomical details of the sky.

The first question that comes to mind when we consider the familiar rhyme of our childhood, "Twinkle, Twinkle, Little Star," is why do the stars twinkle or, rather, why do they appear to twinkle when the planets in our solar system that we see in the sky appear to us as fixed (nontwinkling) light specks? The apparent twinkling of the stars is actually caused by *refraction of the light from star* while passing through different layers (that has different temperature and densities) of Earth's atmosphere. However, Hubble telescope installed in space laboratory outside the Earth's atmosphere do not see any twinkling of stars. As the light from a star (considered to be a point source) passes through our turbulent atmosphere, it undergoes refraction many times. When we finally perceive this light from a star, it is the combination of the light rays that reach us directly and also those that get bent due to refraction in the atmosphere. This results in a quick, apparent dimming and brightening—a star's signature "twinkling." But planets are not seen to twinkle; this is because they are much closer to Earth than any star (except the Sun). They are thus considered to be an extended source of light (i.e., a collection of a large number of point-sized sources of light). Hence, the total amount of light entering our eyes from all the individual point-sized sources of which the planet is approximated to will average out to zero; thereby nullifying the twinkling effect.

In 1838, just 32 years after the rhyme was published, the German mathematician and astronomer Friedrich Bessel found the distance to the stars using the *parallax* method. Parallax is a displacement or difference in the apparent position of an object viewed along two different lines of sight, and it is measured by the angle or semi-angle of inclination between those two lines. Thus, the mysterious white dots, the stars, in the dark background of the night sky now have a measurable distance! The star named 61 Cygni has a distance that is almost a million times that of our Sun—a distance so huge that it would take 11 years for this starlight to reach us (while the sunlight reaches Earth in just 8 minutes). The modern technique presently used to measure the distance of the Moon and planets from Earth is the microwave radar technique. In this approach, the microwave (very short radio waves) signal used in radar beams is sent out into space from Earth and bounces off from a planet such as Venus, and the reflection received is detected and processed. Microwave, an electromagnetic radiation, moves at a known velocity (the velocity of light), and the time that elapses between emission and reception of the microwave signal can be measured accurately. The distance traveled by the microwave beam on the round trip and, therefore, the distance of Venus at a given time, can be determined with greater precision than is possible using the parallax determination method.

When starlight observed by a telescope was investigated through a spectrometer invented by Fraunhofer (who observed the so-called Fraunhofer lines in 1814), it was confirmed that stars are made of roughly the same mixture of gases as our closest star, the Sun! Thus, Giordano Bruno was correct: Stars are distant Suns, similar in both their energy output and content. Hence, in just a decade after the rhyme entitled "The Star" was written by Jane Taylor and published in 1806, the stars went from being inscrutable white dots to being giant balls of hot gas whose chemical composition could be evaluated. The spectrum of the star can also be used to obtain information about its temperature, size, mass, magnetism, and so on.

In 1920, the Indian scientist Meghnad Saha was the first to relate a star's spectrum to its temperature through his *thermal ionization equation,* which he developed by using quantum physics and statistical mechanics. The Saha equation links the composition and appearance of the spectrum with the temperature of the light source and can thus be used to determine either the temperature of the star or the relative abundance of the chemical elements investigated. Saha's work has been foundational in the fields of astrophysics and astrochemistry.

The color and life span of stars depend on their size. The bluish stars are the hottest stars, whereas the reddish stars are the coolest ones. Our Sun has a temperature that is somewhere in between that of the bluish and reddish star and has yellow color. The surface temperature of the hottest stars can be over 40,000°C, while the temperatures of the coolest are as low as 2,500°C. Our Sun's surface temperature is about 6,000°C. It would seem that the big stars with their large mass and store of nuclear fuel might live longer than the small ones, but just the reverse is true. The reason is that the big stars are hungrier, and so they exhaust their fuel much faster than the small stars. While the stars like our Sun can live for billions of years; the life span of truly massive stars is just a few million years. The total life span of the Sun is estimated to be approximately 10 billion years, which is now just the halfway point of its life.

5.4.2 The Life Cycle of Stars—The Stellar Evolution

As explained in the last section, stars were formed after the Big Bang in a similar fashion as the Sun, our closest star. However, the fate of different stars, based on their mass, will be different when they die (i.e., cool down due to the exhaustion of their nuclear fuel), giving rise to the *white dwarf, neutron star,* and *black hole.* The life cycle of stars starting from their birth and subsequent fate after their death, depending on the massiveness of the star, is shown in Fig. 5.11a.

The average stars (like our Sun) with death-time or final mass of the cold star less than or equal to 1.4 times the solar mass first become *red giants* and eventually in the final stage of their life return to their expanded and glowing outer shell of ionized gas to the universe in the form of a *planetary nebula.* A planetary nebula is relatively short-lived and lasts just a few tens of thousands of years. Over time, the enriched material from the planetary nebula gets scattered into space and is used by future generations of stars. (The cosmos began to conserve and recycle long before the mankind learnt to do the same.) In this way, planetary nebulae play a crucial role in the chemical evolution of galaxies. Our Sun is likely to end up forming a planetary nebula via the red giant phase. Until such stars convert 90% of their hydrogen fuel to helium by nuclear reaction, they remain as main sequence stars.

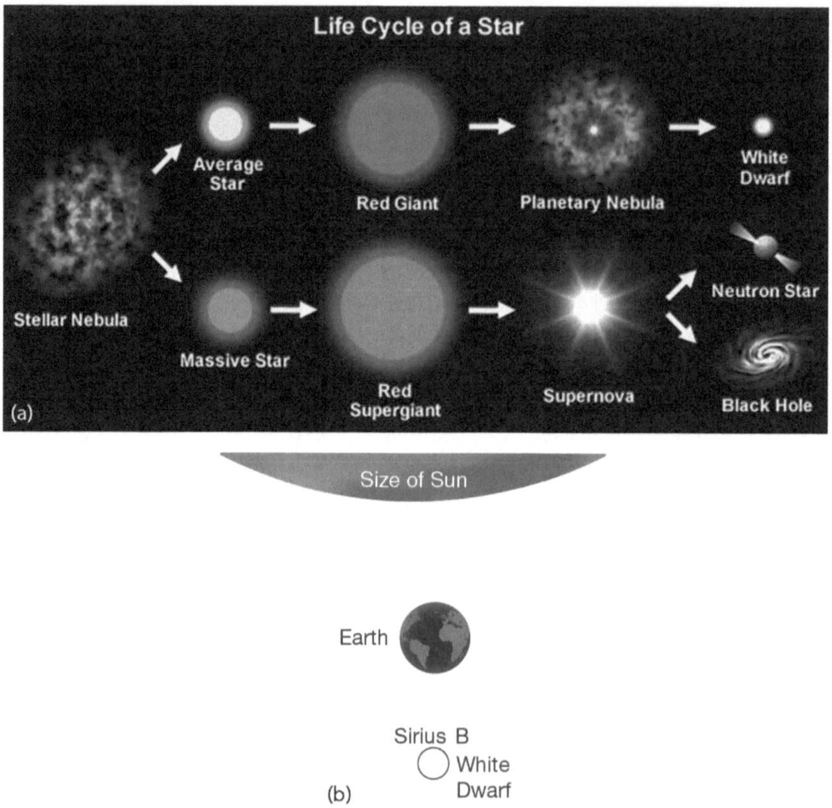

FIGURE 5.11
(a) Life cycle of stars [29]; and (b) relative size of white dwarf (Sirius B) with respect to the Earth and the Sun.

Once all the hydrogen in the core of the star has been burned to helium, energy generation stops and the core begins to contract. This raises the internal temperature of the star and ignites a shell of hydrogen burning around the inert core. Meanwhile, the helium core continues to contract and increase in temperature, which leads to an increased energy generation rate in the hydrogen shell. This causes the star to expand enormously and increase in luminosity. In addition, the star changes color from yellow to red and thus becomes a *red giant*. Our Sun will become a red giant in another 5 billion years or so when with its increased size will engulf our Earth and even the planet Mars!

As the envelope is cast off, the core of the huge star becomes even smaller than the size of the Earth but stupendously dense (see Fig. 5.11b), which is known as the *white dwarf*—a class of low luminosity/faint star and typically white in color. It is given the name white dwarf because it is smaller than Earth. The core of a white dwarf is so dense that one teaspoonful of it on Earth could weigh about 5 tons—as much as a truck. Our closet white dwarf Sirius B is found to have a density of 4 million grams per cubic centimeter. Inside the white dwarf, the tightly compressed electrons produce a quantum repulsive force known as electron degeneracy pressure, which balances the extremely high compressing gravitational force. Thereupon the star again achieves a sort of stability, even though all its nuclear fuel has burned out.

White dwarfs are quite small in size and dim compared to the ordinary stars, and their number is also quite small compared to the huge number of normal stars that decorate the cosmic canvas. Astronomers have observed a number of white dwarf stars, the nearest known white dwarf star is Sirius B, a companion of Sirius. Together, the two form a binary star. Sirius and Sirius B are the sixth and seventh nearest stars to Earth, 8.6 light years away, and Sirius is the brightest star in our sky. Sirius B orbits Sirius just as Earth orbits the Sun, but Sirius B requires 50 years to complete one orbit, while Earth needs only one year to go around the Sun.

The massive stars go through the stages of red supergiant and supernova explosion (details of which is given in Section 5.4.4) and ultimately settle as a neutron star or a black hole (see Fig. 5.11a). In case of the relatively less massive star, but more massive than average stars (with the death-time mass that is typically more than 1.4 but less than 3 solar mass), electronic degeneracy pressure is not sufficient to counterbalance the crushing gravitational force needed to save the star from collapse. As in the case of such stars, the nuclear fuel is exhausted (i.e., when the star reaches its turnoff point), the gravitational crushing force reduces its volume to an exorbitant extent and the density of the star becomes extremely high (10^{11} grams per cubic centimeter). Under such stupendous density of matter in the star and the extreme reduction of its volume, the electron orbits in individual atoms of the matter of the star vanish and the electrons are forced to remain in the crust around a nucleonic core of the star as free electrons. Also, the individual neutrons of atoms find themselves actually "touching" one another, and the star becomes just a solid mass of nucleonic matter. Here again, another quantum degeneracy pressure, known as neutron degeneracy pressure (that prevents two neutrons to be in the same place at the same time, according to Pauli's exclusion principle), comes into play that counterbalances the gravitational crushing force, giving the star a second chance to remain as a stable star known as the *neutron star.*

To imagine the enormous contraction necessary and the exorbitant density that a neutron star assumes, think of compressing the entire mass of the Sun (which is 3,33,000 times that of the mass of Earth) down to the size of a city like Manhattan (with a radius of only 10 miles or so). They are thus much smaller but denser than the white dwarf. At the time they were predicted, there was no way that the neutron star could be observed with optical telescope as they are thousands of lightyears away and with just 20 miles diameter. Their presence was, however, indirectly proved by the discovery of pulsars much later with radio telescope. Pulsars are now known to be the spinning, magnetized neutron star.

Further, when the death-time mass of a massive star is more than 2.5 or 3 times the solar mass, not even neutron degeneracy pressure within the star can save it from its ultimate collapse due to the implosive force of gravity, and it is destined to reach the state of the black hole. The size of black hole is determined by the Schwarzschild radius and its density becomes ideally infinite at the point of singularity (see Fig. 3.5b (i)). For the astronomers of the 1930s–1960s, this was simply beyond all imagination. However, subsequent work on it, especially astronomical observations using the latest generation telescopes in the 1970s, established the practical existence of the black hole.

As noted in Chapter 3, the radius of a black hole that forms out of a star when it dies (after its nuclear fuel is exhausted) is determined with the Schwarzschild radius. The distance from the center of a nonrotating (static) black hole to the event horizon (which is spherical for the static black hole while oblate for rotating black hole) is known as the Schwarzschild radius. The Schwarzschild radius of a black hole can be determined from the formula: $R_{Sch} = 2GM/c^2$ or, equivalently, using the simple formula $R_{Sch} = 3\ M_\odot$, where $M_\odot$ is the solar mass and R is the radius of the event horizon in kilometers. Using this formula, we may note that if our Sun gets converted to a black hole, it will settle with a radius of 3 km, while our Earth will reduce to the dimension of a ping-pong ball (having a radius of 9 mm).

Our nearest black hole is Sagittarius A*, which is at the center of the Milky Way, 25,000 lightyears away from Earth. With its 4 million solar mass, its radius is 12 million km.

Since the 1990s, astronomers have known that the stars which are supermassive at its birth do eject enormous amounts of mass in the form of energy as they age and die. They eject so much, in fact, that most stars born with masses as large as an 8 solar mass lose enough energy to wind up in the white dwarf graveyard. Moreover, those stars born between about 8 to 20 solar mass after loosing enough energy winds up as neutron stars. But supermassive stars heavier than about 25 solar mass, even after loosing energy over time, remain so heavy when they die that their quantum-mechanical (electronic/nucleonic) degenerate pressure provides no protection against a superimplosive gravitational force. Thus, when they exhaust their nuclear fuel and begin to cool, gravity overwhelms all sorts of explosive pressures and they implode to form black holes.

Let us now discuss how a gravitational force known to be much weaker than a nuclear force can overpower nuclear force in a black hole. Real insight into this issue may be obtained as follows. Nuclear force is known to be more powerful than gravitational force when we consider only a few atoms or atomic nuclei at our disposal. But when one has several solar masses' worth of atoms (10^{57}atoms), as in the case of massive stars, then the gravitational force of all the atoms put together can become overwhelmingly more powerful than their nuclear force. This is why it is that when a massive star dies, its huge gravity will overwhelm the repulsion of its atomic nuclei and crush them to form a black hole.

5.4.3 The Chandrasekhar Limit

The fate of a star after its death is determined with the *Chandrasekhar* (mass) *Limit*, a concept named after Subrahmanyan Chandrasekhar, an Indian American astrophysicist and nephew of the great Indian physicist C. V. Raman, who is famous for the *Raman Effect* and who won the Nobel Prize in Physics in 1930. Subrahmanyan Chandrasekhar's mathematical treatment of the stellar evolution, based on the concept of quantum mechanics in addition to relativity theory, yielded many of the best current theoretical models of the later evolutionary stages of all stars, including very massive stars leading to black holes. Chandrasekhar's work conducted during the 1930s was initially ignored by astrophysicists because such a limit would logically require the existence of a black hole, with ideally zero size and with infinite density, and having a stupendous gravitational attraction—all of which were considered scientific impossibilities at that time.

Chandrasekhar's work on the Chandrasekhar Limit aroused controversy, primarily because of the extreme opposition of Arthur Eddington, the renowned British astrophysicist who contributed to understanding the luminosity of stars (*Eddington's Limit*) and speculation regarding the fusion process in stars. Although Niels Bohr, William A. Fowler, Wolfgang Pauli, and other physicists agreed with Chandrasekhar's analysis, owing to Eddington's status in astronomy, they were unwilling to publicly support Chandrasekhar. Eddington argued that Chandrasekhar's theory was just mathematical game-playing and had no relation to the realistic/experimental basis of astrophysics. He claimed that there was no such thing as Chandrasekhar's relativistic degeneracy (which combined relativistic mechanics with nonrelativity, i.e., the quantum theory). He commented that "there should be a law of nature to prevent a star from behaving in this absurd way." Eddington used the full force of his fame and oratorical skills to demolish the young Indian.

It was not until 1972, with the first positive identification of the black hole, that astrophysicists all over the world started appreciating the importance of the Chandrasekhar

Limit. Chandrasekhar was finally awarded the Nobel Prize in Physics in 1983 with William Alfred Fowler, under whom he was a research student while at Cambridge, for "theoretical studies of the physical processes of importance to the structure and evolution of the stars."

Chandrasekhar studied the mathematical theory of black holes during 1971–1983. He also published the treatise, *The Mathematical Theory of Black Holes*, in 1983, the same year in which he received the Nobel Prize. This treatise is considered to be a mathematical handbook for black hole researchers from which they can extract methods for solving any black-hole perturbation problem.

It would be worthwhile to discuss the evolution of Chandrasekhar's work. As an Indian graduate student, he set sail for England in 1928 to study astronomy. During his voyage, he had pondered the possible state of a star when it collapses to a small size under the effect of gravity once its nuclear fuel is exhausted. It may be recollected here that a star is said to be in hydrostatic equilibrium—balancing the implosive gravitational force due to its mass with the explosive gas pressure caused by the heat generated by the thermonuclear reaction of conversion of hydrogen to helium that goes on in the core of a star. When all the hydrogen is converted into helium, the next element in the chain, carbon, forms. The temperature required for hydrogen fusion is about 15 million Kelvin (K), and for helium fusion to carbon it is about 100 million K. Once all the helium burns out into carbon, what is left is an inert carbon core. The temperature required to fuse carbon is 500 million K. However, small to midsized stars do not have the potential to host a full-scale carbon fusion. Thus, what physical process will guarantee the stability of the star, if any? It is here that the Chandrasekhar (mass) Limit enters.

Chandrasekhar was interested in the final states of collapsed stars that is determined by electron degeneracy pressure, as introduced by Ralph Fowler in 1926 in his article "On Dense Matter." In that article, Fowler had attempted to resolve Eddington's Paradox on the final stability of the white dwarf star. When Chandrasekhar began his study by using Eddington and Fowler's work, he soon realized that they had not included relativity in their calculations. When he revised their equations to include relativity in addition to the electron degeneracy pressure of quantum mechanics, he found that above a certain limit there was no solution. His study, based on quantum mechanics and relativity (known as relativistic degeneracy), showed that if a star's death-time mass M is more than or equal to 1.4 $M_\odot$, where $M_\odot$ is the solar mass, then there could be no balance between electron degeneracy pressure and the crushing gravitational force; hence, the star would continue to collapse. Thus, it is that in astronomy, 1.4 $M_\odot$ is known as the Chandrasekhar (mass) Limit, that is, a star with mass more than that value cannot survive as a white dwarf star. Ultimately, however, astronomical studies detected white dwarf stars with mass that can range from 1.2 $M_\odot$ to 1.44 $M_\odot$.

What prompted Chandrasekhar to use relativity in addition to quantum mechanics—in other words relativistic degeneracy—to arrive at the Chandrasekhar (mass) Limit? Matter is said to be degenerate when it is exceptionally dense and the properties of such matter are determined by quantum mechanics. When a star dies due to exhaustion of nuclear fuel, the core of the star becomes heavily compressed due to the unbalanced and implosive gravitational force acting on it. Under this condition, the atoms and electrons in the core get closer and closer to each other. But electrons cannot get closer to each other, violating Pauli's exclusion principle of quantum mechanics. According to that principle, all fermions (i.e., particles like electrons, protons, and neutrons having half-integer spin) must obey the condition that no two particles having the same spin can occupy the same quantum state. Thus, the condition of electrons/neutrons in the degenerate matter of cold stars can be viewed in the following way.

In a highly dense/degenerate star, the volume available for electron to remain in its cell becomes accordingly smaller. Now if we think of the electron as a wave (as per de Broglie's concept), the reduction in volume of the space surrounding the electron means that the wavelength (λ) of the electron becomes smaller to confine it to the smaller volume, making it more energetic (as per Planck–Einstein equation $E = h\nu$, where $\nu = c/\lambda$). Thus, electron will fly about at greater speeds in its cell and by bumping with other particles gives rise to the so-called degeneracy pressure. This pressure is an unavoidable consequence of the laws of quantum mechanics. The issue of the degeneracy pressure can also be understood in terms of the Heisenberg's uncertainty principle. As the electrons in a degenerate white dwarf star are forced to confine in a small space (i.e. Δx becomes very small), the momentum (Δp) of the electron correspondingly increases since $\Delta x \, \Delta p \geq \hbar$ (according to Heisenberg's uncertainty principle), and this contributes to the degeneracy pressure. Thus, when viewed with the de Broglie's concept of electron as wave or with Heisenberg's uncertainty principle, a huge flying-out/repulsive pressure within the white dwarf star (which is of course temperature-independent as is the case with a cold star) is created—this explosive pressure is the so-called electron degeneracy pressure that plays in a white dwarf star and stabilizes it.

When the density of the cold star becomes about 10^7 grams per cubic centimeter due to an implosive gravitational force (as is the case with the average star at cold condition), the erratic speed of the electron reaches even closer to the velocity of light. There is no way to get rid of this quantum mechanical-cum-relativistic degeneracy pressure in a white dwarf star; it is an inevitable consequence of confining the electrons to an extremely small space with the increasing density of the cold star, as discussed earlier. However, in ordinary matter with ordinary densities, the electron degeneracy pressure is so tiny that one never notices it, but at the huge densities of the white dwarf star, it is enormous. It was on the basis of this physical concept of quantum mechanics and relativity applied together to the cold average star that Chandrasekhar worked out his now famous Chandrasekhar (mass) Limit. Subsequently, it was found that similar to electron degeneracy pressure (which counters the gravitational collapse of a white dwarf star), the *neutron degeneracy pressure* works in a neutron star, following the same quantum-mechanical phenomenon (as neutrons are also fermions). Anyway, in the neutron star, the gravitational collapsing force is much higher, making the massive cold star more degenerate or with much higher density of the matter in the cold form of the star which has been discussed earlier.

A question naturally arises why Chandrasekhar used relativity in addition to quantum mechanics to arrive at the Chandrasekhar (mass) Limit though his predecessors Arthur Eddington and Ralph Fowler did not use relativity (and was also not able to come up with such a mass limit for white dwarf star). The reason for this may be understood as follows. Normally the velocity of electron in an atom is nonrelativistic, that is, much less than the velocity of light (as discussed in the Bohr's atomic model, Section 2.7.3, Chapter 2). But because of the extra energy/momentum gained by electrons from compaction/degeneracy in white dwarf star, the electron velocity becomes relativistic and thus the electron degeneracy pressure is relativity-limited (in addition to quantum mechanics)—which leads to a specific limit of mass $M \leq 1.44\ M_{\odot}$ for such stars. If the electron speed remained nonrelativistic in white dwarf star, there would have been no such upper limit of mass of a dead star (the reason why Eddington and also Fowler could not solve the problem of white dwarf star satisfactorily). As mentioned above, the experimental data of astronomical studies on white dwarf star supports the Chandrasekhar (mass) Limit.

5.4.4 Supernovae, Gamma Ray Burst and Novae

5.4.4.1 Supernovae

The supernova is a transient astronomical event that occurs during the last evolutionary stages of a massive star and is caused by the sudden gravitational collapse of a massive star's core releasing gravitational potential energy as a stupendous explosion. It was mentioned earlier that a massive star initially becomes red supergiant, then supernova, and finally settles either as a neutron star or a black hole depending on the death-time mass of the star (see Fig. 5.11a). Once all the hydrogen in a massive star's core fuses to become helium, it becomes a red supergiant that has a helium core surrounded by an expanding shell of gas. Over the next million years, its core shrinks as nuclear reactions progressively create other heavier elements (like carbon, oxygen, neon, magnesium, and silicon), eventually forming a center made of iron where the core temperature rises to 5 billion degrees centigrade or more. (According to a "law of nature," As things get squeezed together, they get hotter!) Anyway, iron cannot burn to create the energy of nuclear fusion. Thus, in this extreme state of the star, the core of the star is left with iron around which the layers of silicon, magnesium, neon, oxygen, carbon, helium, and at the periphery a thin layer of burning hydrogen remains resembling the onion shells. When the iron core of the star exceeds 1.4 solar mass (the Chandrasekhar mass Limit), the electron degeneracy pressure fails to balance the gravitational implosive force of the star and thus the electrons cannot remain in the atomic orbit and take refuge to a crust around the core of the star now made of iron nuclei. This sudden core collapse of the star, with atomic orbits being crushed, by extreme gravitational implosive force, makes the core of the star to reduce from about the size of Earth to about a sphere of about 20 miles in diameter. The core collapse takes place in a split-second when the outer shells of the star ruptures out vigorously and sends out an expanding shockwave in the surrounding interstellar medium that travels at speeds of as much as 30,000 miles per second. Finally, the iron core nuclei with crust of free electrons settle as a neutron star where the neutron degeneracy pressure balances the implosive gravitational crushing force of the star (as mentioned in Section 5.4.2).

The supernova explosion is of mind-boggling magnitude and the energy released during this explosion can be many times larger than the energy produced by our Sun during its entire life of 10 billion years, and the brightness of the light produced can be visible as a bright object in the sky even from 10,000 light years away. Such an exploding star is known as a *supernova*. The word *supernova* was coined by Walter Baade and Fritz Zwicky in 1932. They first propounded the idea of the neutron star and supernova, and they also suggested that cosmic rays are primarily produced from a supernova explosion.

The supernovae explosions are vital events in the development of our universe as it controls the evolution of galaxies. The birth of our solar system is believed to be initiated by the disturbance created when the shockwave from a nearby supernova passed through the lanes of gas and dust molecules in a spiral arm of our Milky Way galaxy. As mentioned earlier, the supernova explosion gives out all forms of elements like carbon, oxygen, neon, magnesium, silicon, etc., that was formed in the shells of the dead massive star which in due course of time get contained in the new star/galaxy formation. This reminds the meaning of the term "universe," derived from Latin term *universum,* whose derivative terms are *unus* (meaning one or unique) and *versus* (meaning to turn). In other words, the universe may be thought to be a unique place where everything reforms and recycles to turn into its universal rhythm of oneness.

It has now been established that the acceleration of particles to high speeds by shock waves in a gas-cloud remnant of supernova explosions, long after the explosions are finished, actually produces cosmic rays. During flare-up, these superluminous novae

FIGURE 5.12
The S. Andromeda supernova explosion in the Andromeda galaxy [30].

(supernovae) shine with a billion times more luminosity than our Sun. However, the debris of the explosion continues to shine very brightly for a few days and remains quite bright for a few months. In Latin, *nova* means "new," referring astronomically to what appears to be a temporary new bright star—a truly spectacular object to be seen in the cosmic canvas. S. Andromeda is a supernova in the Andromeda galaxy (which is much beyond our Milky Way galaxy). It was observed in 1885, outshining all the rest of the Andromeda galaxy. It had brilliance equal to 100,000 ordinary novae, or nearly 10 billion times that of our Sun (see Fig. 5.12). The supernova explosion is so stupendously catastrophic on the Earthly scale that if that had happened in the Sun (of course, there is no chance for such occurrence), Earth's oceans would start boiling immediately and all life would be wiped out instantaneously.

Supernova figures in many myths and legends, including that of the "three wise men" who followed a bright object/star (actually a supernova explosion) to reach the birthplace of Jesus of Nazareth.

The earliest recorded supernova, SN 185, was viewed by Chinese astronomers in 185 A.D. The brightest recorded supernova was SN 1006, which occurred in 1006 A.D. and was described by observers across China, Japan, Iraq, Egypt, and Europe. The widely observed supernova, SN 1054, produced the Crab Nebula, and it was from this supernova that our solar system was born. Supernovae SN 1572 and SN 1604, the latest to be observed with the naked eye in the Milky Way galaxy, had notable effects on the development of astronomy in Europe because they were used to argue against the Aristotelian idea that the universe beyond the Moon and planets was static and unchanging. Kepler began observing SN 1604 at its peak on October 17, 1604 and continued to make estimates of its brightness until it faded from the naked eye a year later. It was the second supernova to be observed in a generation (after SN 1572, which was seen by Tycho Brahe in Cassiopeia).

Anyway, in 1987, a supernova, dubbed as SN 1987A, a type II supernova, was detected in the Large Magellanic Cloud (a dwarf satellite galaxy of the Milky Way), about 1,70,000 light years away from the Earth, became visible from the Earth. SN 1987A is one of the brightest stellar explosions detected since the invention of telescope more than 400 years ago. Neutrinos emitted from the gigantic burst of SN 1987A reached the water tank of physicist Masatoshi Koshiba, of which his research group detected 12 neutrinos, confirming the theories of supernovae. Koshiba was awarded 2002 Nobel Prize in Physics for "pioneering contributions to astrophysics, in particular for the detection of cosmic neutrinos," which he shared with Raymond Davis Jr. and Riccardo Giacconi.

5.4.4.2 Gamma-Ray Bursts (GRB)

The supernovae are believed to be the source of the so-called *gamma-ray bursts* (GRBs). The GRBs—the brightest and most powerful explosions in the universe—are believed to occur when a massive star implodes due to the exhaustion of its nuclear fuel and then goes through the stage of supernova explosion to form ultimately a neutron star or a black hole. Such explosions are so violent that they release as much energy in just a few seconds as the Sun does in its whole lifetime of 10 billion years. The bursts can last from 10 milliseconds to several hours while their longer "afterglow" can be observed even from the Earth. A recently proposed plausible hypothesis on the origin of the bursts is the "magnetic model," which indicates that when the star explodes as supernova, its large magnetic field collapses, unleashing the tremendous energy that powers a GRB. After an initial flash of gamma rays, a longer-lived "afterglow" is usually emitted over a whole range at longer wavelengths in the electromagnetic spectrum (X-ray, ultraviolet, optical, infrared, microwave, and radio wave) from a GRB.

GRBs were first detected in 1967 by the Vela satellites of United States, which had been designed to detect covert nuclear weapons tests; this was declassified and published in 1973. However, the terrestrial gamma-ray flashes (TGFs) were first discovered in 1994 by BATSE, or Burst and Transient Source Experiment, on the Compton Gamma-Ray Observatory, a NASA spacecraft. In the early 2000s, the Ramaty High Energy Solar Spectroscopic Imager (RHESSI) satellite observed TGFs with much higher energies than those recorded by BATSE. Anyway, both the experimental studies on astronomical GRB and their theoretical modelling continued henceforth. NASA's Hubble Space Telescope caught the fading afterglow of GRB 190114C in 2019, across a huge range of frequencies in the electromagnetic spectrum, from a massive star in a galaxy 4.5 billion light years away near the Fornax constellation. This had the highest energy ever observed for a GRB—the afterglow light emitted soon after the burst was found to be a tera-electron volt radiation!

5.4.4.3 Novae

The *Novae*—more popularly known as the exploding stars—are believed to be part of the binary star system. A binary star is formed when a planetary system like ours is not able to form during the making of a star. Under this situation, a pair of stars is created, consisting of a white dwarf and either a main sequence, subgiant, or red giant star, orbiting one another in a cosmic waltz. In course of time, if the orbital period of the white dwarf star falls in the range of several days to one day, it becomes close enough to its companion star to start drawing accreted matter onto the surface of itself, which creates a dense but shallow atmosphere around it. This atmosphere, mostly consisting of hydrogen, is thermally heated by the hot white dwarf and eventually reaches a critical temperature, causing ignition of rapid runaway fusion. Under this condition, the white dwarf star explodes vigorously as a nova (a transient astronomical event that causes the sudden appearance of a bright, apparently "new" star, that slowly fades over several weeks or many months). At its peak, the brightness of a nova becomes 200,000 times that of the Sun. This brightness peak lasts only a few days. The force of the explosion puffs a portion of the star's matter out into space, and with it goes much of its energy. What is left of the star begins to collapse again, and as it becomes like a punctured balloon, it dims. It takes perhaps several months to get back to its pre-nova brightness, after which it continues much as it did before. After a quiescent period, a nova is quite capable of undergoing another explosion with intervals of anywhere from 10 to 100 years. Nova Persei, discovered on February 21, 1901, by the Scottish clergyman Thomas Anderson, exploded in 1901 but brightened again in 1966. The star T Pyxidis in the constellation Pyxis, following its first occurrence in 1890, has been observed to have had five such nova-like peaks of brightness: In 1902, 1920, 1944, 1966, and 2011. This is a reminder of the volcanoes on Earth that occasionally erupt. T Pyxidis is known to be a binary star system, and its distance is estimated at about 15,600 lightyears from Earth.

5.4.5 Classification of Stars

There are trillions and trillions of stars in our observable universe. Astronomers have classified them in various ways but most commonly they are classified in terms of their surface temperature and luminosity. Table 5.3 presents the stellar classification over different temperature ranges, together with their color and luminosity; the mass of stars compared to the solar mass is also included.

There are seven classes of stars (subclasses are also there) that do not follow the alphabetic order but are instead ordered as O, B, A, F, G, K, M; the M-type stars or the reddish stars are the coolest and faintest stars, whereas the O-type stars or the bluish stars are the hottest and brightest stars. The stellar classification sequence may be remembered by using the following easy mnemonic: *O*h *b*e *a* *f*ine *g*irl [or guy], *k*iss *m*e. Our Sun is a G-type star, while Proxima Centauri is an M-type star. Both the Sun and Proxima Centauri are main sequence stars (almost 90% of the stars in the universe are main sequence stars; unlike the protostar, white dwarf, neutron star, and black hole).

In astrophysics, stars are often modeled as black bodies, although it is not always a good approximation. The temperature of a star can be deduced from the wavelength of the peak of its radiation curve (see Fig. 2.5), the black body radiation curve that follows Planck's radiation law. Both the COBE satellite and WMAP have accurately measured the radiation spectrum of stars, which remarkably fits to the black body curve according to Planck's radiation law, from which the color and temperatures of stars have been estimated with great precision.

A question might arise as to why we don't find any green or purple (i.e., violet) stars in the cosmic canvas. The color of a star is linked to its surface temperature. Approximated as a black body, a star emits light over a broad wavelength range. However, the light from the star peaks in one wavelength (color) on a bell-shaped curve governed by Planck's black body

TABLE 5.3
Stellar Classification

Class of Stars	Effective Temperature (K)	Color	Comparison with Solar Mass ($M_\odot$)	Average Luminosity (Sun = 1)
O	≥ 30,000	Bluish	≥ 16	1,400,000
B	10,000–30,000	Light blue	2.1–16	20,000
A	7,500–10,000	Blue white	1.4–2.1	80
F	6,000–7,500	White	1.04–1.4	6
G	5,000–6,000	Yellow	0.8–1.04	1.2
K	3,500–5,000	Orange red	0.45–0.8	0.4
M	2,000–3,500	Reddish	0.08–0.45	0.04

radiation characteristics (see Fig. 2.5b). As the green color falls in the middle of VIBGYOR, if the surface temperature of the star is supposed to peak at the green color, then since green is in the center of the visible light spectrum, the star would appear white—that is, a combination of all colors on the left and right of it. Purple stars are something the human eye won't easily see because our eyes are more sensitive to blue light. Since a star emitting purple light also sends out blue light—the two colors being next to each other on the visible light spectrum—the human eye primarily picks up the blue light and the star appears to be blue.

5.4.6 Measurement of Distance to the Stars and Galaxies

How does one measure the distance of faraway stars (some being tens of light years away) and galaxies (being millions of light years away from us)? There are two basic ways of doing it: (1) By using Cepheid stars, and (2) by using remnants of a supernova. The parallax method that is commonly used for measuring distance to the Moon, planets, and nearby stars will not work for faraway stars and galaxies. Cepheid stars have been used for a long time to obtain accurate distances in such cases.

Cepheids were discovered in 1908 by Henrietta Leavitt based on her study of thousands of variable stars in the small and large Magellanic Clouds in the Milky Way. This discovery permitted knowledge of the true luminosity of a Cepheid by simply observing its pulsation period. This is a star that pulsates radially, varying in both diameter and temperature and producing changes in brightness with a well-defined stable period and amplitude. The term *Cepheid* originated from Delta Cephei in the constellation Cepheus and was identified by John Goodricke in 1784, the first of its type to be so identified.

Cepheids, also called Cepheid Variables (as these stars brighten and dim periodically), are used as cosmic yardsticks (standard candle) out to distances of a few tens of millions of lightyears. A strong direct relationship between a Cepheid variable's luminosity and pulsation period established Cepheids as important indicators of cosmic benchmarks for scaling galactic and extragalactic distances. This is done by using the distance-modulus formula: $m–M = 5 \log(d/10)$. Once both the apparent magnitude, m, and the absolute magnitude, M, of the Cepheid star are known, one can find the value of d, the distance to the star. In 1924, Edwin Hubble announced that he had found a Cepheid variable star in the Andromeda spiral nebula. Since then the Andromeda was revealed to be not a cloud of material lurking in our own galaxy, but rather a separate galaxy, a vast swirl of stars at a tremendous distance. Countless enigmatic nebulae seen by astronomers suddenly revealed their true nature as galaxies, strewn across a very large universe. The photograph of the brightest known Cepheid variable star in the Milky Way galaxy, known as RS Puppis, is shown in Fig. 5.13a.

The Hubble Space telescope was built and launched in 1990 to conduct observations that until then could not be achieved using ground-based telescopes. One of its major initial observational programs, dubbed the Hubble Key Project, sought to refine and calibrate the extragalactic distance scale by observing Cepheids in other galaxies. The goal was to use the recalibrated scale to calculate a more precise value for the Hubble Constant, H, which at that time was measured as somewhere between 50 and 100 km s^{-1} Mpc^{-1} (kilometers per second per megaparsec), depending on the method used. The present value of H as used by the standard model of cosmology is 71 km s^{-1} Mpc^{-1}.

To measure very distant galaxies, however, the luminosity of Cepheids will be practically nondiscernible; thus, type Ia supernovae (see Fig. 5.13b) may be considered to be standard candles as their absolute magnitude reaches about –19 at maximum brightness. Given their extreme luminosity, they can be used to probe much further out into the universe than

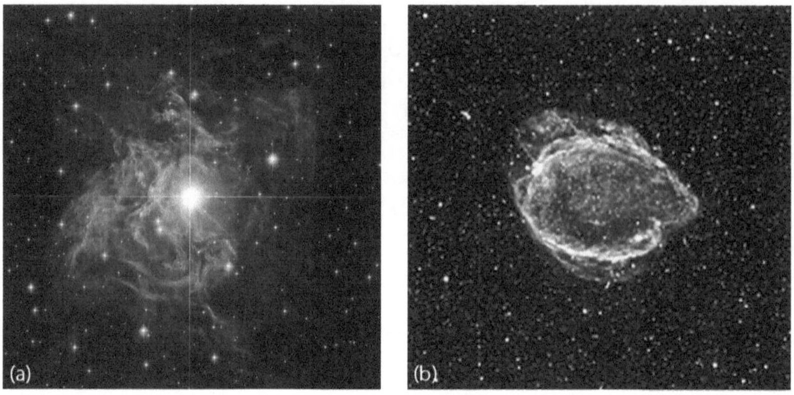

FIGURE 5.13
(a) RS Puppis is one of the brightest known Cepheid variable stars in the Milky Way galaxy [31]. (b) The G299 Type Ia supernova remnant [32].

TABLE 5.4
List of Some Nearer Stars with Their Distance from the Earth

	Distance from Earth	
Name of Star	**Light Years**	**Parsecs**[a]
Proxima Centauri	4.243	1.3001
Barnard's star	5.978	1.8328
Wolf 359	7.795	2.3899
Sirius (binary star)	8.611	2.6401
Ross 248	10.3	3.1580
61 Cygni	11.41	3.4983
Van Maanen 2 (white dwarf)	14.02	4.2985

Note:
[a] *Parsec* is the short form for parallax-second, which is the distance at which a star would have a parallax of one second. 1 parsec = 3.26156 light years = 206,265 astronomical units = 19.17 trillion miles = 30.86 trillion kilometers.

Cepheid stars. Two important projects launched since 1998, the Supernova Cosmology Project and the High-Z SN Search, have observed dozens of supernovae in distant galaxies to help determine very large galactic distances going even into billions of light years. Some of the nearer stars with their distance from the Earth are listed in Table 5.4.

5.5 Nebulae and Galaxies

5.5.1 Introduction

In the 1920s, the distant galaxies were not known as galaxies, and even the idea that other galaxies could exist beyond our own Milky Way galaxy was considered heresy and verging on science fiction. They were thus called *nebulae* (the dictionary meaning being *misty*

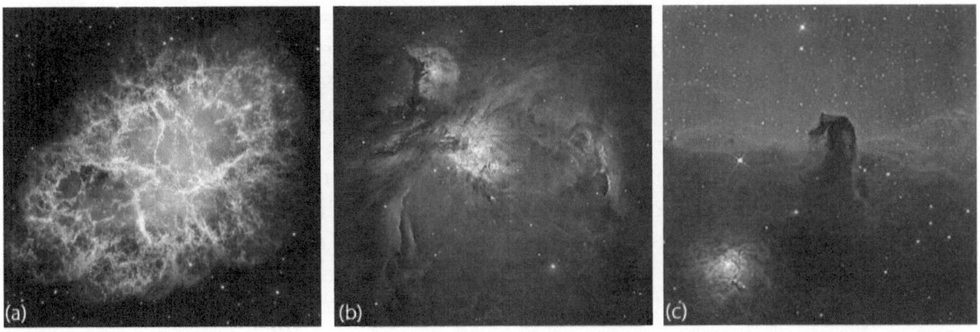

FIGURE 5.14
A few important nebulae: (a) CrabNebula; (b) OrionNebula; and (c) Horse Head Nebula [33].

appearance in the sky made by a group of stars or gaseous matter) because, seen through a telescope of that period, they looked like small, opalescent clouds among the stars. It was not then known that they are in fact distant, immense islands of stars just like our own galaxy. Among the nebulae, Crab Nebula, Orion Nebula, and Horse Head Nebula are three popular ones (see Fig. 5.14).

Galaxies are the largest structures in the universe. They are essentially home to a huge assembly of stars, gas, and dust as well as some other exotic matter, all held together by their mutual gravitational attraction and an exotic matter (yet not fully understandable) dubbed as dark matter. The galaxies are the star-making factory. The Milky Way galaxy (in which our solar system is located) has produced at least 200 billion stars and is still making them.

When the first-generation stars die, they disperse most of their mass into space. The dust and gas of these exploded stars assemble to form the nebulae. Nebulae are the most beautiful and awe-inspiring forms in the universe and are found woven throughout the spiral arms of our galaxy and the most active star-making regions. In fact, our Sun was born from Crab nebula.

5.5.2 Different Types of Galaxies

Until the last century, humans believed that our Milky Way was the only galaxy in the known universe. But with the advanced telescopes we now know that there are two trillion galaxies in the known universe. The smallest of galaxies contain merely a few hundred million stars while the largest galaxies contain up to one hundred trillion stars. Astronomers have classified the galaxies in four main types: spiral, elliptical, peculiar, and irregular.

Spiral galaxies consist of a flat, rotating disc containing stars, gas, and dust, and a central concentration of stars known as the bulge. They are so named because they have spiral structures that extend from the center into the galactic disc. The spiral arms are sites of ongoing star formation and are brighter than the surrounding disc because of the young, hot stars that inhabit them. Our very own Milky Way is a spiral galaxy while our nearest Andromeda galaxy is also a spiral galaxy. As per latest astronomical observations, almost 70% of the galaxies in this universe are the spiral galaxies.

Elliptical galaxies are given their name because of their oblong shape, that is, either shaped like an elongated sphere or cigar-shaped. Most elliptical galaxies are composed of older, low-mass stars, with a sparse interstellar medium and minimal star formation activity, and they are not the dominant type of galaxy in the universe. Because of their age and dim qualities, they are frequently outshone by younger, brighter collection of stars. Their size can range from just a few thousand light years to hundreds of thousand light years, the latter ones are much larger than our Milky Way galaxy. One of the most famous elliptical galaxies is Cygnus A, which is located roughly 600 million light-years from Earth.

A *peculiar galaxy* is a galaxy of unusual size, shape, or composition. Astronomers hypothesize that many peculiar galaxies are formed by the collision of two or more galaxies. About 5–10% of known galaxies are categorized as peculiar.

Irregular galaxies are galaxies that don't fall under any of the three galaxy types mentioned earlier. These galaxies tend to be small, dwarf galaxies that lack any distinguishable shape. Many of these galaxies are companions or satellites to larger galaxies. The Milky Way has dozens of irregular satellite galaxies, the most famous of those is the Large Magellanic Cloud.

5.5.3 The Milky Way Galaxy and Other Nearby Galaxies

The galaxy in which our nearest star, the Sun, is situated and which is our own galaxy is the Milky Way galaxy (Fig. 5.15). We live inside the Milky Way; we are part of it. More precisely, we live about two-thirds of the way out from our galaxy's central hub, on a small planet named Earth orbiting around our closest star, the Sun, one of the 200 billion or more stars of the Milky Way galaxy. By the 1930s, its shape and dimensions were more or less finalized. The Milky Way galaxy has a convex-lens-type shape, which is 80,000–100,000 light years across, with our own solar system about 27,000 light years away from the center and lies in one of the spiral arms of the galaxy. The galaxy is approximately 16,000 light years thick at the center and 3,000 light years thick at the position of our Sun. Our Milky Way galaxy is a spiral galaxy with spiral arms and a huge, bright bulge at the center (Fig. 5.15b). It is in the spiral arms of our galaxy that new stars are made, and it is right on the inside edge of one of these arms, known as the Orion arm, that our solar system resides. With today's latest supertelescopes like the Hubble, the Chandra X-ray Observatory, COBE, and many others located in space well beyond our atmosphere, we can now see far beyond our own galaxy and go billions of years back in time. It is now known that there are at least 125 billion galaxies in the observable universe, with each galaxy containing 100 to 400 billion stars. The Milky Way is one of those galaxies, and the Sun is one of those stars.

Stretching around the sky above through the constellations Orion, Perseus, Cassiopeia, Cygnus, Aquila, Sagittarius, Centaurs, and Carina is a softly luminous band that cuts Earth's equatorial plane at 62° is our Milky Way galaxy. It fades out and becomes invisible in the bright lights of a modern city—but when we go deep into the countryside (where there is no artificial light) we can see in the moonless dark night an awesome

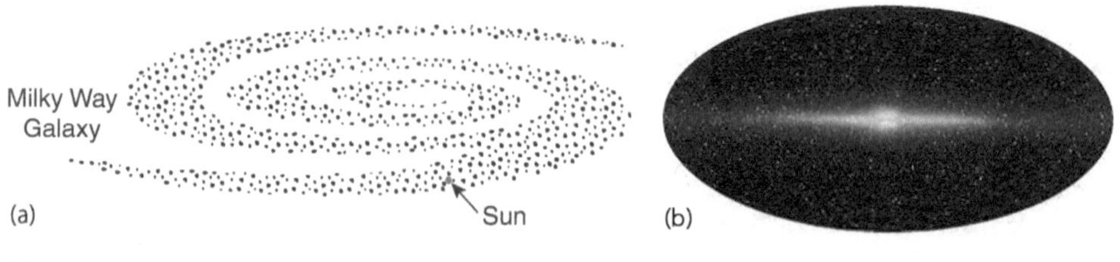

FIGURE 5.15
(a) The spiral-type Milky Way galaxy; and (b) infrared image of the sky showing the Milky Way galaxy (photograph taken from COBE satellite) [34].

glowing river of white light sweeping across the roof of the sky. The ancients called it the "celestial river"—a river made of billions and billions of stars. In 1610, Galileo looked at the Milky Way galaxy through his primitive telescope and found it to be not a featureless luminous band of cloud but a vast collection of very dim stars, as, indeed, some philosophers of pre-telescopic era had speculated it to be. The ancients, who did not possess the dubious blessing (from an astronomer's point of view) of the electric light, were well aware of this luminous band. To them it looked like a milky cloud. The Greeks called it *galaxias kyklos* ("the milky circle"), and the Romans called it *via lactea*, which directly translates into the English as the Milky Way. The English term *galaxy* comes from the Greek version of the name.

Of the estimated 200 billion or more stars in the Milky Way galaxy, we can see only about 5,000 with our eyes unaided. This is because our galaxy is shaped almost like a flat dish, resembling a flying saucer (see Fig. 5.15b) that has a central bulge (where the stars are closely packed together), while becoming sparser as we move to the outskirts. The central stars are old and more yellow than the young, blue stars are found in the arms. Being inside the dish, we can only see the stars that are closest to us. The rest are hidden by the diffuse interstellar gas and dust. When our latest generation telescopes are used to look at the galaxy through infrared light, they penetrate the dust and gas, and we can see deep into the Milky Way galaxy. Even with that, however, we cannot see the far side of the Milky Way galaxy from our place in it. We therefore form our knowledge about its shape by observing other galaxies that seem to behave in the same way as our own galaxy. The distance of the Milky Way galaxy from Earth is 25,000 light years.

Our nearest galaxy is the Andromeda galaxy, which is 2.537 million light years from Earth and has nearly one trillion stars. The farthest known galaxy is MACS0647-JD, which is 13.3 billion light years from Earth and was formed just 420 million years after the Big Bang. Andromeda is also a spiral galaxy and about 2.5 times larger than our Milky Way galaxy, which can be seen with the naked eye as a tiny fuzzy ball of light in the moonless night sky.

Table 5.5 presents some of the naked-eye visible galaxies reported by keen observers in a very dark-sky environment from a high-altitude place, viewed during clear and stable weather.

5.5.4 Far-off Galaxies: Their Possible Distance and the Speed of Recession

Edwin Hubble's astronomical measurements with 100-inch Hooker telescope at Mount Wilson Observatory since 1924 established ultimately in 1929 that nebulae beyond our

TABLE 5.5

List of Naked-Eye-Visible Galaxies from the Earth

	Distance from Earth		
Galaxy	**Light Years**	**Parsec**	**Constellation**
Milky Way	25,000 light years		Sagittarius (at the center)
Andromeda	2.5 million	766 kilo	Andromeda
Triangulum	2.9 million	889 kilo	Triangulum
Bode's galaxy	12 million	3.68 Mega	Ursa Major
Centaurus A	13.7 million	4.2 Mega	Centaurus
Messier 83	14.7 million	4.51 Mega	Hydra

Milky Way galaxy are actually distant galaxies distributed all around the universe. He and his team were the first to map-out how they are distributed in space (known as Hubble sequence) and how they are moving. Hubble provided evidence that the recessional velocity of a galaxy increases with its distance from the Earth (red-shifted), a property now known as Hubble's law (this was also proposed and reported two years earlier by Georges Lemaître based on the then available astronomical data). Hubble-Lemaître's Law (which should be a more fitting name for the law) implies that the universe is expanding. Evidenced by Hubble's observation, Einstein ultimately accepted the view of expanding universe (instead of "static" universe prejudice) that was one of the important conclusions that follows from his own general theory of relativity.

A question naturally arises as to how far-off a galaxy can be as we observe it today and what can be its speed of recession in the expanding universe. If the universe was static (which was the ancient view of the astronomers and even Einstein was initially prejudiced with that view) then the farthest galaxy we could have found is 13.8 billion light years away from us since the age of the universe is known as of today is 13.8 billion years. This is because we know: s (distance) = c (velocity of light) × t (time)—and the light (or any electromagnetic signal like IR, X-ray, etc.) with which we identify them has the highest velocity in the universe (as per the special theory of relativity). But astronomers have detected galaxies as far as 30 billion light years away too (galaxy z8_GND_5296). How is it possible if the age of our universe is only 13.8 billion years? Answer is that, it is possible due to the expanding nature of the universe. This may be understood as follows.

It should be kept in mind that we see a celestial object in space not as it is today, but as it was a finite amount of time ago (as, e.g., in case of Sun it is 8 minutes and for Proxima Centauri, closest star to Earth next to Sun, it is 4.2 years) when it emitted the light that's now striking our eyes. Astronomers often define a quantity called look-back time. It is just the time since the light that we see from an object was actually emitted. For close by celestial object the look-back time is negligibly small (for Sun it is only 8 minutes) because of the stupendously high velocity of light ($c = 3 \times 10^8$ km/s). But as the distance get larger so does this look-back time. When we look at the closest spiral galaxy to our own Milky Way galaxy, the Andromeda galaxy, we see it as it was 2 million years ago (when Humans first began walking on the Earth). Thus the look-back time of Andromeda galaxy is 2 million years. But very distant galaxies in the far reaches of the universe may have look-back time as large as 10–12 billion years. Anyway, due to the expanding nature of the universe, the distance between the electromagnetic signal emitting object and the observation platform continues to increase with time making the original distance between the emitting object and the observation platform smaller than the present distance. If a distant celestial object recedes from us at the Earth with say the velocity of light (c) then the originally 10 billion light years away emitting object when detected presently at the Earth will be found to be 20 billion light years away now (because during 10 billion years of travel of light from the emitting object to reach us the object itself has receded by another 10 billion light years). This makes sense why we can see galaxies even 30 billion light years away though the age of the universe is only 13.8 billion years.

In 1998, it was also found that universe is not only expanding but it is expanding at an accelerated rate and it was also observed that far-off galaxies are even moving away from us with superluminal velocity (i.e., with the velocities more than the velocity of light) which is indicated by their higher than expected red-shift. How is that possible? It appears to contradict the postulate of the special theory of relativity that *nothing in this universe can move faster than the velocity of light.* But in reality, there is no problem with the movement of the very distant galaxies with superluminal velocity when we consider the more broad-based

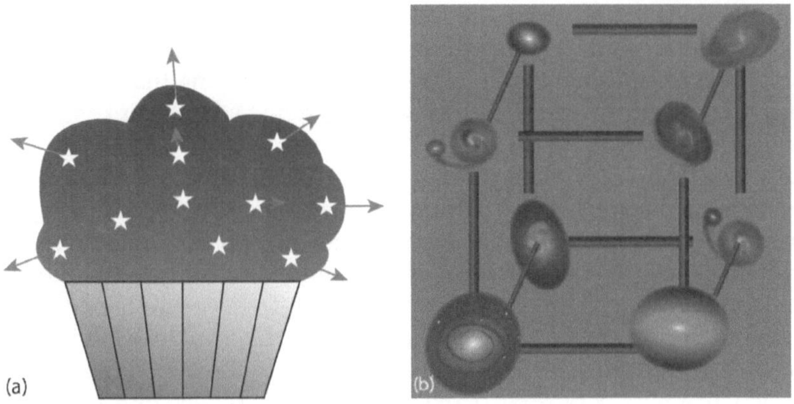

FIGURE 5.16
(a) A rising muffin with chocolate chips (galaxies moving, space not expanding); and (b) space is expanding, but the galaxies are not moving (only their relative distances are redefined).

allowance by the general theory of relativity. This concept may be understood with the help of the model of the expansion of a rising muffin with chocolate chips embedded in it.

Let us look at a rising muffin with chocolate chips (see Fig. 5.16) and assume that the chips are galaxies and the muffin dough is the space of the universe in this model. Because baking powder is mixed with the batter, the muffin fluffs out upon heating. In one viewpoint, we concentrate on moving of the chocolate chips and not on expansion of the dough of the muffin. In other words, we assume that the galaxies are moving and that space is not changing. All chocolate chips (galaxies) move farther apart from all others, and the more widely separated pairs get separated faster. In particular, if one stands on a specific chocolate chip (galaxy), he or she will observe that the motion of all the others relative to him or her obeys Hubble's law. All of them are receding straight away from him or her, and one twice as far recedes at twice the speed (Fig. 5.16a). The same happening may be viewed with concentration fixed on the expansion of the dough (space of the universe), with the chocolate chips (galaxies) not moving relative to the dough. That is, galaxies are not moving through space, but all the distances between them get redefined (on the muffin's surface, the chocolates chip's distance gets redefined from mm to cm, say, Fig. 5.16b).

Thus, in whichever way we consider the concept of the expanding universe (whether in terms of the galaxies moving away from each other or in terms of the space expanding), at the end of the day, it is accepted that our universe is indeed expanding. But Hubble's law, $V = H/D$, implies that galaxies will move away from us faster than the speed of light c if their distance from us becomes greater than $c/H \approx 14$ billion lightyears. We have no reason to doubt that such galaxies exist in our universe. In fact, astronomers have observed experimentally that some galaxies are actually as far as about 30 billion light years (or more) away from us.

Do such galaxies receding faster than the speed of light violate relativity theory (as nothing can go faster than the velocity of light according to the special theory of relativity)? The answer is yes and no: It violates Einstein's special theory of relativity of 1905 but not his general theory of relativity of 1915 (the latter is Einstein's final word on the subject); so we are okay. The special theory of relativity says that no two objects can move faster than light relative to one another *under any circumstances*. The general theory of relativity, however, liberalizes the speed limit and merely insists that they cannot move faster than light relative to one another *when they are in the same place* (in other words, the relative velocity between two objects in space, which are fixed in space, cannot be more than the velocity of light). The galaxies speeding away from us superluminally are actually very far from us. If we think space is expanding, then we can rephrase by saying that nothing is allowed to move faster than light *through space* (when they are fixed in space)—but space itself is free to stretch, however fast it wants to.

In fact, according to the general theory of relativity, the expansion of universe does not subscribe to our commonplace idea of expansion viewed as an explosion-like thing. On the other hand, according to this concept, neither space nor the objects in space moves—it is the metric (which governs the size and geometry of space–time itself; space–time refers to a mathematical model that combines space and time into a single, interwoven medium) that instead changes in scale. Metric expansion is a generic property of the universe we inhabit that can be modelled mathematically with the Friedmann–Lemaître–Robertson–Walker metric based on general theory of relativity of Einstein. However, the model is valid only on very large cosmic scales. According to this model, even the objects within space cannot travel faster than light with respect to each other, but this limitation does not apply to changes in the metric itself. Therefore, at great enough distances from us in the cosmic scale of the expanding universe, the observed speeds of far-off galaxies can exceed even the speed of light. In this context, it may be mentioned that in present understanding, the expansion of universe is a unique phenomenon. All the galaxies in the universe are moving away from each other (as observed by Hubble and his team in 1929), and every region of space is being stretched (as per metric expansion)—but there is no center they are expanding from and no outer edge to expand into anything else; appears to be strange but an acceptable scientific fact as of today.

5.6 Pulsars and Quasars

5.6.1 Radio Astronomy

The history of radio astronomy is quite fascinating. Developments in this subfield of astronomy led to the discovery of the pulsar and quasar, including the unfolding of vast new information about the universe and our own solar system, including the breathtaking imaging of a black hole in April 2019. The probing of different planets of the solar system using microwave signals has revealed a lot of new and refined information about them. The accurate determination of the distance of different planets by measuring the timelapse between emission and echo-return of microwave signals has yielded significantly improved data about the distances of different planets from the Sun when compared with what is known using the parallax method of distance determination. In 1965, microwave reflections indicated the presence of at least two huge mountain ranges on the surface of the planet Venus, one running north and south, the other east and west. The opaque and permanent cloud cover kept Venus's solid surface obscure from the optical telescope, for the optical signal cannot penetrate through clouds. But with the advent of the use of microwaves in the radio telescope, it has been possible to penetrate through it, in as much as the cloud cover is transparent to microwave signal. For example, RF/microwave radar can operate even in cloudy, foggy, and zero visibility night conditions.

Radio astronomy and X-ray astronomy have a special edge over optical astronomy in exploring the mysterious and violent phenomena taking place in the universe. Light, with its wavelength of half a micron, is emitted primarily by hot atoms residing in the atmosphere of stars and planets; thus, optical astronomy has taught us about their atmospheres.

Optical astronomy showed us that the universe is serene and quiescent, dominated by stars and planets that wheel smoothly in their orbits, shining steadily and requiring millions or billions of years to change in discernible ways. But radio waves, with their 10-million-fold greater wavelengths compared to light, are emitted primarily by near-light-speed electrons spiraling in the magnetic fields. Thus, radio astronomy has taught us about the magnetized jets shooting out of galactic nuclei, about the gigantic, magnetized intergalactic lobes that the jet feeds, and about the magnetized beams that emanate from some of the cosmic members like the neutron star.

X-rays, with their 1,000-fold shorter wavelengths than light, are produced mostly by high-speed electrons in the ultra-hot gas accreting into black holes and neutron stars. Thus, they have taught us directly about the accreting gas and indirectly about the black holes and neutron stars. From the 1950s to the 1970s, radio and X-ray observations directed to the cosmos showed us the violent side of our universe: Jets ejected from galactic nuclei, pulsars with intense beams shining off their surfaces and rotating at high speeds, quasars with fluctuating luminosities far brighter than our Milky Way galaxy. The brightest objects seen by optical telescopes were the Sun, the planets, and a few nearby quiescent stars. The brightest objects seen by radio telescopes were violent explosions in the cores of distant galaxies (including ours), powered by gigantic black holes (for our Milky Way galaxy, it is the Sagittarius A^{*}). The brightest objects seen by the X-ray telescope are neutron stars accreting hot gas from binary companions and so forth.

Unlike optical astronomy (that employ visible light/optical waves), radio astronomy uses radio waves in the electromagnetic spectrum received from celestial bodies to characterize their properties. It uses the electronic and electromagnetic technology of wireless communication and radar that has undergone revolutionary developments since World War II. Earth's atmosphere is transparent to visible light but opaque to most other portions of the electromagnetic spectrum; however, it is transparent over a selected band of very short wavelength radio waves known as microwaves. Astronomers were thus offered a second "window" into the so-called Heavens. Any microwave radiation from the celestial bodies could be studied at leisure from the Earth's surface without any need to send instruments aloft in balloons or rockets when resorting to radio astronomical study of celestial bodies. The advantage of using the microwave spectrum of electromagnetic waves is that it works fine even in darkness (at night, with zero optical visibility), in fog, in clouds, and also in rough weather conditions such as snow and hailstorms. Indeed, the use of radio waves, especially microwaves, has led to all sorts of advanced radio aids to electronic navigation of aircraft and ships, including GPS, allowing our seamless movements around the world by air and by sea.

The origin of radio astronomy can be traced back to the work of the American radio engineer Karl Jansky who, in 1931, was engaged in the purely nonastronomical problem of countering the disruptive effects of static in radio communication. There was one source of static that he could not pin down, and that, he concluded at last, must be due to interference from an influx of very short radio waves from outer space. By comparing his observations with optical astronomical maps, Jansky inferred that the radiation was coming from our Milky Way galaxy and was strongest in the direction of the center of the galaxy, in the constellation Sagittarius. Jansky is considered one of the founding figures of radio astronomy. His discovery was of great importance in astronomical studies because the clouds and dust that did not allow the optical telescopes to investigate the center of the galaxy can now be "seen" with radio telescope.

The world's first radio telescope was produced by Grote Reber in 1937 in his mother's backyard. Reber was a radio amateur with a poor science education but good engineering aptitude and a practical streak. Picking up the thread of Jansky's pioneering work in radio astronomy; he made the radio telescope with a 31 ft (9 m) diameter parabolic dish antenna (a type of antenna commonly seen today on roof tops for DTH television signal reception). Through painstaking observations, he made radio maps of the sky in which he could see the central region of the Milky Way and could also faintly observe two radio sources, later called Cyg A and Cas A—A for the "brightest radio sources" and Cyg and Cas for "in the constellations Cygnus and Cassiopeia." He was the first and, for quite a while, the only radio astronomer and he published his first paper on the subject in 1940. However, he was practically unnoticed by astronomers of that era.

Radio astronomers did not take radio astronomy seriously before World War II for two reasons. First, the microwave signal received from the sky had a much shorter wavelength than the signals used for radio communication, so that radiation from space did not ordinarily interfere with radio reception and so astronomers were not bothered at all. But during the War, the microwave radar the British Army used to obtain early warning of the arrival of Nazi Germany's planes suddenly got severely "jammed," making the radar temporarily inoperative by a flood of extraneous microwave radiation. Careful investigation ultimately revealed that the jamming was not the work of the Germans but rather the result of microwave radiation received from a giant solar flare. This finding established that the solar flare sent out radio waves in the microwave region of the electromagnetic spectrum. By the time the war was over, astronomers were ready to turn earnestly to radio astronomy

Roger Jennison and Mrinal Kumar Das Gupta of Lovell's Jodrell Bank Observatory in the United Kingdom, for the first time, in 1953, discovered that Cygnus A is a double radio source (see Fig. 5.17a). They used for their discovery the long-baseline postdetector correlator radio interferometer principle known as the intensity interferometer, which was conceived by their research guide, R. Hanbury Brown, in 1950. In the late 1960s, refined radio-telescopic studies and further studies with a very-large-array (VLA) radio interferometer operating at 5 GHz showed that the radio-emitting lobes were actually powered by jets of gas emerging from a central region of the radio galaxy (Fig. 5.17b). Today, almost all radio galaxies and enigmatic quasars show double structures.

Jennison and Das Gupta's discovery is perhaps the first indirect evidence we have for astronomical black holes, a discovery made at a time when astronomers had not yet accepted the concept of the black hole. The very-long-baseline interferometry used to take the image of black hole M87* in April 2019 with the EHT (discussed in Chapter 3) is a modern version, using state-of-the-art technology, of Jennison and Das Gupta's cruder instrument of 1950s. To realize a large virtual antenna aperture needed to conduct their radio astronomical study, Jennison and Das Gupta kept one antenna array (each consisting of wires in a wooden framework) of the interferometer on site and carried the other one in a truck, moving around the countryside surrounding the city of Manchester.

Das Gupta went from the Institute of Radio Physics and Electronics (IRPE), University of Calcutta (CU), to the Jodrell Bank Observatory in Manchester, England, to complete his PhD and upon completion returned to the IRPE, serving as a teacher there for the rest of his life.

5.6.2 Pulsars

Developments in microwave electronic technology after World War II helped refine radio astronomical studies in subsequent decades. In the 1960s, astronomers observed that some celestial sources changed the intensity of their microwave emission with astonishing

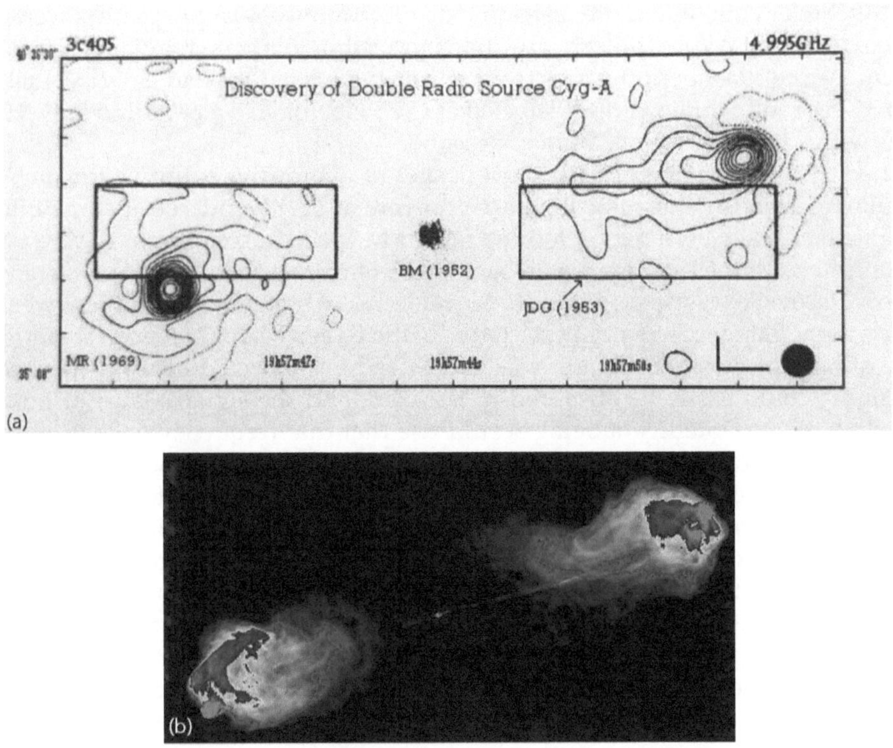

FIGURE 5.17
Picture of the radio emission from the radio galaxy Cygnus A with (a) the long-baseline postdetector correlator radio interferometer [35]; and (b) very-large-array (VLA) radio interferometer [36].

rapidity. It was as if there was a microwave-twinkle! This caused astronomers to design a radio telescope capable of catching very short bursts of microwave energy. They felt this would make it possible to study these fast changes of microwave signals from celestial bodies in greater detail. One astronomer who made use of such a radio telescope along with his coworkers was Antony Hewish at Cambridge University Observatory. Hewish and Martin Ryle were awarded the Nobel Prize in Physics in 1974 "for their pioneering research in radio astrophysics: Ryle for his observations and inventions, in particular of the aperture synthesis technique, and Hewish for his decisive role in the discovery of pulsars."

The Hewish team's first experimental observation showed that the burst of microwave energy received from the celestial body lasted only 1/30th of a second but appeared at regular intervals of 1.33 seconds—that is, in the form of a pulse train with short-duration pulses of microwave energy. The individual bursts (pulses) were very energetic with momentary strong brightness of the celestial object. Thus, they were easy to detect with short-pulse measuring equipment. (All previous radio telescopes were capable of measuring only average power, which was just 3% or less of the maximum brightness and was thus unable to notice them.) Hewish did not have the faintest idea about what this odd microwave phenomenon observed with the celestial body might represent. These newly discovered celestial objects were comparable to pulsed radio signals coming from space, similar to the lighthouse beacon on the seashore. By 1968, when he announced the discovery, Hewish and his team had located four such pulsars. Other astronomers had also begun to search avidly, and more were quickly discovered—40 more pulsars were located in the next 2 years. All the pulsars are characterized by extreme regularity of pulsation, but of course the exact period varies from pulsar to pulsar. One had a period that was as long as 3.7 seconds, while a pulsar detected at Crab Nebula had a period of only 0.0033089 seconds with a pulsing rate of 30 times a second.

Stunned by the reception of such short flashes of microwave energy with such fantastic regularity, Hewish and his fellow astronomers at first wondered if some intelligent life-forms in an alien civilization located far off in space in some galaxy were trying to communicate with the intelligence present in the Earth via these signals. The news after Hewish's discovery was so startling that it remained a state secret for a few weeks. The scientists named their discovery LGM-1, for "Little Green Man." Yes, even scientists have a sense of humor! But that thought was short-lived and rejected because the frequency of the signal could easily be obscured or jammed, and the signal's level of power was extremely high. No intelligent being would use that that frequency or that power level for signaling purposes. Further, detection of more such different sources of a pulsating signal (i.e., pulsar) made it clear that so many different intelligent life-forms in different locations in space cannot just zero-in their signal to us. But something must be producing them; some astronomical body must be undergoing some changes at intervals rapid enough to produce the pulses of microwave signal. Thus, the pulsar was discovered, and we also had our first observational proof of the existence of a neutron star (as the pulsating signal from pulsar are now known to originate from neutron star). This proved a great event for radio astronomers, but the science fiction writers were disappointed that they had lost a great story-making opportunity that would have given them enormous market value.

In this respect, in 2017, a mysterious cigar-shaped object about 800m in length and featuring a dark-red hue, named Oumuamua, zipped through and pointing outside the solar system (for a month or so) puzzled the astronomers, prompting some to think it must represent an "alien technology." Subsequent studies indicated that it was possibly a planetesimal fragment that had been ejected from another star and been thrown into the solar system—that is, it was of natural origin. The object's nongravitational acceleration could be due to a "light sail," a propulsion technology that uses the pressure of photons to move spacecraft. In the canvas of this universe with varieties of cosmic objects like galaxies, stars, and planets, every day and night innumerable mysterious phenomena and incidences are taking place. We, the human beings with rational curiosity, go from the obscurity of ignorance to the effulgence of knowledge in our never-ending thirst for the truth surrounding everything around us.

Although various theoretical propositions were advanced to explain the phenomenon of the pulsar, Thomas Gold proposed the most plausible one. Following the discovery of the pulsar, Gold proposed that these objects were rapidly rotating neutron stars. He argued that because of their strong magnetic fields and high rotational speed, pulsars would emit electromagnetic radiation similar to a rotating beacon. Gold's conclusion was initially not well received by the scientific community; in fact, at the first international conference on pulsars, he was refused permission to present his theory. However, Gold's theory became widely accepted following the discovery of a pulsar in Crab Nebula using the Arecibo radio telescope (see Fig. 5.18a).

Gold pointed out that if his theory was correct, the neutron star would be leaking energy at the magnetic poles and its rate of rotation would subsequently be slowing down. This meant that the faster the period of a pulsar, the younger it was likely to be and the more rapidly it might be losing energy and so it would slow down. The most rapid pulsar known

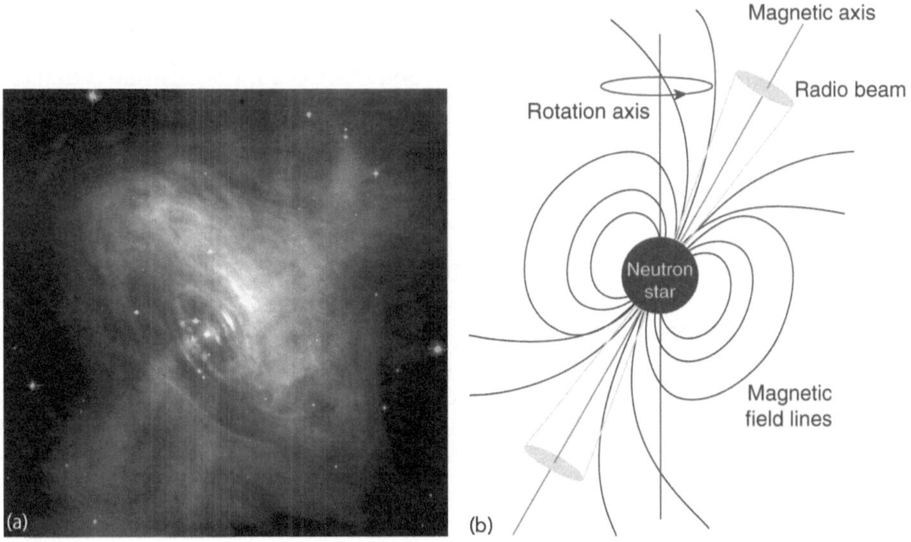

FIGURE 5.18
(a) Pulsar in the Crab Nebula [37]; and (b) magnetic field of a pulsar.

is in the Crab Nebula, and it might well be the youngest too, since the supernova explosion that would have left the neutron star behind took place only a thousand years ago. The period of the Crab Nebula pulsar was studied carefully, and it was indeed found to be slowing down, just as Gold had predicted. The same phenomenon was discovered in other pulsars as well, and the neutron star theory of pulsar proposed by Gold is now an accepted fact.

The neutron stars have intense magnetic fields; they are about a trillion times stronger than Earth's own magnetic field. (The Sun's magnetic field is approximately ten times that of the Earth.) The origin of the magnetic field is as usual due to the rotating electron current. In neutron stars, the electrons are no longer in their orbit of an atom around the nucleus due to the extremely compressed condition of the star caused by the implosive gravitational force. But here the electrons take refuge in the crust of the neutron star as free electrons. However, the proton inside the nucleus balances the charge of the electron ensuring the charge neutrality, but both proton and neutron remain in their domain of the nucleus, with free electrons residing in the crust around the nucleus. The neutron star rotates very fast around its rotation axis (see Fig. 5.18b). Thus, in a neutron star we have a situation of rotating electrons giving rise to its magnetic field. However, the axis of the magnetic field is not aligned with the neutron star's rotation axis, as shown in Fig. 5.18b.

The combination of this strong magnetic field and the rapid rotation of the neutron star produces extremely powerful electric fields, with electric potential in excess of 1 trillion volts. Electrons in the neutron star, which are not in the orbit of the atom but are compressed to the domain of the crust around the nucleus by the stupendous gravity of this star form, are thus accelerated to high velocities by these strong electric fields, producing radiation at microwave frequencies. Because of the neutron star's rotation, the radiation from the pulsar that emerges along the direction of the magnetic poles appears to the external viewers (say from Earth) to be analogous to the light from a lighthouse, which seems to blink as the beam of the pulsar rotates. When the sheaf of microwave signals from the rotating neutron star sweeps past our Earth's direction, then we can detect a short burst of microwave energy once each revolution; thus, received microwave pulses are periodic in nature. However, it must be noted that we would detect only pulsars that happen to rotate in such a way as to sweep one of its magnetic poles in our direction. Some astronomers estimated that only one neutron star out of 100 would do so. Also, they guess that there might be as many as 100,000 neutron stars in our galaxy but that only 1,000 would be detectable from Earth.

5.6.3 Fast Radio Burst

While talking on pulsars, it would be worthwhile to talk a few words about *Fast Radio Burst* or FRB for short. The first FRB was discovered in 2007 by Duncan Lorimer and his students, which was dubbed as the "Lorimar Burst." FRB is a transient radio pulse of length ranging from a fraction of a millisecond to a few milliseconds, caused by some high-energy astrophysical process not yet properly understood. Scientists have dozens of theories about the causes of FRB, from colliding black holes to alien starships, i.e., extraterrestrial intelligence. Many theories suggest the bursts originate from neutron stars, which are corpses of stars that died in catastrophic explosions known as supernovae. Anyway, FRB releases as much energy in a millisecond as the Sun puts out in three days. While extremely energetic at their source, the strength of the signal reaching Earth is very feeble and thus needs very powerful radio telescope to detect them. Most FRBs are extragalactic, but the first Milky Way FRB was detected by the CHIME (Canadian Hydrogen Intensity Mapping Experiment)—an interferometric radio telescope at the Dominion Radio Astrophysical Observatory in British Columbia in April 2020. In June 2021, astronomers reported over 500 FRBs from outer space being detected. In October 2021, astronomers reported the detection of hundreds of FRBs from a single system. From this FRB, named FRB 121102, in just 47 days, the Earth received 1,652 FRBs. To date, this set of signals is considered the largest known to specialists. A possible source of signals from this FRB is understood to be received from a dwarf galaxy.

5.6.4 Quasars

In 1963, Marten Schmidt was studying the radio source 3C 273 with a 200-inch reflector telescope at the Palomar Observatory. (3C is short for Third Cambridge Catalogue for Radio Stars, and the remaining numbers represent the placement of the source on that list.) He noted its star-like appearance—hence the name "quasar," short for quasi-stellar radio source—that has extraordinary luminosity. By studying the spectrum of 3C 273, he found it to have very high red-shift. By 1970, about 150 such strange radio stars were discovered, each one of which showed large red-shifts. The red-shift of all these quasi-stellar radio sources was found to be too large to be caused by a gravitational field. If it had been a gravitational red-shift, the object would have to be so massive and so near to Earth that it would disturb the orbits of planets in the solar system. This consideration suggested that the red-shift was instead caused by the expansion of the universe, which, in turn meant that such celestial objects must be a very long distance away. Moreover, its visibility from such a long distance is possible only if it is extremely bright. That is, it must be emitting a huge amount of energy. The only mechanism that one can think of that would produce such large quantities of energy seemed to be the gravitational collapse not just of a star but of a whole central region of an entire galaxy.

Let us discuss here a few words about the gravitational red-shift or the Einstein shift. It refers to the red-shift of spectral lines that occurs when radiation, including light, is emitted by a source in the gravitational field of a massive body. When the radiation is 'climbing out' from the gravitational field of the body it has to work against the gravitational field and hence looses energy. Thus the radiation frequency must decrease or in other words its wavelength λ, shift by $\delta\lambda$ toward a greater value. The gravitational red-shift is given by: $\delta\lambda/\lambda = GM/c^2R$, where G is the universal gravitational constant, M and R the mass and radius of the massive body and c is the speed of light. Gravitational red-shift was predicted by Einstein's general theory of relativity. It is a very small quantity for the Sun and Earth—but quite significant for highly massive bodies in the far reaches of the universe. In the spectra of several stars including the Sun and also that of Earth the gravitational shift has been measured and the predicted and measured values agree very closely.

NASA's Hubble telescope equipped with Wide Field and Planetary Camera 2, which can see images in the visible, near-UV, and near-infrared parts of the electromagnetic spectrum, permitted the best image of the bright and closest (2.5 billion light years from Earth) quasar 3C 273 (Fig. 5.19a) in 2013. The farthest quasar was ULAS J1120+0641, which is 12.9 billion light years from Earth. The quasar 3C 273 resides in a giant elliptical galaxy in the constellation Virgo (The Virgin). Its light has taken some 2.5 billion years to reach us. But despite this great distance, it is still one of the closest quasars to Earth.

Quasars are the most distant celestial object in the universe; today they can be said to be at the boundaries of the universe. Quasars are capable of emitting hundreds or even thousands of times the entire energy output of our own Milky Way galaxy, making them some of the most luminous and energetic objects in the entire universe. Of these very bright objects, 3C 273 is the brightest in our skies. If it was located 30 lightyears from our own planet—roughly seven times the distance between Earth and Proxima Centauri, the nearest star to us after the Sun—it would still appear as bright as the Sun in the sky.

Quasars are the intensely powerful centers (nuclei) of distant, active galaxies, powered by a huge disc of particles called accretion discs surrounding a supermassive black hole. The accretion disc grows when the supermassive black hole at the heart of the active galaxy pulls in materials such as gas clouds, dust, planets, and stars from its surroundings,

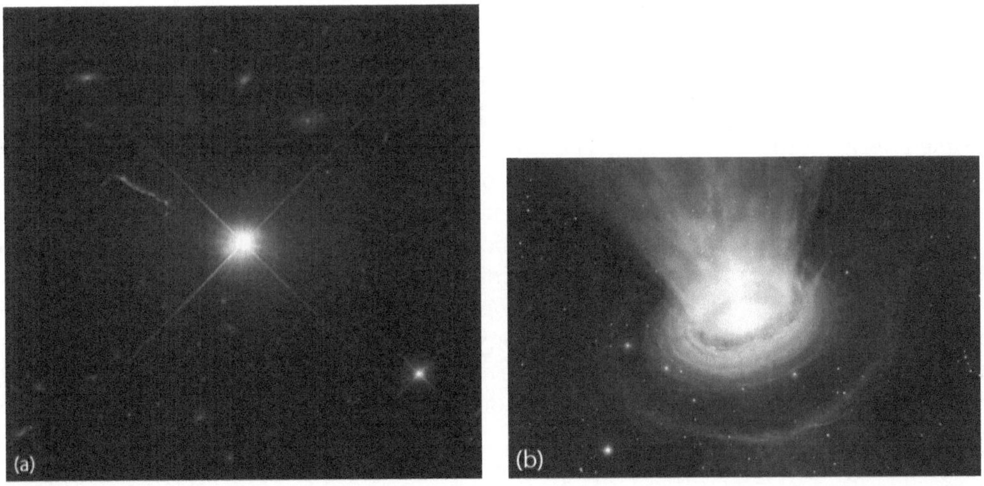

FIGURE 5.19
(a) Brilliant quasar 3C 273 that resides in a giant elliptical galaxy in the constalation Virgo [38]; and (b) quasar firingoff a superfast jet from the center of a supermassive black hole at the nuclei of an active galaxy NGC 3783 [39].

tearing it to shreds and heating it up to tremendous temperatures before swallowing it up. As material from this disk falls inward, the quasars fire off superfast jets into the surrounding space (see Fig. 5.19b)—some sort of implosion—a catastrophe on the galactic scale (creating a small "Bang," so to say), just like a supernova in the stellar scale. As quasars are understood to be powered by giant black holes in active galaxies (which are still in the evolution stage), its extraordinary luminosity, which may be a trillion times that of our Sun, derives from the compact accretion region of the supermassive black hole (see, Fig. 3.5a). Quasars might only occur once in the lifetime of a galaxy, and if it does, it only lasts a few million years, while the black hole works through all the backed-up material. Once the black hole has finished its "stuff buffet," the accretion disc disappears, and the light from the quasar shuts off.

The quasar's energy requirement is so great that if one were to convert all the mass of 10 million (10^7) Suns into pure energy with 100% conversion efficiency, it might be able to meet that energy budget. Thus, beginning in the mid-1960s, astrophysicists tried to examine all conceivable sources of power for quasars in search of a viable explanation. With *chemical power* (the burning of gasoline, oil, coal, etc.), which meets a person's energy requirement, that is absolutely impossible. It will need $10^7 \times 10^9$ ($=10^{16}$) solar masses of chemical fuel, or 100,000 times more fuel than is contained in our entire Milky Way galaxy (since the chemical conversion efficiency for converting mass into energy is only one part in a billion, i.e., 1 part in 10^9). *Nuclear power,* the basis of the hydrogen bomb and the source of power for the Sun and all other stars, has a mass-to-energy conversion efficiency of hardly 1% (i.e., 1 part in 10^2). Thus, a quasar would need $10^7 \times 10^2$ ($=10^9$), or one billion solar masses of nuclear fuel, to energize its radio-emitting lobes if the process of converting this nuclear energy to the magnetic and kinetic energy were 100%. This is never the case, however, and such a conversion process in nature is not known. Thus, nuclear fuel was a possibility but not a likely one.

The *annihilation of matter with antimatter* can give 100% conversion of mass to energy, so 5 million solar masses of antimatter annihilating with 5 million solar masses of matter could satisfy a quasar's energy needs. But it is a known fact that antimatter does not exist in the present universe except for a tiny amount which is created in accelerators or by nature through the collision of high-energy particles in space. Moreover, such annihilation of antimatter with matter would produce very-high-energy gamma rays rather than the magnetic energy and electron kinetic energy required to power the central engine and hence the radio-emitting lobes of a quasar. Thus, matter–antimatter annihilation is not a possible energy source of quasars. The only remaining possibility is *gravity*—the implosion of a massive star to form a neutron star or a black hole; that conceivably converts 10% of stars' mass into magnetic and kinetic energy. The massive star needs to be equivalent to $10^7 \times 10$ ($=10^8$) normal stars, that is, 10 million Suns. In the early 1960s, the concept of the black hole was poorly understood, and the idea of the implosion of a massive star to form a black hole that might energize a quasar was just a radical idea. However, by early 1980, through advanced radio telescope technology and good understanding of black holes, it was now understood that the central engine of the quasar, including the jets and its radio-emitting lobes, was powered by the accretion disc of the black hole, which can have the energy of 100 million Suns.

In 1969, Roger Penrose made a marvelous discovery: He found that a spinning black hole stores huge amounts of rotational energy in the swirl of space around itself. Since the swirl's energy is outside the hole's event horizon and not inside it, this energy can be extracted and used to power things. A decade following Penrose's discovery, physicists discovered that nature is clever enough to harness that energy as a black hole "machine" to power the gigantic jets and hence quasars. Penrose received the 2020 Nobel Prize in Physics "for the discovery that black hole formation is a robust prediction of the general theory of relativity."

A final question about quasars: If the radio galaxies and quasars are powered by the same kind of black hole engine, then what makes quasars look so different? The light of a radio galaxy seems to come from a Milky Way-like assemblage of stars, 100,000 light years across, while the light of a quasar appears to originate from an intensely luminous, star-like object, only one light month in size or less. The quasar is also a radio galaxy but an active one with an active black hole embedded in it whose accretion disc shines so brightly that its optical brilliance is hundreds or thousands of times greater than that of all the other stars taken together in the surrounding galaxy. Thus, the whole galaxy becomes visible as a tiny but intense point of light—as if a superbrilliant star (blotting out all other stars of the galaxy)—and hence given the name quasi-stellar or in short quasar. But in a nonquasar-type radio galaxy the central accretion disc presumably is rather quiescent and produces low luminosity, so that the disc shines much less brightly than the rest of the galaxy. Hence, the disc is not visible separately, but the whole galaxy with billions of stars becomes visible to the optical telescope. However, the disc, the spinning hole, and magnetic fields threading through the hole together produce strong jets, and those jets shootout through the galaxy and into intergalactic space, where they feed energy into the galaxy's huge radio-emitting lobes as usual.

Since our Milky Way is already a middle-aged galaxy, its quasaring days are probably long over. But possibly a monster is chasing the beauty of our galaxy—the Andromeda galaxy is going to cuddle with the Milky Way galaxy, disrupting the cores of both galaxies. During this colossal event, the supermassive black holes in the center of these two galaxies (for the Milky Way galaxy, this is known as Sagittarius A*) will interact, messing with the orbits of stars, planets, gas, dust, and so on. Some will be thrown out into space, while others will be torn apart and fed to the hungry black holes. And if enough material piles up, perhaps our Milky Way galaxy will become a quasar again. This possibility need not worry us for the moment, for it will take another 4 billion years from now and by that time the Sun may swallow the Earth, including the planet Mars. The collision of galaxies is not a rare phenomenon in the vast platform of the universe and, in fact, such collision of galaxies make old galaxies to disappear, paving the way for new ones to be born.

5.7 Few Unsung Female Scientists

Although Antony Hewish was rightly awarded the Nobel Prize in 1974 for discovering the pulsar, the contribution of his PhD student, Jocelyn Bell Burnell, was not recognized in any way. Her exclusion from the Nobel Prize (jointly with Hewish) is widely seen as one of the biggest Nobel snubs in history. In reaction, some of her contemporaries called it the "No-Bell prize." As for Burnell herself, however, she doesn't hold any grudge against the Nobel committee. Here is her comment: "I was delighted when I heard the news. I knew it created a very important precedence. There is no Nobel in astronomy, and until then, no astronomer had received a Nobel Prize. I was proud that it was my stars (pulsars) that had convinced the committee that there is good physics in astrophysics. I wasn't too bothered not to get it. The Nobel committee didn't normally recognize students. It didn't even know how many students there were, what gender they were, what their names were. Students were in the noise." In 2018, a half century after her first publication on pulsars, Burnell was, however, honored with the Special Breakthrough Prize in Fundamental Physics. (The Breakthrough Prize in Fundamental Physics is awarded by the Fundamental Physics Prize Foundation, a not-for-profit organization dedicated to awarding physicists involved in fundamental research. The foundation was founded in July 2012 by Russian physicist and Internet entrepreneur Yuri Milner.)

In this context, we want to call attention to another woman physicist, Bibha Chowdhuri, one of the earliest particle and cosmic ray physicists in India. She discovered a new subatomic particle, the pi-meson (pion), while doing experiments at Darjeeling, India, during 1938–1942, with her research guide, D.M. Bose. The results were published in *Nature*. Chowdhuri and Bose could not access full-tone photographic plates because World War II was raging at the time. Using the same technique, but with high-quality full-tone photographic plates, British physicist Cecil Frank Powell identified the pion at least four years later and won the Nobel Prize in Physics in the year 1950. Powell, however, acknowledged the work of Chowdhuri and Bose. Anyway, in 2019, the International Astronomical Union (IAU) renamed a yellow-hued star HD 86081 in the equatorial constellation of Sextans located 340 light years away from Earth as Bibha; which means "a light beam," to honor the contribution of this pioneering Indian female scientist.

When talking about the heroes of physics and astronomy, we tend to breathlessly utter names like Newton, Einstein, Bohr, Heisenberg, Dirac, and Feynman. They are definitely the stalwarts in their respective field of research, but we often forget to honor many of the heroines of physics and astronomy, which is very unfortunate. But there are a number of women Nobel laureates in science including those in physics and astrophysics (https://en.m.wikepedia.org) who have also made significant contributions to the development of science. We all know the name of the great woman scientist Marie Curie, but how many of us are aware of the contributions of other women scientists?

For example, there is Maria Goeppert Mayer, a German-born American, who was the second woman to win the Nobel Prize in Physics (1963), for proposing the nuclear shell model of the atomic nucleus. During the Great Depression (that caused worldwide economic downturn) of 1930s, no university would think of employing the wife of a professor. But Maria kept working, "just for the fun of doing physics." She received the Nobel Prize when she was just 57 years old and did most of her Nobel Prize-awarded research for free.

In addition, there is Donna Strickland, a Canadian physicist who is a pioneer in the field of pulsed lasers and the winner of the Nobel Prize in Physics in 2018 for the invention of chirped pulse amplification.

Andrea M. Ghez, an American astronomer who is known for her research on the Milky Way, won the Nobel Prize in Physics in 2020 jointly with R. Genzel and Roger Penrose.

Irène Joliot-Curie, daughter of Marie Curie, received the Nobel Prize in Chemistry (1935) along with her husband for the discovery of artificial radioactivity. Their discovery paved the way for innumerable medical advances, especially in the fight against cancer.

Another woman whose name practically remained unknown in the history of astrophysics is Williamina Fleming, who discovered the "first dead star"—the so-called white dwarf—in the universe. She was also the one who discovered the horse head nebula and cataloged more than 10,000 stars, which would later be compiled in the Henry Draper Catalogue, where she received no credit for her work.

Yet another woman we should not miss to mention is Lise Meitner, an Austrian–Swedish physicist who spent most of her scientific career in Germany, made significant contribution in the discovery of nuclear fission but did not share the 1944 Nobel Prize in Chemistry though she received many other awards and honors. The chemical element 109 meitnerium (Mt), an artificially produced element belonging to the transuranium group, was named in 1997 after her name. She was praised by Albert Einstein as the "German Marie Curie." Her work in the group, who developed the nuclear fission process, can be fittingly credited for the development of nuclear reactor used for the generation of electricity and also the nuclear weapons during World War II.

We mentioned earlier that another female scientist who was denied proper recognition, even though she provided the first evidence of dark matter, was Vera Rubin, an American astronomer. Further, it was also mentined that Henrietta Leavitt, who worked in Hubble's team of researchers, credited for the discovery of *Cepheids* (used as cosmic yardstick for measuring distance to far off galaxies) did not even get the due recognition for such an important discovery by her. Last but not the least, how many of us have heard the name of Emmy Noether, the genius who made unmatched contributions to mathematical sciences (that has immensely helped physics too), despite all the hardships that she had to face because of her gender? But she was the one who helped and resolved elegantly with her so-called Noether's theorem, when two of the world's top male mathematicians, David Hilbert and Felix Klein, in 1915, were stumped with a serious problem that cropped up in Albert Einstein's new theory of gravity, general relativity. Not only this, but Noether's theorem can also be used to calculate the entropy of stationary black holes. Physicists had known about the conservation of momentum, angular momentum, and energy long before Noether's theorem came along. They are foundational precepts of classical mechanics. But it was not known that these hallowed laws shared a common origin, each bound to a particular symmetry. This new insight, which sprang from Noether's work, is a guiding principle that has immensely helped in the development of many issues of modern physics. Noether's theorem states that every "continuous" symmetry in nature has a corresponding conservation law and vice versa. This theorem uncovered a hidden relationship between two basic concepts—symmetries and conserved quantities—that until then had been treated separately. The theorem provides an explicit mathematical formula for finding the symmetry that underlies a given conservation law and, conversely, finding the conservation law that corresponds to a given symmetry. When Noether died in 1935, Einstein wrote in the New York Times, "Noether was the most significant creative mathematical genius thus far produced since the higher education of women began."

5.8 The Observable Universe and Multiverse

How do we know that everything that we can see is everything there is? A number of suggestions have been offered in response to this question, ranging from the idea of cosmic inflation (hyperexpansion) to string theory to the revival of Einstein's cosmological constant (though with a different interpretation of it with today's perspective). Some of the notions that have been presented based on these lines of evidence are pretty counterintuitive and somewhat weird—yet that doesn't worry physicists and astronomers. The idea of multiple universes, or a multiverse, poses the notion of a Grand Universe for the cosmos existing like a giant sponge, with each bubble representing a distinct universe, like ours, but others could exist separated by giant voids (Fig. 5.20a).

But what evidence has been offered for a multiverse? First, measurements of distant supernovae suggest that the expansion of the cosmos is accelerating (understood in terms

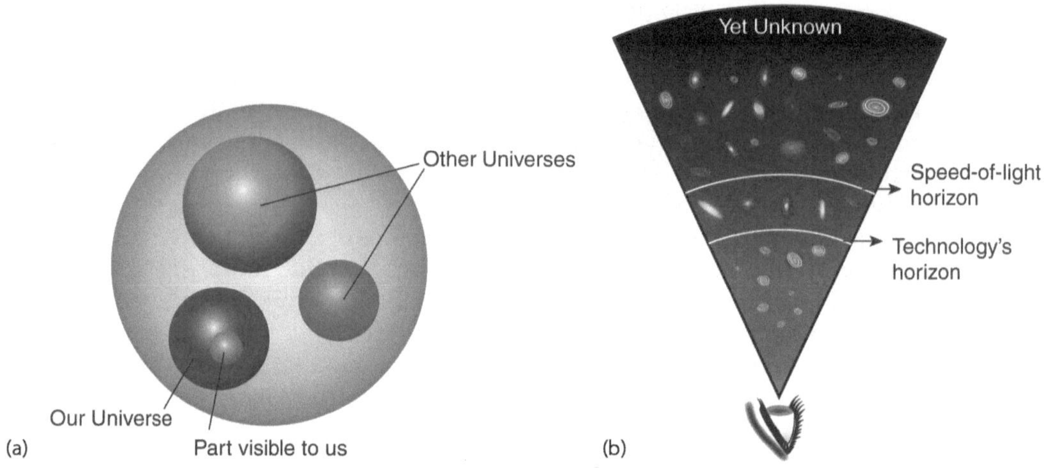

FIGURE 5.20
Artist's impression of (a) a grand universe/multiverse; and (b) the observable universe with technology's horizon and speed-of-light horizon and yet unknown domain of it.

of the revival of Einstein's cosmological constant). Second, more and more evidence support the inflationary scenario of the early history of the universe. Third, ideas about inflation suggest many Big Bangs may have occurred. Fourth, recent notions about string theory suggest that many different types of universes have formed in the past. Altogether, these notions suggest it was possible, if not probable, that multiple universes of different types did arise in the past and that they coexisted with the familiar cosmos that we observe.

At the beginning of this chapter, we mentioned that our universe is 92 billion light years across (i.e., it may be possible for us to see as far as 46 billion light years in all directions), but the age of the universe is known to be only 13.8 billion years. How is it possible that the size of our universe is much larger than the size that can be calculated from its age (when light cannot travel more than the velocity of light as per special theory of relativity) and we can see celestial objects lying much beyond 13.8 billion light years? The answer to this question may be understood as follows. If the universe had been a static one, it would have been possible for us to see the light coming from objects maximum to the extent of 13.8 billion light years away. But Einstein's theory of general relativity and Hubble's astronomical measurement on the motion of galaxies established that the universe is an "expanding universe." In an expanding universe, we are expected to be able to see celestial objects twice as far away compared to what we could see in the case of a static universe if the celestial body moves away from us with the velocity of light. That is, objects as far as 27.6 billion light years away can be seen (assuming their light just reaches us now and they speed away from us at *almost* the speed of light). If, at the time of the Big Bang, only radiation energy was present, then the maximum limit to our astronomical observation (i.e., our observable universe; see Fig. 5.20a) would have been 27.6 billion light years (the speed-of-light horizon shown in Fig. 5.20b), though the "technology's horizon" is much less than that today.

But it is known that at the time of the Big Bang, the universe was extraordinarily hot and dense, and subsequently the universe became filled with a tremendous diversity of energy sources: Radiation, matter, and the intrinsic energy to space itself (dark energy?). In the very early stages of the universe, for the first few thousand years, it was dominated by radiation, mostly in the form of photons and neutrinos. After that, a transition took place, and matter—both normal (like protons, neutrons, and electrons) and dark matter, including dark energy—became the most important component for billions of years. And then very recently, even after the formation of our solar system and Earth, dark energy became important enough to dominate, accelerating the expansion of our universe. Our astronomical data show that outer galaxies exhibit red-shift larger than normal, indicating that space is expanding at a stupendous speed, even at superluminal speed. Thus, according to general relativity, the radius of the universe would be limited to 41.4 billion light years (the relativistic derivation of that figure is $R = 3ct = 3 \times c \times 13.8$ when the universe is endowed with radiation and normal matter). But in a universe with dark energy, the accelerating universe gets pushed out to an even greater number: 46 billion light years for the mysterious dark energy that our cosmos is believed to possess.

A further question now arises: Does the more than normal red-shift of the outer galaxies of the universe observed by astronomical experiments indicate that expansion of space is taking place with a velocity greater than the velocity of light? The answer is yes, and thankfully, it does not violate the assumption of special relativity. This is because special relativity is true for a gravity-free frame of reference, and when the question concerns expansion of space, celestial objects in the frame of reference are not moving with respect to each other. They remain in place (Fig. 5.16), but the expansion of space due to its energy content constantly continues to create new space and results in an apparent increase of the distance between celestial bodies at the far reaches of the universe. So, yes, objects in the universe can travel faster than c (velocity of light) away from us due to the expansion of the universe, and the universe itself can be much larger than expected given its age and the speed of light.

Our current instruments have allowed us to see galaxies, pulsars, quasars, the black hole, and other celestial objects out to the technology's horizon. We can't observe a galaxy past the "speed-of-light horizon," however, and hence it is marked as "yet unknown." But it might even become visible in the future if the universe's expansion decelerates or if we can devise a time machine that will allow us to travel through a wormhole. All this of course is part of our meandering with a science-fiction-like-fantasy thought.

So, we come again to the central question: Do we live in a multiverse? Daydreamers and science fiction writers have pondered parallel universes for as long as scientists have described our own universe. What if our universe isn't the only one? What if alternate universes are humming along undetected, right next to ours? Cosmologists call this idea the multiverse/parallel universe, and there's good reason to consider the concept. Indeed, many of the best scientific models describing the origin of our universe actually depend on the existence of the multiverse. It is believed that the multiverse, or the parallel universe, could come in different flavors—which may be qualitatively new and different from our own universe, having totally different fundamental laws of physics or those might have same fundamental laws of physics, but have started with different initial conditions and so on.

Physicists and astrophysicists remain sharply divided about the existence of the multiverse. Many scientists dismiss the very idea on the basis of one simple fact: If you can't leave your own universe, then there's no way to prove that any other universe exists. However, not everyone agrees with that premise. My view is that science and technology have been marching forward together since the dawn of civilization, and we have progressed from the Stone Age to the age of nanotechnology, from the geocentric model of the universe through the heliocentric universe to today's mind-boggling concept of a multiverse. Let us hope that our journey to explore the unknown through the rational approach of science and technology continues with indomitable spirit so that one day we may emboss our footprints to other universes.

Appendix I

Some Important Physical Constants of Physics and Astrophysics

- Universal gravitational constant: $G = 6.67428 \times 10^{-11}$ N m^2/kg^2
- Velocity of light in vacuum: $c = 299{,}792{,}458$ m/s
- Boltzmann Constant: $k_B = 1.3806504 \times 10^{-23}$ J/K
- Planck's Constant: $h = 6.62607015 \times 10^{-34}$ J-s
- Reduced Planck's Constant: $\hbar$ (=h/2π) = $1.054571818 \times 10^{-34}$ J-s
- Planck length: $l_p = 1.616 \times 10^{-35}$ m
- Hubble's Constant: $H = 71$ km/sec/mega parsec
- Mass of electron: $m_e = 9.1 \times 10^{-31}$ kg
- Mass of proton: $m_p = 1.67 \times 10^{-27}$ kg (i.e., 1835 m_e)

Multiple and Submultiple SI Prefixes (Powers of Tens)

Power	Symbol	Prefix (UK)	Prefix (US)	Feel the Number
10^{24}	Y	Yotta	Septillion	1000000000000000000000000
10^{21}	Z	Zeta	Sextillion	1000000000000000000000
10^{18}	E	Exa	Quintillion	1000000000000000000
10^{15}	P	Peta	Quadrillion	1000000000000000
10^{12}	T	Tera	Trillion	1000000000000
10^{9}	G	Giga	Billion	1000000000
10^{6}	M	Mega	Million	1000000
10^{3}	k	Kilo	Thousand	1000
10^{2}	h	hecto	Hundred	100
10^{1}	da	deca	Ten	10
10^{0}			One	1
10^{-1}	d	deci	Tenth	0.1
10^{-2}	c	centi	Hundredth	0.001
10^{-3}	m	milli	Thousandth	0.0001
10^{-6}	μ	micro	Millionth	0.0000001
10^{-9}	n	nano	Billionth	0.0000000001
10^{-12}	p	pico	Trillionth	0.0000000000001
10^{-15}	f	femto	Quadrillionth	0.0000000000000001
10^{-18}	a	atto	Quintillionth	0.0000000000000000001
10^{-21}	z	zepto	Sextillionth	0.0000000000000000000001
10^{-24}	y	yocto	Septillionth	0.0000000000000000000000001

Appendix II

The photograph of a galaxy of scientists of 20th century (17 of the 29 attendees were or became Nobel Laureates). 5th Solvay Conference, 1927; Theme: *Electrons and Photons* (Institute International de Physique Solvay in Leopold Park, Brussels, Belgium). [Photograph taken from: https://rarehistoricalphotos.com/solvay-conference-probably-intelligent-picture-ever-taken-1927/]

Name of the scientists in the photograph at different rows:

Back row: Piccard, A.; Henriot, E.; Ehrenfest, P.; Herzen, E.; de Donder, T.; Schrödinger, E.; Verschaffelt, J.E.; Pauli, W.; Heisenberg, W.; Fowler, R.H.; and Brillouin, L.

Middle row: Debye, P.; Knudsen, M.; Bragg, W.L.; Kramers, H.A.; Dirac, P.A.M.; Compton, A.H.; de Broglie, L.; Born, M.; and Bohr, N.

Front row: Langmuir, I.; Planck, M.; Curie, M.; Lorentz, H.A.; Einstein, A.; Langevin, P.; Guye, Ch.-E.; Wilson, C.T.R.; and Richardson, O.W.

Copyright Acknowledgements

Some of the images used in this book have been downloaded from public domain websites, the list of which is given below:

List of Referred Illustrations

1. https://www.space.com/23017-weightlessness.html
2. https://news.stanford.edu/news/2011/may/gravity-probe-mission-050411.html
3. https://en.wikipedia.org/wiki/Gravitational_lens
4. http://mentalfloss.com/article/32288/11-great-geeky-math-tattoos
5. https://www.britannica.com/science/periodic-table
6. https://iveybusinessjournal.com/the-big-bang-theory-of-disruption/
7. https://www.nasa.gov/feature/making-sense-of-the-big-bang-wilkinson-microwave-anisotropy-probe
8. https://www.eso.org/public/images/black-holes-infographic-v2/
9. https://www.bbc.com/news/science-environment-47873592
10. https://directory.eoportal.org/web/eoportal/satellite-missions/content/-/article/gravitational-waves
11. https://sites.google.com/site/gss12m33/which-theory-is-better-supported
12. Evolution of the Early Universe: phys.libertexts.org
13. https://astrobites.org/wp-content/uploads/2016/04/Dark_Energy.jpg
14. https://home.cern/news/news/physics/higgs-boson-one-year
15. https://www.geekwire.com/2018/physicists-lhc-confirm-happens-higgs-bosons-die/
16. https://www.secretsofuniverse.in/future-circular-collider/
17. http://www.webexhibits.org/causesofcolor/14.html
18. https://www.spaceanswers.com/solar-system/why-is-mars-red/
19. https://www.eyeem.com/search/pictures/sahara%20desert
20. https://theconversation.com/curious-kids-why-is-the-sky-blue-and-where-does-it-start-81165
21. https://esahubble.org/images/opo9545b/
22. https://theplanets.org/solar-system/
23. https://www.space.com/19878-halleys-comet.html
24. https://www.news18.com/news/tech/spotting-neowise-in-india-skywatchers-these-apps-can- help-you-see-the-rare-comet-2727799.html

25. https://en.wikipedia.org/wiki/Mercury, Venus, ... Pluto
26. https://en.wikipedia.org/wiki/Near_side_of_the_Moon#/media/File:Moon_nearside_LRO.jpg
27. https://en.wikipedia.org/wiki/Far_side_of_the_Moon#/media/File:Back_side_of_the_Moon_AS16-3021.jpg
28. http://www.tp.umu.se/space/Proj_10/Amir_A-10.pdf
29. https://www.schoolsobservatory.org/learn/astro/stars/cycle
30. https://www.space.com/21854-andromeda-galaxy-m31-photos-gallery.html
31. http://www.spacetelescope.org/images/heic1323a/
32. https://en.wikipedia.org/wiki/Type_Ia_supernova
33. https://en.wikipedia.org/wiki/Crab_Nebula, Orion_Nebula, Horsehead_Nebula
34. https://apod.nasa.gov/apod/ap000130.html
35. https://www.thestatesman.com/ [Subal Kar: Journey into Space, The Statesman, Monday, 26 Dec. 2005, Kolkata
36. https://en.wikipedia.org/wiki/Cygnus_A#/media/File:3c405.jpg
37. https://en.wikipedia.org/wiki/Crab_Pulsar
38. https://www.nasa.gov/content/goddard/nasas-hubble-gets-the-best-image-of-bright-quasar-3c-273/#.YH-6J-4zbIU
39. https://www.discovermagazine.com/the-sciences/quasars-shine-light-on-universe%27s-mysterious-energy-signals

Further Readings

1. Hawking, Stephen. *A Brief History of Time: From the Big Bang to Black Holes*. London: Bantam Dell Publishing Group, 1988.
2. Thorne, Kip S. *Black Holes and Time Warps: Einstein's Outrageous Legacy*. New York: W. W. Norton, 1994.
3. Rovelli, Carlo. *Reality Is Not What It Seems: The Journey to Quantum Gravity*. New York: Penguin Random House, 2018.
4. Tegmark, Max. *Our Mathematical Universe: My Quest for the Ultimate Nature of Reality*. New York: Penguin Books, 2015.
5. Feynman, R. P., Leighton, R. B., and Mathew, S. *The Feynman Lectures on Physics*. Vols. 1, 2, & 3). New Delhi: Narosa Publishing House, 2008.
6. Einstein, Albert. *Relativity: The Special and General Theory*. New Delhi: General Press, Revised Edition 2020.
7. Wheeler, J. A. *A Journey into Gravity and Spacetime*. New York: Scientific American Library (A Division of HPHLP), 1990.
8. Kaku, Michio, and Thompson, Jennifer. *Beyond Einstein: The Cosmic Quest for the Theory of Universe*. New York: Anchor Books, Doubleday, 1995.
9. Syrovatkin, Sergei. *Einstein and the Philosophical Problems of 20th-Century Physics*. Moscow: Progress Publishers, 1979.
10. Susskind, L, and Friedman, A. *Quantum Mechanics: The Theoretical Minimum*. New York: Basic Books, 2014.
11. Weinberg, Steven. *First Three Minutes: A Modern View of the Origin of the Universe*. North Lincolnshire: Fontana Paperbacks, 1993.
12. Shapiro, Stuart L., and Teukolsky, Saul A. *Black Holes, White Dwarfs, and Neutron Stars: The Physics of Compact Objects*. New York: John Wiley, VCH Verlag GmbH & Co KGaA, 1983.
13. Smolin, Lee. *The Trouble with Physics: The Rise of String Theory, the Fall of a Science and What Comes Next*. New York: Houghton Mifflin Harcourt, 2006.
14. Lewin, W., and Goldstein, W. *For the Love of Physics: From the End of the Rainbow to the Edge of Time—A Journey Through the Wonders of Physics*. New York: Free Press, 2011.
15. Sagan, Carl, and Druyan, Ann. *The Demon-Haunted World: Science as a Candle in the Dark*. New York: Ballantine Books, 1997.
16. Sagan, Carl. *Cosmos*. New York: Random House, 1980.
17. Rovelli, Carlo, and Carnell, Simon (Trans.). *Seven Brief Lessons on Physics*. New York: Riverhead Books, 2014.
18. Varvoglis, Harry. *History and Evolution of Concepts in Physics*. New York: Springer, 2014.
19. Glasby, John S. *Boundaries of the Universe*. Routledge, 2013.
20. Clegg, Brian. *Dark Matter and Dark Energy: The Hidden 95% of the Universe*. Hot Science, 2019.

Subject Index

Scientist Index

S

T

W

Y